THE X-RAY STANDING WAVE TECHNIQUE

Principles and Applications

Series on Synchrotron Radiation Techniques and Applications
(ISSN: 2010-2844)

Published

Vol. 1 Synchrotron Radiation Sources — A Primer
H. Winick

Vol. 2 X-ray Absorption Fine Structure (XAFS) for Catalysts & Surfaces
ed. Y. Iwasawa

Vol. 3 Compact Synchrotron Light Sources
E. Weihreter

Vol. 4 Novel Radiation Sources Using Relativistic Electrons: From Infrared to X-Rays
P. Rullhusen, X. Artru & P. Dhez

Vol. 5 Synchrotron Radiation Theory and Its Development: In Memory of I. M. Ternov
ed. V. A. Bordovitsyn

Vol. 6 Insertion Devices for Synchrotron Radiation and Free Electron Laser
F. Ciocci, G. Dattoli, A. Torre & A. Renieri

Vol. 7 The X-Ray Standing Wave Technique: Principles and Applications
eds. J. Zegenhagen & A. Kazimirov

Series on Synchrotron Radiation Techniques and Applications – Vol. 7

THE X-RAY STANDING WAVE TECHNIQUE

Principles and Applications

Editors

Jörg Zegenhagen
European Synchrotron Radiation Facility, France

Alexander Kazimirov
Cornell University, USA

World Scientific

NEW JERSEY • LONDON • SINGAPORE • BEIJING • SHANGHAI • HONG KONG • TAIPEI • CHENNAI

Published by

World Scientific Publishing Co. Pte. Ltd.
5 Toh Tuck Link, Singapore 596224
USA office: 27 Warren Street, Suite 401-402, Hackensack, NJ 07601
UK office: 57 Shelton Street, Covent Garden, London WC2H 9HE

British Library Cataloguing-in-Publication Data
A catalogue record for this book is available from the British Library.

Series on Synchrotron Radiation Techniques and Applications — Vol. 7
THE X-RAY STANDING WAVE TECHNIQUE
Principles and Applications

ISBN 978-981-227-900-7

Typeset by Stallion Press
Email: enquiries@stallionpress.com

Printed in Singapore by Mainland Press Pte Ltd.

DEDICATION

Boris W. Batterman (August 25, 1930–December 14, 2010)
Founder of the X-ray standing wave technique

Boris W. Batterman, Bob as he was known, attended Brooklyn Technical High School and Cooper Union College in New York before earning his undergraduate degree at Massachusetts Institute of Technology in 1952 and his PHD in Physics there in 1956 under the supervision of Bertram E. Warren. He spent the academic year of 1953–1954 as Fulbright Scholar at the Technische Hochschule in Stuttgart, Germany. Bob was a member of Technical Staff at AT&T Bell Labs, Murray Hill, NJ from 1956 until 1965, when he moved to Cornell University. He was awarded both a Guggenheim Fellowship and a Fulbright-Hayes Fellowship from 1971–1972. As co-founder, he became the first director of the Cornell High Energy Synchrotron Source (CHESS) in 1978 and he remained it until his retirement in 2001. In 1983, he received the Humboldt Research Award of the German Alexander von Humboldt Foundation and in 1985, he was named the Walter S. Carpenter Jr. Professor of Applied and Engineering Physics at Cornell. Many current leaders in the synchrotron X-ray field, now working at synchrotrons and a host of universities and laboratories around the world, lived and learned under Bob's tenure as Professor at Cornell University and director of CHESS.

PREFACE

The X-ray standing wave (XSW) technique has been established almost 50 years ago. It has found widespread use in many field of condensed matter research. However, an experimentalist who wants to employ it but is not familiar with its requirements in theory and practice is facing some difficulties. Until now, there is not a single book which one can refer to in order to learn about its basics, technical details and possible applications. So far, all necessary information is scattered in many different papers and some review articles. To improve this situation and to try to present the most important basic information comprehensively in one volume was the motivation for this book.

It is in fact a hybrid consisting of two parts of different style. The first part is more like a textbook, providing much of the necessary framework of the XSW method. The second part comes along like a review of recent work on the XSW technique presenting a collection of papers with up-to-date highlights. It is meant to wet the appetite of the reader for his/her own application. We also decided to ask some of the (very) early practitioners, experts and furtherers to provide their personal view of the early time of the development of the XSW method. We are extremely fortunate that Boris W. Batterman, Walter M. Gibson, Gene A. Golovchenko, Seishi Kikuta, and Gerhard Materlik responded positively to this request. Standing in awe in front of perfect theories and marvelling at brilliant experiments, it is often forgotten how many faults and meanders had been necessary to arrive at that point. We believe that in particular for young scientists, it is not only important to know science but also to know how science is done, how it happened.

From its very beginning, it took a while to arrive at the current perception of the XSW technique. For some time, the dynamical theory of X-ray diffraction (DTXD) was considered as its theoretical foundation and its knowledge a necessity for any XSW practitioner. Indeed, in the beginning, the X-ray standing wave was exclusively produced by dynamical diffraction from single crystals. Nowadays, knowledge of the DTXD is considered useful but not necessary. In fact, there are various ways to generate the XIF: Bragg diffraction, diffraction from multilayer, reflection from mirror surfaces. With highly coherent light sources such as ultimate storage ring light sources and free electron lasers, other methods may be devised. Already until now, the availability of highly brilliant synchrotron radiation has simplified the generation of the necessary plane wave(s) enormously.

However, producing the XIF is only one part of the theoretical framework of the XSW technique. The other part, and nowadays maybe even the more complex one, is the inelastic scattering of the X-ray photon interference field. With more extensive use of the excited photoelectrons, it was realized that the interference extends to the scattering process itself, to the complex matrix elements describing the photo excitation. Thus, in particular polarization effects, limits of the dipole approximation and multipole effects in the photo absorption need to be understood for the increasingly popular case of photoelectron emission by the standing wave. Last but not least, the data analysis and interpretation requires a solid foundation, for instance the equivalence of XSW and Fourier analysis for structural studies.

We would like to express our gratitude to Donald Bilderback. After we had edited a small selection of papers on the XSW technique for Synchrotron Radiation News, he suggested collecting a more comprehensive set for World Scientific. After giving this some thought, we came up with the idea for this volume. The work on this book took much longer than originally scheduled and we are immensely grateful for the patience the publisher displayed with us. We sincerely hope that the result finds your approval and that you find the reading useful and enjoyable.

Alexander Y. Kazimirov and Jörg Zegenhagen
Ithaca and Grenoble, July 2011

ACRONYMS

AA	Anomalous absorption
AES	Auger electron spectroscopy
AFM	Atomic force microscope
APS	Advanced Photon Source
ARXPS	Angle-resolved X-ray photoelectron spectroscopy
BESSY	Berliner Elektronen-Speicherring Gesellschaft für Synchrotronstrahlung
B-H	Bonse-Hart
BNL	Brookhaven National Laboratory
BO	Bridging oxygen
CAE	Constant analysis energy
CC	Channel-cut
CCD	Charge-coupled device
CERN	Conseil Européen pour la Recherche Nucléaire
CESR	Cornell Electron Storage Ring
CHA	Concentric hemisphere analyzer
CHESS	Cornell High-Energy Synchrotron Source
CMA	Cylindrical mirror analyzer
COM	Center of mass
CRL	Compound refractive lens
CRR	Constant retard ratio
CS	Compton scattering
CSA	Cylindrical sector analyzer
CXDI	Coherent X-ray diffraction imaging
DA	Dipole approximation
DAS	Dimer adatom stacking fault
DESY	Deutsches Elektronen Synchrotron
DFT	Density functional theory
DMFT	Dynamical mean field theory
DORIS	Doppel Ring Speicher
DOS	Density of states
DPPC	Dipalmitoylphosphatidylcholine

DPPE	Dipalmitoylphosphotidylethanolamine
DTXD	Dynamical theory of X-ray diffraction
DW	Debye-Waller
EDC	Energy distribution curve
EDC	electron-distribution curve
ERL	Energy recovery LINAC
ESCA	Electron spectroscopy for chemical analysis
ESRF	European Synchrotron Radiation Facility
EXAFS	Extended X-ray absorption fine structure
FC	Front coupling
FEL	Free electron laser
FET	Field effect transistor
FFT	Fast Fourier transform
FOM	Figure of merit
FT	Fourier transform
FWHM	Full width at half maximum
GAXSW	Grazing-angle X-ray standing wave
GE	General Electric
GID	Grazing incidence diffraction
GIXRD	Grazing incidence X-ray diffraction
GMR	Giant magnetoresistance
HASYLAB	Hamburger Synchrotronstahlungs Labor
HAXPES	Hard X-ray photoelectron spectroscopy
HVPE	Hydride vapor phase epitaxy
ID	Insertion device (wriggler or undulator)
IR	Infrared radiation
IS	Inelastic scattering
IXS	Inelastic X-ray scattering
KFA	Kernforschunganlage
KXSW	Kinematical X-ray standing waves
LB	Langmuir-Blodgett
LCAO	Linear combination of atomic orbitals
LCLS	Linac coherent light source
LCP	Left circular polarized
LDA	Local density approximation
LEED	Low energy electron diffraction
LEEM	Low energy electron microscopy
LINAC	Linear accelerator

LLL	Laue-Laue-Laue (reflections)
LSM	Layered synthetic microstructure
MBE	Molecular beam epitaxy
MCD	Magnetic circular dichroism
MIT	Massachusetts Institute of Technology
ML	Monolayer
MO	Molecular orbital
MOCVD	Metal organic chemical vapor deposition
MOVPE	Metal organic vapor phase epitaxy
MPI-FKF	Max-Planck-lnstitut für Festkörperforschung
MTJ	Magnetic tunnel junction
NIST	National Institute of Standards and Technology
NIXSW	Normal incidence X-ray standing waves
NSLS	National Synchrotron Light Source
pDOS	Partial density of states
PE	Primary extinction (Chapter 1), photoelectron emission (Chapter 11)
PEEM	Photoelectron emission microscopy
PES	Photoelectron spectroscopy
PHA	Pulse height analysis
PM	Post-monochromator
PML	Periodic multilayer
PMMA	Polymethyl methacrylate
PSD	Photon stimulated desorption
PTCDA	1,4,5,8-perylene-tetracarboxylicacid-dianhydride
RBC	Resonant beam coupling
RBS	Rutherford backscattering spectrometry
RCP	Right circular polarized
RIE	Reactive ion etching
RIXS	Resonant inelastic X-ray scattering
RNA	Ribonucleic acid
RÖMO	Röntgen monochromator
SAM	Self-assembled monolayer
SEM	Scanning electron microscope
SEXAFS	Surface extended X-ray absorption fine structure
SIMS	Secondary ion mass spectrometry
SPEAR	Stanford Positron-Electron Accelerating Ring
SPLEEM	Spin polarized low energy electron microscopy

SP-PEEM	Spin polarized photoelectron emission microscopy
SR	Synchrotron radiation
SRS	Synchrotron Radiation Source (Daresbury, UK)
SRY	Secondary radiation yield
SSRL	Stanford Synchrotron Radiation Laboratory
STM	Scanning tunnelling microscope
STXM	Scanning transmission X-ray microscope
SUNY	State University of New York
SW	Standing wave
SXRD	Surface X-ray diffraction
SXSW	Soft X-ray standing waves
TDS	Thermal diffuse scattering
TMR	Tunnel magnetoresistance
TO	Terminal oxygen
TOF	Time of flight
TR	Total reflection
TR-XSW	Total reflection X-ray standing waves
UHV	Ultra high vacuum
UPS	Ultraviolet photoelectron spectroscopy
VUV	Vacuum ultraviolet
WG	(X-ray) waveguide
XANES	X-ray absorption near edge structure
XES	X-ray emission spectroscopy
XESD	X-ray induced electron stimulated desorption
XFEL	X-ray free electron laser
XIF	X-ray interference field
XM	X-ray microscope
XPS	X-ray photoelectron spectroscopy
XPSD	X-ray photon stimulated desorption
XRD	X-ray diffraction
XRF	X-ray fluorescence
XRO	X-ray optical
XSW	X-ray standing waves
XSW-PSD	X-ray standing wave photon stimulated desorption
YPF	Yield probability function

CONTENTS

Part I

Chapter 1

X-RAY STANDING WAVES IN A NUTSHELL

JÖRG ZEGENHAGEN

European Synchrotron Radiation Facility (ESRF),
BP 220, F-39043, Grenoble, France

ALEXANDER KAZIMIROV

Cornell High Energy Synchrotron Source (CHESS),
Cornell University, Ithaca, NY 14853, USA

We begin with a brief historical overview and then present the basic principles of the X-ray standing wave (XSW) technique, including the basics of elastic and inelastic scattering, photoelectron absorption, dipole approximation, and the nature of the structural information provided by the XSW method. How do we create XSWs? What signal should we detect? Do we need perfect crystals? One can find answers to these and other questions, which will be discussed in depth in the following chapters.

1.1. Introduction

About 50 years have passed since the X-ray standing wave (XSW) technique was conceived by Boris W. Batterman.[1] Meanwhile, the method has found widespread applications and is used in different fields of scientific research. It has diverse applications such as in surfaces, soft condensed matter, biophysics, semiconductors, and electrochemical interfaces. Some useful review papers on the XSW technique are available; however, they all deal with some rather specific applications of the method. It seems quite astonishing that no comprehensive description of XSW's general principles or an overview of its many applications has been published up to now. In our opinion, this is due to the very nature of the technique, since it requires knowledge from different fields, which we would like to stress from the very beginning.

XSW is formed by the superposition of (two) coherent plane waves. Already the generation of one (sufficiently) coherent plane X-ray wave, i.e. sufficiently collimated and monochromatic, is not trivial. However, by using modern third-generation synchrotron sources, it has become much easier. The formation of a second coherent wave is based on the elastic (i.e., coherent) scattering of X-rays for which several possibilities already exist. However, our main interest rests with the process of the inelastic scattering from the stationary X-ray wave, i.e., from the spatially modulated X-ray intensity. Here, several inelastic scattering processes of the X-ray photon need to be considered. Consequently, several spectroscopic techniques may be used. Thus, the XSW technique requires specific knowledge from technical and scientific areas which are largely represented by different communities. A comprehensive description of the XSW technique thus requires ingredients from different fields and thus needs necessarily some room.

The XSW technique adds spatial resolution to many traditional X-ray spectroscopy techniques. Instead of a transient X-ray wave, it utilizes an X-ray interference field (XIF), i.e., a stationary X-ray wave field or standing wave, as a source of excitation. Several other techniques employ the interference of X-rays, but what distinguishes the XSW technique is the utilization of the inelastic scattering excited by an XIF. Within the wave field, the X-ray intensity is spatially modulated. Since the strength of the inelastically scattered X-rays and the X-ray photo-absorption is proportional to the X-ray intensity, the intensity of these scattering signals depends on whether the scattering objects (atoms, electrons, etc.) are located at the maximum (antinodes) or the minimum (nodes) of the XSW (Fig. 1.1). Thus, if one is able to manipulate the wave field, i.e., shift it in space, information about the location of the scatterers with respect to the XSW planes can be obtained. Particularly useful information can be obtained if the entity of the scatterers is distributed inhomogeneously with respect to the XSW pattern.

The XSW technique is in fact a Fourier technique, as will be explained in detail further below. However, in simple cases, the result of an XSW measurement can be immediately interpreted in real space, because of the phase information provided by the technique. Being a Fourier technique, in terms of structural analysis, some analogy with diffraction techniques exists. Indeed, employed for structural investigations by recording the photoelectric scattering (such as photoelectrons, X-ray fluorescence) out of the interference field, the two fundamental XSW parameters determined

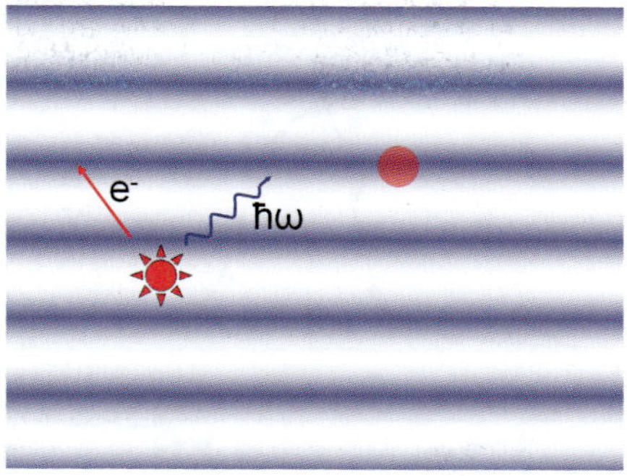

Fig. 1.1. Atoms experiencing maximum (left) and minimum (right) of a spatially modulated X-ray intensity.

by an XSW experiment, i.e. coherent fraction f and coherent position P, are closely related to the X-ray structure factor. Yet, it is important to stress already at this point some important differences with diffraction techniques. In principle, and in contrast to diffraction, information about the location of a *single* scatterer can be obtained using XSWs. (In practice, the weakness of the signal will most likely prohibit such a measurement.) The reason behind this unique feature is that the XSW technique probes the correlation of the scatterer(s) with the XIF, whereas diffraction techniques probe the autocorrelation of the scatterers. Moreover, diffraction techniques suffer from the so-called phase problem since only the magnitude of the (complex) structure factor can be recorded and the phase information is lost. (For completeness we should note that there are some ways to regain this information.[2]) Furthermore, element-specific information is difficult to obtain by diffraction, whereas the XSW method allows to analyze the structure of materials element specific, even with chemical sensitivity and with respect to electronic properties when using high-resolution spectroscopic techniques.

1.2. Historical Background

In 1964, Boris W. Batterman published a paper[1] entitled "*Effect of dynamical diffraction in X-ray fluorescence scattering.*" The experiment that Batterman reported provided evidence that an (= XSW) is formed inside a crystal by the superposition of an incident and Bragg reflected X-ray beams. The existence of the X-ray interference wave field and the shift of the position of its antinodal planes (intensity maxima) from being between

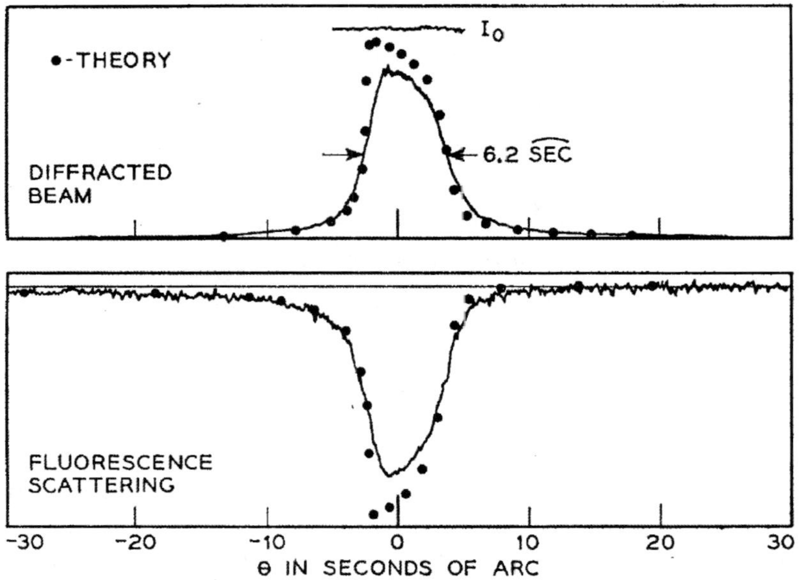

Fig. 1.2. First XSW experiment scanning the Ge(220) reflection curve (top) using Mo K_α radiation and recording simultaneously the Ge K_α radiation from the sample. The maxima of the XSW are in-between the atomic planes on the left-hand side of the rocking curve and on the atomic planes on the right-hand side, proven by a slight asymmetry of the Ge K_α fluorescence, being a bit stronger on the right-hand side. From Ref. 1.

to being on the crystal lattice planes while traversing the Ge(220) Bragg reflection was proven by the slightly asymmetric profile of the recorded K_α fluorescence of the atoms of the germanium crystal. The Ge K fluorescence was noticeably a bit weaker on the low angle and a bit stronger on the high angle side of the simultaneously recorded Ge(220) reflection curve (Fig. 1.2). Under the experimental conditions chosen by Batterman, this asymmetry is quite faint because of the extinction effect; under the Bragg reflection condition, the incident beam is reflected from a shallow layer limited by the extinction length and thus it does not penetrate deep into the crystal (it is "extinct" within the bulk of the crystal). Consequently, fewer crystal atoms are photo-excited when the beam is Bragg reflected, and the X-ray fluorescence signal from these atoms will be strongly reduced. Batterman understood this and, in a subsequent experiment, reported in the same publication, could enhance the asymmetry in the Ge fluorescence by monitoring the signal at a grazing exit angle, thus limiting the "escape depth" of the fluorescence radiation, as shown in Fig. 1.3.

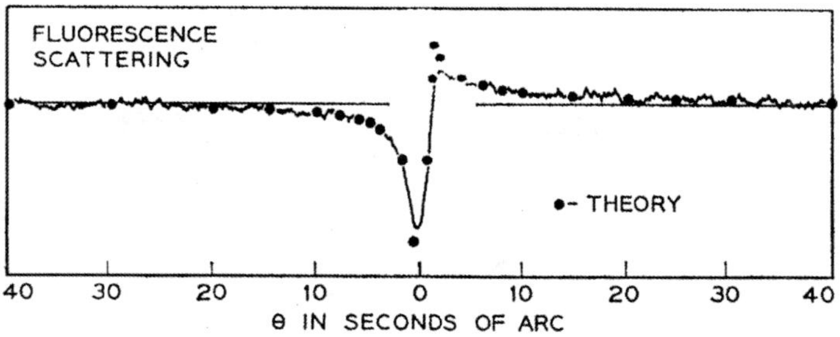

Fig. 1.3. Ge K fluorescence recorded while passing the Ge(220) rocking curve under the same conditions as in Fig. 1.2 except that the fluorescence radiation is recorded at grazing exit, thus limiting its escape depth and minimizing the effect of extinction. In this way, the enhancement of the fluorescence intensity at the high angle side, when the maxima of the XSW coincide with the Ge atomic planes, is shown more clearly. From Ref. 1.

Batterman was not the first to realize that an XIF, i.e., an XSW, is generated during Bragg reflection. It had been recognized much earlier, but the proof was considered difficult. With respect to XSW, which he calls "X-ray beats" (Röntgenschwebungen), Max von Laue writes in his book *Röntgenstrahl-Interferenzen*[3]: "... weil man außer dem Kossel–Effekt kein Mittel kennt, Dasein und Lage der Röntgenschwebungen im Raumgitter unmittelbar nachzuweisen; ..." ("... since except for the Kossel effect there is no means to prove existence and location of the X-ray beats within the crystal space lattice ..."). The Kossel effect[4] can in fact be understood as the space inversion of the XSW method: the fluorescence radiation emitted from the atoms within the lattice is enhanced or diminished along certain angular directions depending on whether the elastic scattering of the characteristic radiation by the atomic planes leads to constructive or destructive interference. This leads to the beautiful spatial pattern of Kossel lines, frequently with a characteristic light/dark fine structure. The other early manifestation of the XSW inside a crystal was the Borrmann effect, i.e. the anomalous transmission of X-rays through thick absorbing crystal in the so-called Laue case. Owing to the specific boundary condition in the Laue case, Bragg diffraction of a linearly polarized incident wave gives rise to two wave fields inside the crystal. One wave field, which has the nodes on the atomic planes, is anomalously strongly transmitted, whereas the other wave field with the antinodes on the atomic planes is anomalously absorbed.[5] While Kossel and Borrmann effects proved the existence of the interference field in crystals, Batterman's idea to use the characteristic X-ray fluorescence radiation to track the location and the movement of the wave field during Bragg reflection was certainly particularly elegant. Batterman describes his motivation, how he conceived the XSW idea, and the early experiments in detail in the Appendix of this volume.

Scientists in Japan were alerted by the first experiment of Batterman and started immediately to study the Compton scattering and thermal diffuse scattering excited by an XSW.[6] Both signals are not only affected by the movement of the XSW but also suffer strongly from the extinction affect. An account of the early activities related to the XSW technique, which led to important activities in Japan in the field of XSW over the years, is given by Seishi Kikuta, also in the Appendix of this volume. First experiments on photoelectron emission excited by the XSW in perfect crystals and, later, from crystals with distorted surface layers were performed in Russia.[7] Since the escape depth for photoelectrons is much smaller than the extinction length, the angular dependence of the movement of the XSW field was observed unhidden by the extinction effect. This work stimulated the further development of the theory of the XSW in distorted crystals[8] (see also Chapter 3), in particular, with respect to applications of the XSW technique to study the structure of thin surface layers.

The observation of the characteristic asymmetrical shape of the Ge fluorescence with rocking angle by Batterman was initially only a demonstration that XSW exists. He had been stimulated to do his first experiment after studying the dynamical theory of X-ray diffraction[3] (DTXD). A highly cited review of this theory was published by Batterman and Cole in 1964.[9] Batterman had been additionally inspired by a paper published in 1956 by Knowles,[10] who had performed a neutron standing wave experiment, detecting the γ-rays originating from a calcite crystal when rocking through the (211) Laue reflection. Realizing the potential usefulness of his experimental approach, Batterman continued his work and published in 1969 another paper entitled "*Detection of foreign atom sites by their X-ray fluorescence scattering*".[11] From carefully studying the DTXD, he had learnt that the wave field intensity and its location with respect to the lattice planes of a crystal can be calculated exactly at every point of the single crystal reflection curve during Bragg reflection. Thus, the characteristic profile of the intensity of the X-ray fluorescence from an impurity atom, in that experiment arsenic in silicon, recorded over the angular range of Bragg reflection, contains the information on the position of this atom with respect to the crystal atomic planes. The result of his experiment was in agreement with the known fact that the (majority of) As atoms reside on substitutional sites. However, since he used a bulk sample uniformly doped by As, the characteristic shape of the As fluorescence was again strongly masked by the extinction effect. Only several years later in 1974, using a silicon substrate crystal with arsenic diffused into the shallow surface region instead of homogeneously distributed in the bulk, thus minimizing the extinction effect, the possibility to locate impurity atoms using XSWs was convincingly demonstrated by Golovchenko, Batterman, and Brown.[12] A personal view on his involvement in the early days of the XSW technique is given by Jene Golovchenko in the Appendix of this volume. It is complemented by some remembrances, also about XSW work at SUNY Albany, by Walter Gibson, who had been Golovchenko's thesis supervisor at Bell Labs.

In 1980, Cowan, Golovchenko and Robbins demonstrated that XIF is not restricted to the interior of the crystal but extends into the region above the surface.[13] This laid the foundation of the further use of the XSW technique in surface science, which is presently the most prominent area of its application. In their paper, Cowan *et al.* showed that the XSW technique is able to determine the position of an adsorbate on a surface. The example chosen was bromine on the Si(110) surface. Since X-ray physicists at that time were not typically UHV experts, the XSW had been performed with Br adsorbed on a silicon surface from a liquid. However, such an XSW measurement for Br on Si(110) using X-rays from a laboratory source lasted many hours. It required the application of intense synchrotron radiation (SR) a few years later,[14] to achieve reasonable measuring times and impressively low detection limits. A personal view of the development of the XSW technique using synchrotron radiation in Hamburg is given by Gerhard Materlik in the Appendix of this volume. Because of the high value of the XSW technique for the study of surface adsorbate systems, the possibility of performing measurements in ultra-high vacuum (UHV) was another important step forward.[15]

As soon as measuring times had become reasonable by using SR and the usefulness of the method in surface science had been demonstrated, the XSW method became attractive and the number of publications increased sharply. In the following years, XSW experiments ceased being merely demonstrations and were applied to an increasing number of problems in surface physics and material science. While about 20 papers had been published in the first 20 years of its existence, if we date the birth of the XSW method back to 1964, roughly 200 papers were published in the following decade until about 1994. In the beginning of 2010, "googling" "X-ray standing waves" produces about 300,000 hits (compared to about eight million for "X-ray diffraction"). Searching ISI Science Web for "X-ray standing wave" yields almost 900 hits. In the last 15 years, about 500 papers were found. The growth in the 1980s and early 1990s has led to a small user community and now about 40 papers are published yearly (compared to about 200 per year for "grazing incidence X-ray diffraction"). Thus, the XSW method is established but has not become a mainstream technique. One simple reason for this is the fact that the technique is of very limited value without SR. The other reason is at least as important: since the XSW technique needs the combination of elastic and inelastic scattering (or absorption), it requires knowledge about experimental areas that are typically represented by different communities. Furthermore, for many years, the DTXD had been considered essential for the XSW technique, which represented a stumbling stone since the DTXD is not usually part of a graduate student education. Knowledge of the DTXD is very useful when applying the XSW technique, but the DTXD is not essential in understanding the basics of the XSW technique. The DTXD is needed (1) if the XSW technique is used with single crystals and (2) for understanding perfect crystal optics to prepare the incident (plane) wave for the XSW experiment.

The XSW technique up to now has been used most intensively for structure investigations by employing the photoelectric "scattering" of X-ray photons. However, as mentioned above, the XSW technique has also been applied to Compton scattering (see contribution by V. Bushuev, Chapter 10) and thermal diffuse scattering.[16] More recently, XSW has been used to gain information about the valence band electronic structure of materials (see contributions by J. Woicik, Chapters 12 and 26).

1.3. The Basic Principle of the XSW Technique

Many descriptions or reviews of the XSW technique, in particular, older ones, start with a more or less extensive treatment of DTXD in perfect crystals. The DTXD is needed to exactly quantify the elastic scattering of X-rays in crystals under Bragg diffraction conditions (see Chapters 2 and 3) and allows to calculate the properties (such as intensity, phase, and extinction) of the wavefield created during Bragg reflection. For many newcomers interested in the XSW technique, the seeming necessity to understand the DTXD first represents an unpleasant hurdle. However, while Boris Batterman conceived the XSW technique based on his knowledge of the DTXD employing Bragg reflection from a crystal, it is important to keep in mind that this represents only the way to produce an X-ray standing wave. Thus, it was pointed out already in 1993[17] that the principle of the XSW technique can be understood completely independent of the DTXD.

The XSW technique is in essence an X-ray interference technique. Knowing the exact pattern and position of an X-ray interference field and being able to manipulate it in space, information about the real space distribution of inelastically scattering objects can be obtained. In case of atomic distributions, which is the most important application of the XSW technique, the photo-absorption or subsequent decay channels (such as photoelectrons, Auger electrons, and X-ray fluorescence) serve as marker signals upon shifting the wave field in space. A strong signal indicates that the maxima of the wave field are on atomic positions. A weak signal would indicate that the minima of the wave field intensity mostly coincide with atomic positions. This principle is based on the fact that the probability of the photoelectric process reflects the wave field intensity at the center of the atom. It furthermore requires the ability to manipulate XIF, i.e., to move its position relative to atoms in a controlled way and to switch the interference effect on and off. In case of a planar wave field, i.e. the superposition of two coherent plane waves, which is the most common case for the XSW technique, we can analyze atomic positions on the length scale of the wave field spacing d_H. More accurately, utilizing a planar X-ray wave field with spacing d_H for the XSW analysis of an atomic distribution, we obtain a Fourier component of the atomic distribution corresponding to the wave field spacing, i.e. the **H**th Fourier component with $|\mathbf{H}| = 1/d_H$ with the vector **H** being normal to the wave field planes. This equivalence of XSW and Fourier analysis had been pointed out first by Hertel, Materlik and Zegenhagen in 1985[18] and is proven further below.

We will restrict ourselves in the following to the production and utilization of a planar X-ray wave field, which is created by the superposition of two plane X-ray waves with the electric field vectors $\mathbf{E}_0$ and $\mathbf{E}_H$:

$$\begin{aligned} \mathbf{E}_0 &= \mathbf{e}_0 E_0 \exp[2\pi i(\nu_0 t - \mathbf{K}_0\mathbf{r})] \quad \text{and} \\ \mathbf{E}_H &= \mathbf{e}_H E_H \exp[2\pi i(\nu_H t - \mathbf{K}_H\mathbf{r})] \end{aligned} \tag{1.1}$$

as shown schematically in Fig. 1.4. Here, $\mathbf{e}_0$ and $\mathbf{e}_H$ are polarization vectors and $\mathbf{K}_0$ and $\mathbf{K}_H$ are the X-ray propagation vectors with

$$|\mathbf{K}_0| = |\mathbf{K}_H| = |\mathbf{K}| = K = \lambda^{-1}, \tag{1.2}$$

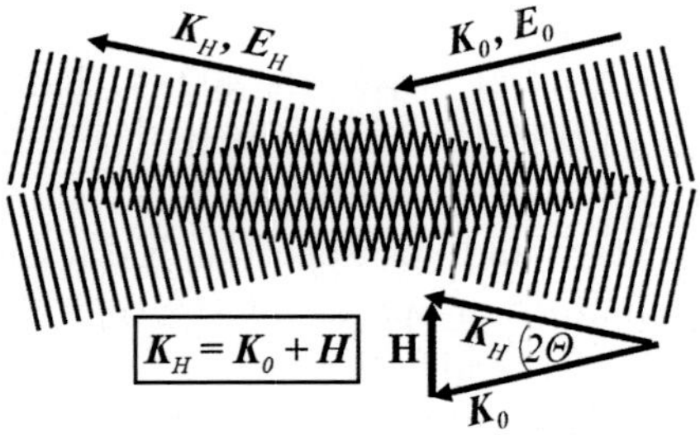

Fig. 1.4. Snapshot at fixed time of two coherent plane waves traveling in different directions leading to the formation of a standing wave (interference field) in the overlap region.

where λ is the X-ray wavelength. Since $E_\gamma = \hbar\omega = h\nu$ and $c = \lambda\nu$, X-ray energy E_γ and λ are related via $\lambda E_\gamma = hc = 1.239842\,\text{keV}\cdot\text{nm}$. The two waves are coherent and thus their frequencies are identical, i.e.

$$\nu_0 = \nu_H = \nu. \tag{1.3}$$

Thus one can relate the two E-field vectors by an amplitude and a phase factor via

$$E_H = \sqrt{R}E_0 \exp{(iv)} \tag{1.4}$$

with

$$R = \frac{I_H}{I_0} = \frac{|E_H|^2}{|E_0|^2}. \tag{1.5}$$

Furthermore, the propagation direction of the two plane waves is related by

$$\mathbf{K}_H = \mathbf{K}_0 + \mathbf{H} \tag{1.6}$$

as shown in Fig. 1.4.

From Fig. 1.4 it is easy to see that $\sin(\theta) = (H/2)/K$ with $H = |\mathbf{H}|$. The XIF (i.e., XSW) in the overlap region of the two plane waves is expressed as

$$|\mathbf{E}| = |\mathbf{E}_0 + \mathbf{E}_H| = (\mathbf{e}_0 \cdot \mathbf{e}_H)[E_0 e^{2\pi i(\nu t - \mathbf{K}_0\mathbf{r})} + \sqrt{R}E_0 e^{iv} e^{2\pi i(\nu t - \mathbf{K}_H\mathbf{r})}], \tag{1.7}$$

which can be written as

$$|\mathbf{E}| = (\mathbf{e}_0 \cdot \mathbf{e}_H)E_0 e^{2\pi i(\nu t - \mathbf{K}_0\mathbf{r})}(1 + \sqrt{R}e^{i(v - 2\pi\mathbf{H}\mathbf{r})}), \tag{1.8}$$

and using

$$\mathbf{E} \cdot \mathbf{E}^*/[(\mathbf{e}_0 \cdot \mathbf{e}_H)^2 E_0^2] = I \tag{1.9}$$

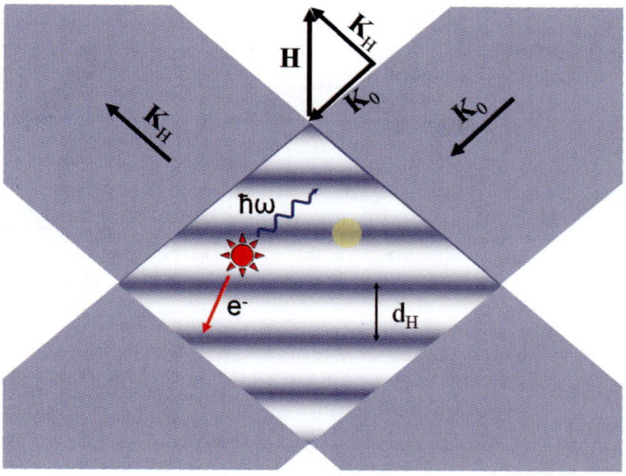

Fig. 1.5. Averaged over time, the intensity of the two traveling, coherent, plane X-ray waves $\mathbf{E}_0$ and $\mathbf{E}_H$ is uniform along $\mathbf{K}_0$ and $\mathbf{K}_H$, respectively. In the overlap region a standing wave is created, i.e., the intensity is spatially modulated in the direction of $\mathbf{H}$ with the spacing d_H. For an atom within the range of the interference field, high intensity will lead to pronounced scattering and photo-absorption and consequently the atom will radiate (electrons or fluorescence) strongly. An atom located at the minimum of the wave field intensity will radiate minimal. The wave field position can be varied by manipulating the phase of one of the traveling waves. Changing the relative phase υ by π rad will exchange the position of maxima and minima of the wave field.

it can be transformed to

$$I = |1 + \sqrt{R}e^{i(\upsilon - 2\pi\mathbf{Hr})}| \tag{1.10}$$

and finally, using the cosine theorem of vector algebra, into the more convenient form:

$$I = 1 + R + 2\sqrt{R}\cos(\upsilon - 2\pi\mathbf{Hr}). \tag{1.11}$$

The scalar product $\mathbf{Hr}$ describes the spatial modulation of the wave field intensity. In the direction normal to $\mathbf{H}$ the wave field intensity is constant. In the direction along $\mathbf{H}$, which we denote as z, the wave field intensity is periodically modulated with the spacing $d_H = H^{-1}$ with $H = |\mathbf{H}|$, as shown in Fig. 1.5.

1.4. How to Create a Suitable XSW

Creating a stationary XIF requires an X-ray source of sufficiently large coherence length, i.e., sufficiently small $\Delta\lambda/\lambda$. Furthermore, we need a beam splitter and subsequently some proper guidance of the two beams to create

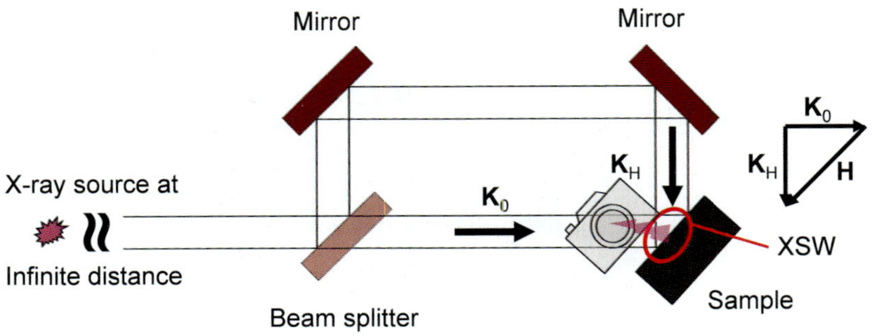

Fig. 1.6. Principle experimental set-up of an XSW experiment. A plane X-ray wave is divided by a beam splitter. With the help of mirrors both beams arrive on the sample, creating a standing wave field. Inelastic radiation from the sample is detected. The wave field planes are oriented normal to **H**. The phase can be adjusted by, e.g., adjusting the traveling distance of $\mathbf{K}_H$ to the sample by moving one of the mirrors. The scattered radiation from within the overlap region of the two waves is recorded by a proper detector.

a region of overlap. We will restrict our considerations to a standing wave created by two plane waves. The sample to be studied is exposed to XSW. In principle, an XSW experiment can be performed with a set-up for a typical interferometer experiment, as shown in Fig. 1.6.

The experimental set-up shown in Fig. 1.6 is impractical in reality. It would require an exceptionally long coherence length of the X-rays of at least the path-length difference of the two beams. The mirrors would need to be adjusted with a pm resolution and the relative positions of beam splitter, mirrors, and sample would need to be stable to the same degree. Nevertheless, an XSW experiment had been realized, which had used an interferometric set-up, which was based on this principle, and is shown in Fig. 1.7.[19] However, in this case, the beam splitter, the mirrors and the sample are all cut from one piece of silicon single crystal and therefore rigidly connected. The phase v of one of the two interfering waves is changed by placing a medium (a phase shifter) of certain refractive index and thickness into the beam path, using the fact that the wavelength of the X-ray in the material differs from the wavelength in air. The phase velocity of X-rays $\nu_p = c/n$. Here, n is the refractive index with $n = 1 - \delta + i\beta$. Neglecting β, which describes absorption, the path difference Δs produced by the phase shifter is given by $\Delta s = \delta \cdot D$, where D is the thickness of the shifter the wave has to penetrate. Assuming that for the material $\delta = 10^{-6}$ and an X-ray with $\lambda = 0.1\,\text{nm}$, $D = 50\,\mu\text{m}$ will shift the wave by $\lambda/2$, i.e., the phase will be shifted by π. Such a Bonse–Hart (BH) interferometer with a similar phase shifter arrangement had been used to determine the

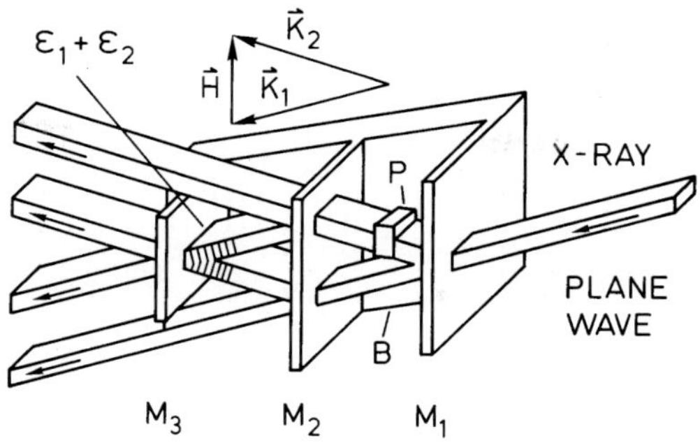

Fig. 1.7. Bonse–Hart interferometer employing Laue reflections to create an XIF. The interferometer is cut from a single piece of a perfect crystal, e.g. silicon or germanium. The first "mirror" M_1 splits the beam by Laue reflection. The second mirror M_2 splits each of the two beams again and two beams arrive at the third mirror M_3, creating in their overlap region an XSW with its planes normal to the surface. A low absorption material with a refractive index $n < 1$ (phase shifter P) is placed in one of the beams, creating a phase shift, since the phase velocity of the X-rays in the medium is $> c$. For example, by rotating the phase shifter P, the path-length of the X-ray in the medium and thus the phase of the X-ray wave can be varied.

in-plane lattice position of Br on a Si(111) surface.[20] The authors used a phase shifter made of Lucite and detected the Br K_α fluorescence excited by the interference field as a function of the phase shift. The proof of principle had been demonstrated earlier by Funke using a BH interferometer produced from a Ge single crystal.[21] Even though in the BH interferometer mirrors and sample are rigidly connected, such an XSW experiment is quite difficult, since minor imperfection and strain in the crystal as well as minor temperature gradients can destroy the phase relationship.

As early as 1890, Otto Wiener, by recording a light/dark pattern on a photographic plate, had proven that an interference field is created by visible light reflected from a silver mirror surface.[22] An XIF can be created in the same way as shown schematically in Fig. 1.8 (see Chapter 5). For X-rays, the index of refraction is smaller than unity, and thus the beam entering the medium is refracted toward the surface. Eventually, at sufficiently small glancing angle, the so-called critical angle θ_c, the beam cannot enter the medium any longer. Since the refractive index of X-rays is close to unity, the critical angle θ_c for the incident X-rays is small. An XIF is formed for the whole range of total reflection, i.e., for glancing angles $\theta < \theta_c$. The critical angle can be expressed as $\theta_c = 2\sqrt{\delta}$. Since θ_c is small, typically a few mrad, i.e., a few 10^{-3}, the magnitude H of $\mathbf{H}$ is small as well. Consequently, the wave field spacing $d_H = H^{-1}$ is comparably large.

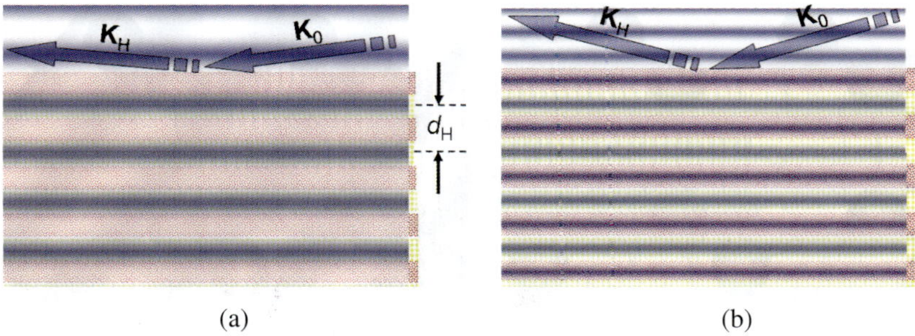

Fig. 1.11. Schematic representation of the formation of an XSW in a multilayer; (a) first-order reflection, $m = 1$; (b) second-order reflection, $m = 2$.

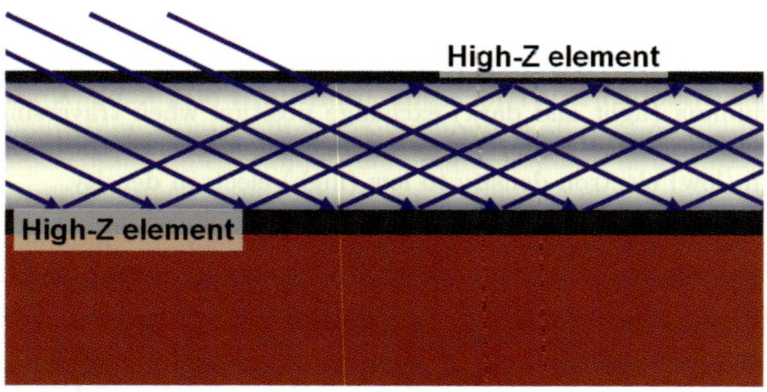

Fig. 1.12. XSW in a waveguide structure. A low absorption material of a few ten nanometers thickness is sandwiched between two layers of high absorption material, all deposited on a flat substrate. The standing wave and its intensity distribution in the interior of the waveguide is indicated.

than 2–3 nm and the wave field periodicity is accordingly significantly larger than in the case of Bragg reflection from single crystals.

Finally, there is a variant of the total reflection method to produce an XIF. Instead of producing a standing wave by a single reflection, the wave field is resonantly coupled into a waveguide structure, i.e., the X-ray wave is reflected back and forth from two confining surfaces/layers made of high-Z material and thus guided in the interior of a low-Z material, as shown schematically in Fig. 1.12 (see Chapter 9). For introducing the X-ray wave into the confined area, different "coupling" schemes exist. Resonant coupling is achieved by adjusting energy or angle of incidence in such a way that the wave field exhibits nodes at the confining surfaces. Different "modes" m can be excited by varying angle/energy such that $m = 1, 2, 3, \ldots$ maxima of the wave field occur in the low-Z material of the guiding layer. Since a wave of macroscopic dimension is compressed into a microscopically small channel, the interference enhancement of the wave field intensity in the antinodes can be much larger than four, which is otherwise the maximum enhancement for the overlap of two equally strong waves.

Very recently, Snigirev *et al.*[24] demonstrated a novel way of creating an XSW by employing the large degree of coherence provided by a third-generation storage ring X-ray source. The interference field is created by the overlap of two X-ray waves emitted from two closely spaced compound refractive lenses (a bi-lens) placed into the monochromatic X-ray beam from an undulator. The spacing d_S of the interference pattern is, in this case, a function of the distance z_1 from the bi-lens to the observation point, i.e. $d_S = \lambda z_1/d_L$, with d_L being the separation of two lenses, which needs to be smaller than the transverse coherence length $l_{\mathrm{coh}-T} = \lambda z_0/S$. Here z_0 is the distance to the X-ray source and S is the size of the source, being typically several 10 m and several microns, respectively.

1.5. X-Ray Scattering

In principle, we distinguish between elastic scattering and inelastic scattering. The elastic scattering is typically employed to generate the XSW field. For the elastic scattering events, the energy of the photon is conserved, while the phase of the E-field may change and the **k**-vector of the individual photon may change direction. The process of interest here is charge scattering, which is for a free charge carrier classically described by the Rayleigh/Thomson cross-section $\sigma_T = 8\pi r_0^2/3$ with the classical electron radius $r_0 = e_0^2/(4\pi\varepsilon_0 mc^2)$, where e_0 is the elementary charge and m is the mass of the charge carrier. Since the nucleus is more than three orders of magnitude heavier than an electron, the elastic scattering of the electron cloud is more than six orders of magnitude stronger than the nuclear scattering and thus it is the only way to create a strong interference field. For the XSW technique, the elastic scattering is only employed to create the interference field.

The inelastic scattering represents the signal of interest to be detected during an XSW experiment. In the energy region ranging from the soft X-ray (several hundred eV) to the hard X-ray (a few ten keV) range, which is most useful for the XSW technique, some different inelastic scattering processes of an X-ray photon can occur:

(1) Nuclear scattering, especially Mössbauer absorption.
(2) Raman scattering.

(3) Compton scattering.
(4) Photo-absorption.

Photon-absorption by the nucleus requires that the excitation energy matches exactly the transition energy with small deviation determined by the lifetime broadening of the excited nuclear state. Since nuclear energy levels are very sharp, X-ray/γ-ray absorption happens only within a very small band of excitation energy of typically μeV to neV width range. Thus, recoil of a free atom upon the absorption of an incoming photon will typically prevent the appropriate matching of excitation and transition energy. However, appreciable nuclear absorption can be observed when the atom is contained in a solid which takes the photon momentum such that the nuclear transition is happening recoil free (resonant absorption, i.e. Mössbauer effect).

Raman scattering is a "photon in, photon out" process and can generally be characterized by $E'_\gamma = E_\gamma \pm \Delta E$. Here, E'_γ is the energy of the photon after the scattering process and the energy ΔE is transmitted to (or received from) the scattering medium which is driven from the state $|i\rangle$ to the state $|f\rangle$. The states involved in the scattering process can be associated with phonons ($\Delta E \approx 10\,\mathrm{meV}$) or electrons or, can be a collective excitation such as plasmons ($\Delta E \approx 10\,\mathrm{eV}$) or specific electronic transitions. The cross-section for Raman scattering is rather small, but with the help of brilliant synchrotron radiation, the process is increasingly exploited in condensed matter research, in particular, since the cross-sections are considerably enhanced at specific energies (at resonance). The technique is commonly called inelastic X-ray scattering (IXS) or resonant IXS (RIXS). It will be an exciting perspective for the future to increasingly utilize IXS and RIXS, i.e. Raman scattering, in combination with an XSW.

The Compton effect is the scattering of a photon by a free electron.[25] The photon transmits kinetic energy and momentum to the electron. In the classical approximation, this process can be viewed as hard sphere collisions of photons with electrons (billiard of photon with electrons). However, when $E_\gamma > E_B$, Compton scattering can happen with a bound electron as well. Here, E_B denotes the binding energy of the electron. The angular profile of the Compton scattered photon is strongly anisotropic in intensity and energy. The final state of the Compton process is a photon with $E'_\gamma < E_\gamma$ and a free electron with a certain amount of kinetic energy.

The most important process for the XSW technique is photo-absorption, which we consider also as a scattering process. The photon is absorbed and the energy is transmitted to an electron, which is either excited to an unoccupied bound state or emitted as a free electron with kinetic energy $E_{\text{kin}} = E_\gamma - E_B - \Phi_E$. The latter process corresponds to the classical photoelectric effect,[26] where Φ_E is the work function of the photo-excited medium. For energy and momentum conservation, absorption of a photon by a force-free electron is not possible and in case of photo-absorption, the Coulomb field of the heavy nucleus is essential.

In case of the first three processes discussed above, it is clear which signals will be detected: the outgoing photon in case of nuclear scattering and inelastic scattering and the photon and/or electron in case of Compton scattering, since these signals are characteristic of the scattering event.

However, in comparison with photo-absorption, all three scattering types have found very limited application for the XSW technique and further down in this book besides the photo-effect, only the Compton effect will be considered (Chapter 10).

1.6. Photo-Excitation and Dipole Approximation

The inelastic scattering channel which is by far most frequently employed for the XSW technique is the photo-effect. It is the key to the excellent structural resolution of the technique. A photon with kinetic energy $E_\gamma = \hbar\omega$ is absorbed by an atom, exciting an electron from the bound state $|i\rangle$ to the final state $|f\rangle$ with the wave functions $\psi_i = \psi_i(\mathbf{r}_e)$ and $\psi_f = \psi_f(\mathbf{r}_e)$, respectively. The final state is usually an ionized atom and a free electron, i.e. an electron wave is emitted from the atom. The transition probability $|i\rangle \rightarrow |f\rangle$ for the photo-effect is proportional to the transition matrix element $\mathbf{M}_{fi}$, which in first-order perturbation theory can be written as

$$\mathbf{M}_{fi} \sim \int d^3\mathbf{r}_e \psi_f^* \mathbf{A} \cdot \mathbf{p} \psi_i. \tag{1.13}$$

In our case, $\mathbf{A}$ is the vector potential of the interference field, i.e.

$$\mathbf{A} = \mathbf{A}_0 e^{-i2\pi\mathbf{K}_0\mathbf{R}} (\mathbf{e}_0 + \mathbf{e}_H \sqrt{R} e^{iv} e^{-i2\pi\mathbf{H}\mathbf{R}}) \tag{1.14}$$

where $\mathbf{e}_0$ and $\mathbf{e}_H$ are the polarization vectors of the two plane waves. We can decompose the vector $\mathbf{R}$ into the position vectors $\mathbf{r}_e$ and $\mathbf{r}$, which are related via $\mathbf{R} = \mathbf{r} + \mathbf{r}_e$, where $\mathbf{r}$ is the position vector of the center of the atom and $\mathbf{r}_e$ is the position of the excited electron relative to the center of the atom. Thus, one can write the vector potential as

$$\mathbf{A} = \mathbf{A}_0 e^{-i2\pi\mathbf{K}_0\mathbf{r}} e^{-i2\pi\mathbf{K}_0\mathbf{r}_e} (\mathbf{e}_0 + \mathbf{e}_H \sqrt{R} e^{iv} \cdot e^{-i2\pi\mathbf{H}\mathbf{r}} e^{-i2\pi\mathbf{H}\mathbf{r}_e}). \tag{1.15}$$

The exponential functions of $\mathbf{r}_e$ can be expanded in a Taylor series (multipole expansion), yielding

$$e^{-i2\pi\mathbf{K}_0\mathbf{r}_e} = 1 - i2\pi\mathbf{K}_0\mathbf{r}_e + 2\pi^2(\mathbf{K}_0\mathbf{r}_e)^2 - \cdots \tag{1.16a}$$

$$e^{-i2\pi\mathbf{H}\mathbf{r}_e} = 1 - i2\pi\mathbf{H}\mathbf{r}_e + 2\pi^2(\mathbf{H}\mathbf{r}_e)^2 - \cdots . \tag{1.16b}$$

In the dipole approximation (DA), only the first term of the multipole expansion is considered, which is justified if $\mathbf{K}_0\mathbf{r}_e$, $\mathbf{H}\mathbf{r}_e \ll 1$. Therefore, with $e^{-i2\pi\mathbf{H}\mathbf{r}_e} = e^{-i2\pi\mathbf{K}_0\mathbf{r}_e} = 1$, the vector potential $\mathbf{A}$ can be written as

$$\mathbf{A} = \mathbf{A}_0 e^{-i2\pi\mathbf{K}_0\mathbf{r}}(\mathbf{e}_0 + \mathbf{e}_H\sqrt{R}e^{iv}e^{-i2\pi\mathbf{H}\mathbf{r}}), \tag{1.17}$$

which is no longer a function of $\mathbf{r}_e$, the position of the electron relative to the atom, but only of $\mathbf{r}$, which is the position of the center of the atom. Thus, we can write for the transition matrix element

$$\mathbf{M}_{fi} \sim \mathbf{A}\int d^3\mathbf{r}_e\psi_f^*\mathbf{p}\psi_i. \tag{1.18}$$

The photo-absorption cross-section σ is in this case (approximated by) the dipole cross-section σ_D and the absorption probability is now modulated in space by the interference field $\mathbf{A}$. Consequently, we can write for the photoelectron absorption probability and thus also for the photoelectron yield Y_H

$$\begin{aligned} Y_H &\sim |\mathbf{M}_{fi}|^2 \sim |\mathbf{A}|^2 \sim |1 + \sqrt{R}e^{i(v-2\pi\mathbf{H}\mathbf{r})}|^2 \\ &= 1 + R + 2\sqrt{R}\cos(v - 2\pi\mathbf{H}\mathbf{r}). \end{aligned} \tag{1.19}$$

The position of the atom $\mathbf{r}$, but not the position of the electron, appears in the interference term, and thus the spatial extent of the electron cloud is neglected and the absorption probability depends only on the position of the center of the atom relative to the maxima or minima of the XIF.

Astonishingly, the dipole approximation of the photo-absorption is pretty good even for harder X-rays. The fact that the photon is absorbed virtually at the center of the atom can be understood qualitatively. In the matrix element there also appears the momentum operator $\mathbf{p} = (\hbar/i)\nabla$ and thus a large matrix element requires a strong gradient of the wave function, which is the case close to the positively charged center of the atom. As mentioned above, the photo-effect is not possible with a free electron. The reason behind that is simple. Energy conservation in the absorption process is not sufficient but also the momentum needs to be conserved. This requires participation of the heavy nucleus via its Coulomb potential.

The influence of higher-order multipole terms on the total cross-section (or total yield) remains small. However, they can significantly influence the angular profile of the photoelectron emission. Thus higher-order terms (mostly quadrupole) cannot be ignored for XSW photoelectron spectroscopy experiments when detecting differential yields (cf. Chapter 11 and chapters thereon).

1.7. Photo-Excitation and Decay Channels: Which Signal to Detect

When an X-ray photon is absorbed by an atom, an electron is excited and may be emitted as a "photoelectron." The probability for the photo-excitation process is (in the DA) directly proportional to the strength of the X-ray wave field at the center of the atom. Monitoring the intensity of the photoelectron signal as a function of the movement of the XSW provides information about the structural arrangement of the emitting atoms. After the initial excitation, the excited atom relaxes rapidly, leading to a cascade of subsequent emission processes, as shown in Fig. 1.13. In case of a photo-excited core hole, the first step is the filling of the hole by an electron from an upper shell. The de-excitation energy can be carried away by a photon ($\hbar\omega_F$) or by an electron kicked out from another shell with lower binding energy. This so called Auger electron is emitted (to a good approximation), with the kinetic energy $E_{\text{kin}} = E_{B1} - E_{B2} - E_{B3}$, where E_{B1} is the binding energy of the photo-emitted electron, E_{B2} is the binding energy of the electron, which fills this hole, and E_{B3} is the binding energy of the electron

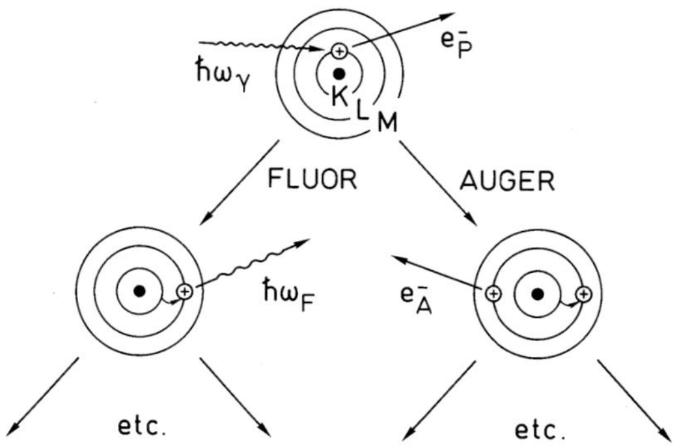

Fig. 1.13. Photo-excitation and radiative decay of an atom.

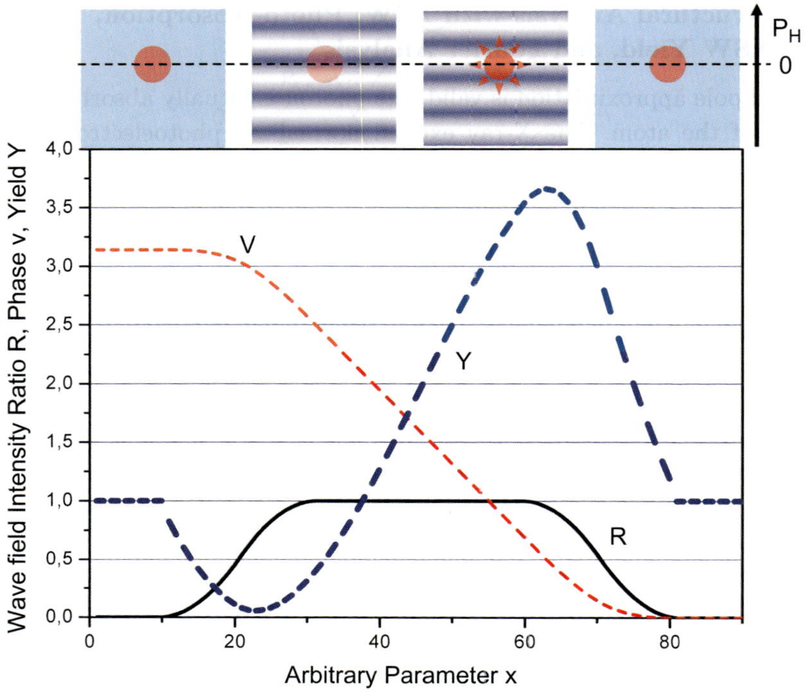

Fig. 1.15. The scattering yield Y from a single atom at $P^H = 0$ according to Eq. (1.20) as a function of an arbitrary parameter x (which could be energy or angle in case of Bragg reflection). On the left side, at $x = 0$, only one wave field is present and the atom is exposed to an isotropic X-ray intensity, as shown schematically on the top. At $x = 10$, a second, coherent wave (traveling in a different direction) is slowly "switched on," the phase υ of which is initially out of phase by π rad at the position of the atom. Thus, they are destructively interfering at the position of the atom, and the corresponding scattering signal Y is minimal. With the phase changing from π to 0, the wave field intensity becomes maximal at the atomic position, leading to a maximum in the scattering yield, before the intensity of the second wave goes to zero again at the right-hand side.

It is clear that the summation over cosine function will yield again a cosine function. However, if the phases, i.e $\mathbf{Hr}_{Aj} = P_{Aj}$ of the individual cosine functions varies, the resulting cosine function will exhibit a reduced amplitude f_A^H and a averaged phase P_A^H, i.e.

$$Y_H = 1 + R + 2\sqrt{R} f_A^H \cos(\upsilon - 2\pi P_A^H) \tag{1.22}$$

with

$$P_A^H = H z_{AH} = \mathbf{Hr}_A \tag{1.23}$$

where $\mathbf{r}_A$ and z_{AH} represent the "average" position and the $\mathbf{H}$-projected "average" position, respectively, of the entity of atoms, which, however, does not correspond to the arithmetic average of the r_{Aj} or z_{Aj}.

Because the number N_A of atoms is in practice very large, Eq. (1.21) can be rewritten by defining a distribution function $G(\mathbf{r})$ by using the properties of the delta function via

$$G(\mathbf{r}) = N_A^{-1} \sum_j^{N_A} \delta(\mathbf{r} - \mathbf{r}_j) \tag{1.24}$$

and exchanging the summation over specific atoms of type A with an integral, we can write

$$Y_H = \int_{\mathbf{r}} d\mathbf{r} G(\mathbf{r}) \{1 + R + 2\sqrt{R} \cos(\upsilon - 2\pi \mathbf{H}\mathbf{r})\}. \tag{1.25}$$

The function $G(\mathbf{r})$ can be interpreted as a quasi-continuous, normalized particle density function of the particular atomic species under consideration. The two parameters f_A^H and P_A^H (coherent fraction and coherent position) in Eq. (1.22) are in fact the Fourier transform of $G(\mathbf{r})$, i.e.

$$G(\mathbf{H}) = G^H = f^H e^{-i2\pi \mathbf{H}\mathbf{r}} = f^H e^{-i2\pi \mathbf{P}^H}. \tag{1.26}$$

This can be easily shown by converting the cosine function into exponentials

$$\cos(\upsilon - 2\pi \mathbf{H}\mathbf{r}) = \cos(\upsilon - 2\pi P^H) = [e^{-i(\upsilon - 2\pi \mathbf{P}^H)} + e^{i(\upsilon - 2\pi \mathbf{P}^H)}]/2. \tag{1.27}$$

Using $\int_{\mathbf{r}} d\mathbf{r} G(\mathbf{r}) = 1$ we convert Eq. (1.25) into

$$\begin{aligned} Y_H &= \int_{\mathbf{r}} d\mathbf{r} G(\mathbf{r}) \{1 + R + \sqrt{R}[\exp(i(\upsilon - 2\pi \mathbf{H}\mathbf{r}) + \exp(-i(\upsilon - 2\pi \mathbf{H}\mathbf{r})]\} \\ &= 1 + R + \sqrt{R} \\ &\quad \cdot \left[e^{i\upsilon} \int d\mathbf{r} G(\mathbf{r}) \exp i(-2\pi \mathbf{H}\mathbf{r}) + e^{-i\upsilon} \int d\mathbf{r} G(\mathbf{r}) \exp i(2\pi \mathbf{H}\mathbf{r}) \right] \\ &= 1 + R + 2\sqrt{R} \cdot \mathrm{Re}[e^{i\upsilon} G(\mathbf{H})], \end{aligned} \tag{1.28}$$

and consequently we obtain Eq. (1.22), i.e., $Y_H = 1 + R + 2\sqrt{R} f^H \times \cos(\upsilon - 2\pi P^H)$.

As mentioned in the very beginning of this introduction, the parameters f^H and P^H, coherent fraction and coherent position, are related to the XRD structure factor F_H, which represents (neglecting dispersion corrections) the Fourier transform of the electron density function $\rho_e(\mathbf{r})$. The amplitude

f_A^H and phase P_A^H are amplitude and phase of the (complex) Fourier coefficient of a particle density function, where the specific element A is selected by spectroscopy. Thus, measuring f_A^H and phase P_A^H for enough **H** values allows to create an image of the analyzed distribution by Fourier back transformation. However, one has to keep in mind that the available **H** values are restricted. In particular, using Bragg reflection long-range structural information is typically not accessible by an XSW measurement. The analogy and the differences of information obtained by XSW and XRD had been pointed out already earlier.[17]

1.9. Simple Structural Analysis in Case of an XSW Excited by Bragg Reflection

Most frequently used is the XSW technique employing Bragg reflections from a single crystal. While a thorough Fourier analysis employing sufficient diffraction vectors/planes is the best (and the only) way to obtain structural information without any *a priori* assumptions or models, a simpler approach is often used to exploit XSW data (e.g., obtained from a surface adsorbate).

The intensity I_H of electron emission or X-ray fluorescence from N_A atoms of a particular element A within the range of an XSW excited by a reflection characterized by the diffraction vector **H** can be written (neglecting multipole contributions and extinction effect) as

$$I_H = I_0 \left\{ 1 + R + 2\sqrt{R} N_A^{-1} \left[\sum_{j=1}^{N_A} \cos(\upsilon - 2\pi \mathbf{H} \mathbf{r}_{Aj}) \right] \right\}. \tag{1.29}$$

The proportionality constant I_0 is called the off-Bragg yield since it characterizes the yield for $R = 0$. It is a linear function of parameters such as atom coverage/density, cross-section, and emission probabilities for the chosen element and emission line, the incident X-ray intensity, as well as detector solid angle and efficiency. As introduced before, $\mathbf{H}\mathbf{r}_{Aj} = P_j^H$ because of the periodic nature of the wave field and the above equation can thus be rewritten as

$$I_H = I_0 \left\{ 1 + R + 2\sqrt{R} \left[\sum_{i=1}^{n} D_i^H \cdot c_i \cos(\upsilon - 2\pi P_i^H) \right] \right\} \tag{1.30}$$

if we assume that the element A under consideration occupies a small number n of discrete coherent positions P_i^H with respect to the wave field. Furthermore, the Debye–Waller factor $D_i^H \leq 1$ takes into account thermal vibrations (or some static disorder considering that N_A is a very large number). In the harmonic approximation, $D_i^H = \exp(-2\pi^2\sigma^2)$ with $\sigma^2 = \langle u^2 \rangle / d_H^2$, where $\langle u^2 \rangle$ is the mean square vibration amplitude of the considered atom and d_H is the spacing of the used diffraction planes. The factor c_i describes the relative population of the element on the position P_i^H with $\sum_{i=1}^n c_i = 1$. If we assume that the Debye–Waller factor is the same for all atoms of the element A independent of the particular coherent position P_i^H we can simplify further and write for the normalized yield Y_H

$$\begin{aligned} Y_H = I_H/I_0 &= 1 + R + 2\sqrt{R} \cdot D_A^H \sum_{i=1}^{n} c_i \cdot \cos(\upsilon - 2\pi P_i^H) \\ &= 1 + R + 2\sqrt{R} \cdot f_A^H \cos(\upsilon - 2\pi P_A^H). \end{aligned} \tag{1.31}$$

Here, the coherent position P_A^H is now the "average" coherent position of all atoms A excited by the XSW and the coherent fraction $f_A^H = D_A^H \cdot F_A^H$ describes their distribution around this mean position. The parameters P_A^H and F_A^H can be expressed via

$$F_A^H = [(G_c^H)^2 + (G_s^H)^2]^{1/2} \tag{1.32a}$$

and

$$P_A^H = (2\pi)^{-1} \tan^{-1}(G_s^H / G_c^H) + x \tag{1.32b}$$

with $x = 0.5$ if $G_c^H < 0$ and $x = 0$ otherwise.

We denote here the "geometric coherent fraction" as F_A^H, which should not be confused with the X-ray structure factor F_H. The two variables G_c^H and G_s^H are simple functions of the n population factors c_i and the n discrete positions P_i^H via

$$G_c^H = \sum_{i=1}^{n} c_i \cdot \cos(2\pi P_i^H) \tag{1.33a}$$

and

$$G_s^H = \sum_{i=1}^{n} c_i \cdot \sin(2\pi P_i^H). \tag{1.33b}$$

These relationships allow a simple modeling of the XSW results in case one has some information or knowledge about the possible sites which may be occupied by the particular element under study. For more details see Ref. 17. Frequently, f^H and P^H, i.e., coherent fraction and coherent position, are considered as expression of the order and mean position of the contemplated species. However, some caution is advised here since a particular coherent fraction f^H (i.e., a particular Fourier component) of the particle distribution being zero does not prove that there is no order. For a well-ordered system, a particular coherent fraction may by chance be zero. A simple example would be that two specific sites $\mathbf{r}_{A1}$ and $\mathbf{r}_{A2}$ are occupied such that for the chosen diffraction vector $\mathbf{Hr}_{A2} = P_{A2} = \mathbf{Hr}_{A1} + 0.5 = P_{A1} + 0.5$. The XSW yield is then given by $Y_H = 1 + R + 2\sqrt{R}f^H \cos(\upsilon - 2\pi P^H) = 1/2\sum_{j=1}^{2}\{1 + R + 2\sqrt{R} \cdot \cos(\upsilon - 2\pi P_{Aj})\}$, and adding two cosine functions which are out of phase by π leads consequently to $f^H = 0$.

1.10. XSW Yield from the Bulk

The equations for the normalized yield discussed in the previous sections are valid only for the secondary radiation originating from a very thin surface layer. For the yield observed during Bragg reflection from species deeper below the surface in the bulk of a crystal, two processes have to be taken into account additionally: (1) attenuation of the E-field intensity as it propagates from the surface to the depth z; and (2) attenuation of the secondary radiation on its way from the depth z to the surface. In general, this means that deeper layers are contributing less to the yield than shallower layers. Any X-ray wave will of course be attenuated by absorption when entering a medium. However, during Bragg reflection from a crystal, the wave can be much stronger attenuated with depth because the wave is reflected and thus the penetration depth is minimal when the reflectivity is maximal, an effect which is coined primary extinction (PE). Furthermore, the strength of absorption depends on the wave field position relative to the lattice atoms and, therefore, depends on angle/energy within the range of Bragg reflection. Thus, the X-ray wave experiences enhanced anomalous absorption (AA) when the maxima of the wave field intensity coincide with the atomic planes.

Because of these reasons, the normalized wave field intensity I from Eq. (1.11) has to be complemented by a factor $e(E_\gamma, \theta, z)$, which is in fact a function of depth in the crystal and of energy E_γ or Bragg angle θ within the range of Bragg reflection. The intensity is thus expressed as

$$I = e(E_\gamma, \theta, z)[1 + R + 2\sqrt{R}\cos(\upsilon - 2\pi\mathbf{Hr})]. \tag{1.34}$$

The pre-factor can be written as

$$e(E_\gamma, \theta, z) = \exp[-z(\mu_\theta + \mu_e)\sin^{-1}\theta], \tag{1.35}$$

where μ_e is the linear coefficient of absorption for the incident X-rays in the material of the crystal (or multilayer), whereas μ_θ accounts for the decay of the wave field inside the crystal or multilayer due to the primary extinction and anomalous absorption, characterized by μ_{PE} and μ_{AA}, respectively. The latter two absorption effects are strongly dependent on angle (or energy) within the range of Bragg diffraction. They can be expressed as

$$\mu_{AA} = -(2\pi/\lambda)\chi''_H \operatorname{Re}(E_H/E_0) \tag{1.36a}$$

and

$$\mu_{PE} = -(2\pi/\lambda)\chi'_H \operatorname{Im}(E_H/E_0), \tag{1.36b}$$

with $\chi_H = \chi'_H + \chi''_H$ being the complex susceptibility of the wave field generator for the chosen X-ray energy E_γ and reflection $\mathbf{H}$. As an example, the ratio of μ_θ/μ_e is shown in Fig. 1.16. For more details about the wave field formation in the context of Bragg diffraction and for calculating intensities and other associated parameters, the reader is referred to the chapter on dynamical diffraction by Authier (Chapter 2).

The yield from the bulk of a material is further influenced by the absorption of the signal on the way to the surface. Originating from a

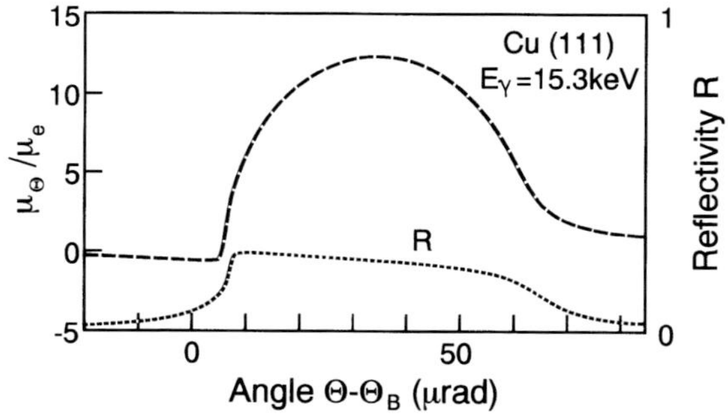

Fig. 1.16. The ratio of μ_θ/μ_e for a Cu(111) Bragg reflection at 15.3 keV as a function of (relative) angle, where θ_B marks the "Bragg angle" as given by Bragg's law. It is interesting to note, that at the left-hand side of the total reflection range $\mu_\theta < 0$. The absorption is anomalously low because $v = \pi$ and the maximum of the wave field intensity is located between the lattice planes. Within the range of strong Bragg reflection, μ_θ is becoming more than 10 times stronger than μ_e and the penetration of the wave field is strongly reduced. Outside of the Bragg reflection region, μ_θ is slowly approaching 0 and the incident X-rays experience "normal" absorption μ_e.

depth z within a crystal, it is attenuated by a factor $e_a(z) = \exp(-\mu_a l)$. Here μ_a is the coefficient of linear absorption for the corresponding signal in the material and l is the length the signal travels through the material, $l = z \sin^{-1} \alpha$, where α is the angle at which the signal is emitted with respect to the surface plane ($l = z$ for $\alpha = 90°$). Thus, the scattering intensity $I_S(\mathbf{r})$ recorded from a distribution of atoms $G(\mathbf{r})$ within the bulk of the crystal/multilayer is expressed by

$$I_S(\mathbf{r}) = I_0 \int_{\mathbf{r}} d\mathbf{r} G(\mathbf{r}) e(E_\gamma, \theta, z) e_a(z)[1 + R + 2\sqrt{R} \cos(v - 2\pi \mathbf{H}\mathbf{r})]. \quad (1.37)$$

The integration has to be carried out over the X-ray exposed range of $\mathbf{r}$. Since $e(E_\gamma, \theta, z)$ and $e_a(z)$ tend to zero for large z, the integration over z can simply be carried out from zero to infinity. In practical applications, the integral has to be evaluated for each specific distribution function $G(\mathbf{r})$.

We will show here the result for two specific cases. In both cases we assume that the considered scatterers A have a specific distribution in the unit cell of the crystal, which is not a function of depth. For the first case we assume that the density of scatterers is constant in the volume of the crystal. The integration can be first carried out over x, y in which case $e(E_\gamma, \theta, z)$ and $e_a(z)$ are constant. Then, the integration is carried out over z in which case $G(\mathbf{r})$ is constant. The integral can in this way be evaluated easily, and the scattered intensity I_S is given by

$$I_S = I_0 d_\theta [1 + R + 2\sqrt{R} f_A^H \cos(v - 2\pi P_A^H)]. \quad (1.38)$$

The parameter d_θ is the effective thickness from which the scatter signal appears to originate and which varies strongly within the range of Bragg reflection. It is given by

$$d_\theta = \frac{1}{(\mu_\theta + \mu_e) \sin^{-1} \theta + \mu_a \sin^{-1} \alpha}. \quad (1.39)$$

The other specific case, which is frequently encountered, is the situation that the species of interest are located at a certain depth z_A below the surface. Only integration over x, y is necessary, yielding the distribution function $G(z_A)$. The intensity I_S is then given by

$$I_S = I_0 e(E_\gamma, \theta, z_A) e_a(z_A)[1 + R + 2\sqrt{R} f_A^H \cos(v - 2\pi P_A^H)]. \quad (1.40)$$

With respect to $e(E_\gamma, \theta, z)$ and $e_a(z)$, one can distinguish two extreme cases: (1) If $e(E_\gamma, \theta, z)$ is much smaller than $e_a(z)$, the scattering yield will be dominated by the extinction effect. This is typically the case when detecting (hard) X-ray fluorescence from deep within the bulk; (2) If $e_a(z)$ is much smaller than $e(E_\gamma, \theta, z)$, the scattering yield is not affected by the extinction effect and the scattering yield is practically described by Eq. (1.22). This is typically the situation when detecting Auger or photoelectrons or soft X-ray fluorescence at shallow angle α.

More complex structures such as epitaxial films and multilayers, buried interfaces, bi-crystals, crystals with nonuniform crystalline layers (e.g. produced by diffusion or ion implantation), and partially disordered or amorphous layers require a more elaborate theory. Computational algorithm based on recurrent equations for the E-field and the secondary radiation yield in multilayer crystals consisting of many layers characterized by their own set of parameters is described by Victor Kohn in Chapter 3.

1.11. Preview

In part one of this book, the material will be presented in greater detail, which allows one to understand and use the XSW technique, i.e. the formation of the XSW field, the scattering from the XSW, and experimental arrangements. We invited contributions to cover all most important aspects. However, not all aspects are treated in depth but only to an extent as we deemed it necessary for the XSW technique, keeping the size of this volume within reasonable limits. An exhaustive treatment of some of the chapters would have otherwise justified book-sized treatments on its own. On the other hand, we did not attempt to completely avoid redundancies in the presentations. In fact, most contributions are largely self-consistent and can be understood to a good degree on their own without referring to other chapters.

This part of the volume starts after the present introduction with a concise treatment of the X-ray wave field formation in perfect crystals within the framework of the dynamical theory of X-ray diffraction. The second chapter is written by André Authier, who recently published a 600-plus-page book on DTXD. Afterward, in Chapter 3, Victor Kohn, one of the leading theorists on different X-ray issues such as scattering and optics, elaborates the X-ray wave field formation in complex and distorted crystals. This treatment is quite important for widening the application of the XSW technique since many crystals cannot be produced and prepared with the perfection necessary for an XSW experiment. However, the requirements for the crystalline perfections can also be relaxed if the XSW is produced by reflection at very high Bragg angles close to 90°. This will be described in Chapter 4 by D. Phil Woodruff, who is one of the first to have used this technical approach, which is, meanwhile, quite extensively applied for studying adsorbates on metal surfaces.

it simplifies the description of the polarization states of the field inside the crystal; χ is the dielectric susceptibility, or polarizability, and is given by:

$$\chi = \chi_n + \chi_a + i\chi_i = \chi_r + i\chi_i,$$

where χ_a and χ_i are the so-called anomalous dispersion corrections and $\chi_n(\mathbf{r})$ is classically given by:

$$\chi_n(\mathbf{r}) = -\frac{\rho(\mathbf{r})R\lambda^2}{\pi},$$

where $\rho(\mathbf{r})$ is electron density and R the classical radius of the electron. The electron density of the crystal is triply periodic and so is the polarizability, which, along with its real and imaginary parts, can therefore be expanded in Fourier series:

$$\chi = \sum_{\mathbf{h}} \chi_h \exp(2\pi i\mathbf{h}\cdot\mathbf{r})$$

$$= \sum_{\mathbf{h}} \chi_{rh} \exp(2\pi i\mathbf{h}\cdot\mathbf{r}) + i\left[\sum_{\mathbf{h}} \chi_{ih} \exp(2\pi i\mathbf{h}\cdot\mathbf{r})\right] \tag{2.2}$$

where $\mathbf{h}$ is a reciprocal-lattice vector, the summation is extended over all reciprocal-lattice vectors, and

$$\chi_h = -R\lambda^2 F_h/(\pi V), \tag{2.3}$$

where V is the volume of the unit cell and F_h is the structure factor related to the electron density through:

$$F_h = \int_{\text{unit cell}} \rho(\mathbf{r}) \exp[-2\pi i\mathbf{h}\cdot\mathbf{r}]d\tau$$

$$= |F_h| \exp(i\varphi_h) = \sum_j (f_j + f_j' + if_j'') \exp[-M_j - 2\pi i\mathbf{h}\cdot\mathbf{r_j}], \tag{2.4}$$

where f_j is the form factor of atom j, f_j' and f_j'' are the anomalous dispersion corrections and $\exp(-M_j)$ is the Debye–Waller factor.

In absorbing crystals, Friedel's law does not hold ($F_h^* \neq F_{\bar{h}}$, $\chi_h^* \neq \chi_{\bar{h}}$) and the normal absorption coefficient is given by:

$$\mu_o = -2\pi k\chi_{io}. \tag{2.5}$$

The quantities $\sqrt{\chi_h\chi_{\bar{h}}}$ and $\sqrt{\chi_h\chi_{\bar{h}}}/\chi_{\bar{h}}$ that come in the expressions of the Pendellösung distance and of the amplitude of the reflected waves, respectively, are complex and can be written as:

$$\sqrt{\chi_h \chi_{\bar{h}}} = \sqrt{|\chi_h \chi_{\bar{h}}|}\exp(i\beta) \tag{2.6}$$

$$\frac{\sqrt{\chi_h \chi_{\bar{h}}}}{\chi_{\bar{h}}} = -\sqrt{|\chi_h/\chi_{\bar{h}}|}\exp(i\beta'), \tag{2.7}$$

where the phase angles β and β' are related by $\beta + \beta' = \varphi_h$ (φ_h phase of the structure factor).

2.2.2. *Fundamental equations of dynamical theory*

The solutions of the propagation equation (2.1) with a triply periodic polarizability have the same periodicity and are of the form, in the two-beam case:

$$\begin{aligned} \mathbf{D} &= \mathbf{D_h}\exp(-2\pi i \mathbf{K_h}\cdot\mathbf{r}) + \mathbf{D_o}\exp(-2\pi i \mathbf{K_o}\cdot\mathbf{r}) \\ &= \exp(-2\pi i \mathbf{K_o}\cdot\mathbf{r})[\mathbf{D_h}\exp(2\pi i \mathbf{h}\cdot\mathbf{r}) + \mathbf{D_o}] \end{aligned} \tag{2.8}$$

where $\mathbf{K_h} = \mathbf{K_o} - \mathbf{h}$ and $\mathbf{h} = \mathbf{OH}$ is a reciprocal-lattice vector. The wave vectors $\mathbf{K_o} = \mathbf{OP}$ and $\mathbf{K_h} = \mathbf{HP}$ are oriented *from* the reciprocal-lattice points *toward* their common extremity, P, called *tiepoint* (Fig. 2.1).

The index of refraction of the medium for X-rays is

$$n = \sqrt{1+\chi} \approx 1 + \frac{\chi_o}{2} = 1 - \frac{R\lambda^2 F_o}{2\pi V}$$

The waves propagating inside the crystal have as wave number $nk = (1 + \chi_o/2)k$. Far from the diffraction condition, one wave only propagates in the crystal, of wave vector $\mathbf{K_o}$ or $\mathbf{K_h}$. Its extremity lies on a sphere of radius nk and centered at the reciprocal-lattice points O and H respectively (Fig. 2.1). When two waves propagate inside the crystal, the common extremity of their wave vectors lies on a connecting surface, called the *dispersion surface.* Its intersection with the scattering plane is represented schematically on the figure. The point where the two spheres cross in the scattering plane is

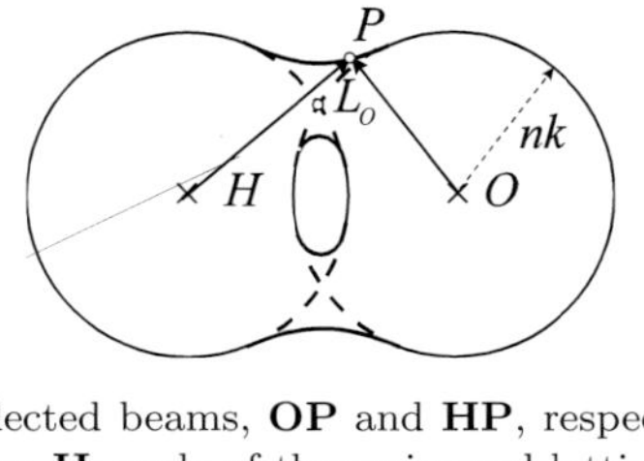

Fig. 2.1. Incident and reflected beams, **OP** and **HP**, respectively, in reciprocal space. **O**: origin of reciprocal space; **H**: node of the reciprocal lattice.

called the *Lorentz point*, L_o. The choice of vector $\mathbf{K_o}$ and the values of the amplitudes $\mathbf{D_o}$ and $\mathbf{D_h}$ are determined by the boundary conditions.

By substituting the expansions (2.2) and (2.8) into the propagation equation (2.1), one obtains the *fundamental equations* of the dynamical theory:

$$\begin{aligned} \mathbf{D_o}(K_o^2 - k^2) &= K_o^2(\chi_o \mathbf{D_{o[o]}} + \chi_{\bar{h}} \mathbf{D_{h[o]}}) \\ \mathbf{D_h}(K_h^2 - k^2) &= K_h^2(\chi_h \mathbf{D_{o[h]}} + \chi_o \mathbf{D_{h[h]}}) \end{aligned} \tag{2.9}$$

where the indices $[o]$ and $[h]$ design the components of the corresponding vectors on the planes normal to $\mathbf{K_o}$ and $\mathbf{K_h}$, respectively. The fundamental equations are solved separately for the two directions of polarization, normal and parallel to the scattering plane:

(1) *σ-polarization.* $\mathbf{D_o}$ and $\mathbf{D_h}$ are perpendicular to the scattering plane and:

$$\mathbf{D_{o[h]}} = \mathbf{D_o}; \quad \mathbf{D_{h[o]}} = \mathbf{D_h}. \tag{2.10}$$

(2) *π-polarization.* $\mathbf{D_o}$ and $\mathbf{D_h}$ are parallel to the scattering plane and:

$$\mathbf{D_{o[h]}} = D_o \cos 2\theta_B \frac{\mathbf{D_h}}{D_h}; \quad \mathbf{D_{h[o]}} = D_h \cos 2\theta_B \frac{\mathbf{D_o}}{D_o}. \tag{2.11}$$

If we introduce the polarization coefficient, $C = 1$ for σ-polarization and $C = \cos 2\theta_B$ for π-polarization (θ_B Bragg angle), the set of fundamental equations (2.9) can be written, after projection on $\mathbf{D_o}$ and $\mathbf{D_h}$, respectively, and to a small approximation:

$$\begin{aligned} [K_o^2 - k^2(1+\chi_o)]D_o - k^2 C\chi_{\bar{h}} D_h &= 0 \\ -k^2 C\chi_h D_o + [K_h^2 - k^2(1+\chi_o)]D_h &= 0. \end{aligned} \tag{2.12}$$

For the fundamental equations to have a nontrivial solutions, the determinant of Eq. (2.12) must be set equal to zero. This is the equation of the *dispersion surface*, which is the locus of the tiepoint P:

$$X_o X_h = k^2 C^2 \chi_h \chi_{\bar{h}}/4 \tag{2.13}$$

where

$$\begin{aligned} X_o &= [K_o^2 - k^2(1+\chi_o)]/2k \\ X_h &= [K_h^2 - k^2(1+\chi_o)]/2k. \end{aligned} \tag{2.14}$$

2.2.3. *Dispersion surface in the infinite medium*

2.2.3.1. *Non-absorbing crystals*

The dispersion surface consists of two sheets, one for each direction of polarization ($C = 1$ or $\cos 2\theta_B$); each sheet is a fourth-degree surface, connecting the spheres of radius $nk = k(1 + \chi_o/2)$ centered at O and H. The tiepoints of the wave fields excited in a particular situation are determined by the boundary conditions. Let $\mathbf{K_o^{(a)}} = \mathbf{OM}$ be the wave vector of an incident plane wave; its extremity, M, lies on a sphere of radius $OM = k = 1/\lambda$. The intersection of that sphere and of the dispersion surface with the scattering plane is represented schematically in Fig. 2.2.

According to the condition of the continuity of the tangential components of wave vectors, the extremities, M of the incident wave vector and P_j of the wave vectors excited inside the crystal, must lie on a common normal (**Mz** in reciprocal space, **Az** in direct space — insets of Figs. 2.2(a) and 2.2(b)) to the crystal surface. The figure shows two examples for a given direction of polarization (σ or π): Fig. 2.2(a) for the Laue, or transmission geometry, and Fig. 2.2(b) for the reflection, or Bragg geometry. The normal **Mz** intersects the dispersion surface at four points, P_1 to P_4. If the angle of incidence is smaller than the critical angle for specular reflection, three of them only usually have a physical significance. The wave vector of the specularly reflected beam is $\mathbf{K_o^{(s)}} = \mathbf{OM_s}$ (Fig. 2.2). The extremity, N, of the wave vector of the reflected beam $\mathbf{K_h^{(a)}} = \mathbf{HN}$ lies on the sphere of center H and radius k. The intersection, in the scattering plane, of the two spheres of radius k and of centers O and H, respectively, is called the *Laue point* (Fig. 2.2). An incident plane wave of wave vector $\mathbf{OL_a}$ satisfies exactly the Bragg condition of reflection, according to the kinematical theory, namely neglecting refraction.

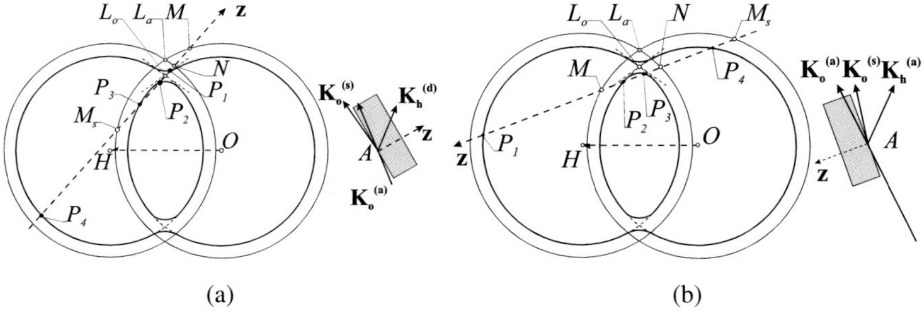

Fig. 2.2. Tiepoints excited inside the crystal. (a) Laue geometry. (b) Bragg geometry.

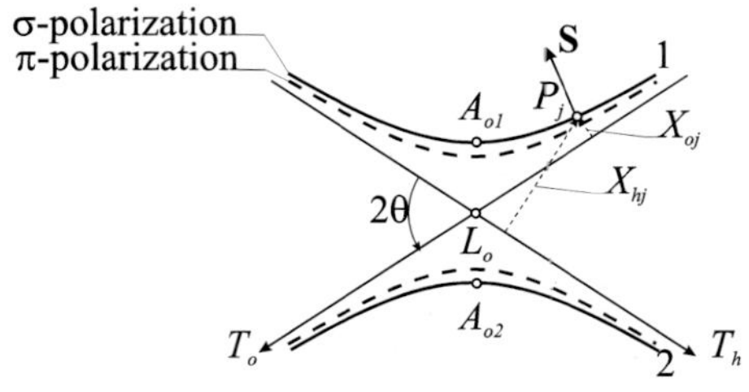

Fig. 2.3. Dispersion surface.

We shall limit ourselves here to the case where the incident or emergence angles are not too small and the Bragg angle is not too close to $\pi/2$.[a] In that case, the two spheres of radius nk can be replaced in the vicinity of the Lorentz point by their tangential planes whose intersections with the scattering plane are denoted T_o and T_h (Fig. 2.3) and there are at most *two* tiepoints excited in the crystal. The quantities X_o and X_h can then be approximated by:

$$\begin{aligned} X_o &\approx K_o - k(1 + \chi_o/2) \\ X_h &\approx K_h - k(1 + \chi_o/2), \end{aligned} \tag{2.15}$$

and can be interpreted as the differences between the wave numbers $K_o = OP$ and $K_h = HP$ on the one hand and the radius of these two spheres and therefore as the distances of the tiepoint P to T_o and T_h. They are called *resonance errors*.

Equation (2.13) is then that of a hyperbola whose asymptotes are T_o and T_h. Its diameter $A_{o2}A_{o1} = k|C|\sqrt{\chi_h \chi_{\bar{h}}}/\cos\theta_B$ is called the *Bragg gap*. The larger the Bragg gap, the stronger the reflection. The dispersion surface corresponding to σ-polarization is represented on the figure by a full line and that corresponding to π-polarization is represented by a dashed line. The propagation direction of a wave field inside the crystal is that of the normal to the dispersion surface, **S**, at the tiepoint (Fig. 2.3).

2.2.3.2. *Absorbing crystals*

When the crystal is absorbing, the dielectric susceptibility and the quantities X_o and X_h are complex, and so is Eq. (2.13) of the dispersion surface. It can be split up into two equations, one for the real part and one

[a]For such situations, see Ref. 25 and the references therein.

for the imaginary part. The former is:

$$X_o^r X_h^r = X_o^i X_h^i + \frac{k^2}{4} C^2 \mathcal{R}e(\chi_h \chi_{\bar{h}}),$$

where X_o^r, X_o^i and X_h^r, X_h^i are the real and imaginary parts of X_o and X_h, respectively.

If the ratio of the imaginary to the real part of χ_h is small enough, the equation of the real part of the dispersion surface may be approximated by:

$$X_o^r X_h^r = \frac{k^2}{4} C^2 \mathcal{R}e(\chi_h \chi_{\bar{h}}).$$

The direction of propagation of the wave fields may then be approximated by the normal to the real part of the dispersion surface. This is, however, no longer true when the imaginary part is large.[26]

2.2.4. *Determination of the tiepoints*

The tiepoints lie at the intersection of the normal to the crystal surface, **Mz**, drawn from the extremity of the incident wave vector, $\mathbf{OM} = \mathbf{K}_\mathbf{o}^{(\mathbf{a})}$, with the dispersion surface (Fig. 2.4). The two spheres of centers O and H and of vector k can also be approximated by their tangential planes whose intersections with the scattering plane are denoted T_o' and T_h' (Fig. 2.4).

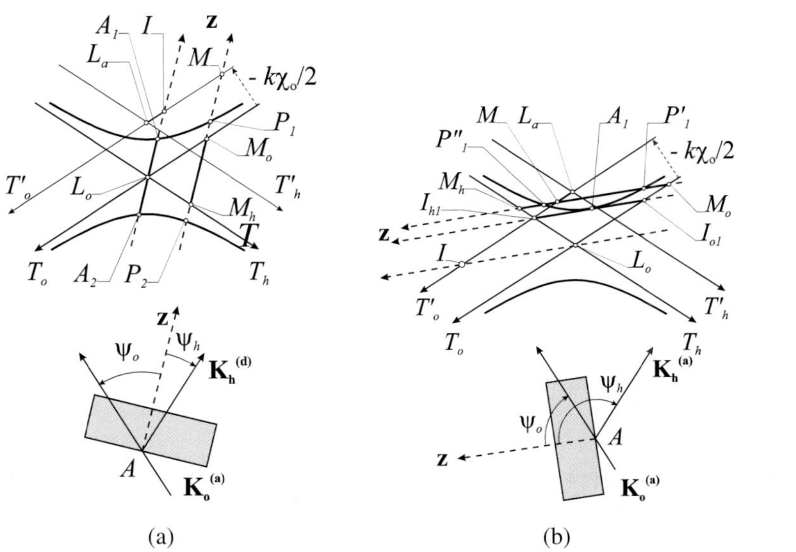

Fig. 2.4. Tiepoints in the two-beam case. (a) Laue geometry. (b) Bragg geometry.

The intersection of T'_o and T'_h is the Laue point, L_a. The distance between T_o and T'_o, T_h and T'_h is $1 - n = -k\chi_o/2$. The orientation of **Mz** is defined by its direction cosines relative to the incident and reflected directions, respectively:

$$\gamma_o = \cos\psi_o; \qquad \gamma_h = \cos\psi_h.$$

Their ratio, $\gamma = \gamma_h/\gamma_o$, is called the *asymmetry ratio.* It can be seen from the figures that γ_h, γ_o and γ are always positive in the Laue geometry (Fig. 2.4(a)), while γ_h and γ are always negative in the Bragg geometry (Fig. 2.4(b)). The equation of **Mz** is:

$$X_{oj}/\gamma_0 - X_{hj}/\gamma_h = \overline{M_oM_h} \tag{2.16}$$

where M_o and M_h are the intersections of **Mz** with T_o and T_h.

Let us set:

$$\Lambda_o^{-1} = \frac{k|C|\sqrt{\chi_h\chi_{\bar{h}}}}{\sqrt{\gamma_o|\gamma_h|}} \tag{2.17}$$

$$\eta = \Lambda_o S(\gamma_h)\overline{M_oM_h} \tag{2.18}$$

where $S(\gamma_h)$ means sign of γ_h, equal to $+1$ in the Laue geometry and -1 in the Bragg geometry. Using these equations and Eqs. (2.13) and (2.16), one obtains for the coordinates of the tiepoints:

$$X_{oj} = \frac{\gamma_o S(\gamma_h)}{2\Lambda_o}[\eta \pm \sqrt{\eta^2 + S(\gamma_h)}] \tag{2.19}$$

$$X_{hj} = \frac{|\gamma_h|}{2\Lambda_o}[-\eta \pm \sqrt{\eta^2 + S(\gamma_h)}] \tag{2.20}$$

where the top sign corresponds to branch 1 ($j = 1$) and the bottom one to branch 2 ($j = 2$).

2.2.5. *Deviation parameter*

Parameter η, defined by Eq. (2.18), is called the *deviation parameter.* It is related to the deviation from Bragg's incidence $\Delta\theta = \overline{LaM}/k$ by noting (Figs. 2.4(a) and 2.4(b)) that:

$$\overline{M_oM_h} = \overline{L_oM_o}\sin 2\theta_B/\gamma_h = \overline{IM}\sin 2\theta_B/\gamma_h = [\overline{L_aM} - \overline{L_aI}]\sin 2\theta_B/\gamma_h$$

where I is at the intersection of T_o and the normal to the crystal surface, $\mathbf{L_oz}$, drawn from the Lorentz point. It can be shown (see Ref. 25, Sec. 4.7.4) that the wave vector of the incident plane wave corresponding to the middle of the reflection domain is **OI**. The departure from the kinematic Bragg incidence of that wave vector is:

$$\Delta\theta_{os} = \overline{L_a I}/k = -\chi_o(1-\gamma)/(2\sin 2\theta_B).$$

It follows that:

$$\overline{M_o M_h} = k(\Delta\theta - \Delta\theta_{os})\sin 2\theta_B/\gamma_h.$$

$$\eta = (\Delta\theta - \Delta\theta_{os})\frac{\Lambda_o}{\lambda}\frac{\sin 2\theta_B}{|\gamma_h.|} = \frac{\Delta\theta\sin 2\theta_B + \chi_o(1-\gamma)/2}{|C|\sqrt{|\gamma|}\sqrt{\chi_h\chi_{\bar{h}}}}. \quad (2.21)$$

It can further be shown that the geometrical interpretation of the quantity Λ_o is:

(1) Laue geometry (Fig. 2.4(a)):

$$\Lambda_o = \left(\overline{A_2 A_1}\right)-1$$

where A_2 and A_1 are the intersections of $\mathbf{L_o z}$ with the two branches of the dispersion surface. It is the *Pendellösung distance.*

(2) Bragg geometry (Fig. 2.4(b)):

$$\Lambda_o = \left(\overline{I_{oj} I_{hj}}\right)-1$$

where I_{oj} and I_{hj} are the intersections with T_o and T_h of the tangent to the dispersion surface parallel to the normal to the crystal surface (at A_1 on the figure). It is the *extinction distance.*

The wave vectors of the reflected and refracted waves excited in the crystal are, respectively:

$$\begin{aligned}\mathbf{K_{oj}} &= \mathbf{OP_j} = \mathbf{OM} + \mathbf{MP_j} = \mathbf{K_o^{(a)}} + \mathbf{MP_j}\\ \mathbf{K_{hj}} &= \mathbf{HP_j} = \mathbf{OP_j} - \mathbf{OH} = \mathbf{OM} + \mathbf{MP_j} - \mathbf{OH} = \mathbf{K_o^{(a)}} + \mathbf{MP_j} - \mathbf{h}\end{aligned} \quad (2.22)$$

with, using Eq. (2.19):

$$\overline{MP_j} = \left(X_{oj} + \frac{k\chi_o}{2}\right)\Big/\gamma_o = \frac{S(\gamma_h)}{2\Lambda_o}[\eta \pm \sqrt{\eta^2 + S(\gamma_h)}] + \frac{k\chi_o}{2\gamma_o}$$

$$= k\frac{\Delta\theta\sin 2\theta_B}{2|\gamma_h|} + k\frac{\chi_o}{4}\left(\frac{1}{\gamma_o} + \frac{1}{\gamma_h}\right) \pm \frac{S(\gamma_h)}{2\Lambda_o}\sqrt{\eta^2 + S(\gamma_h)}. \quad (2.23)$$

In an absorbing crystal, the quantities Λ_o, $\Delta\theta_s$, η, X_{oj}, X_{hj}, and $\overline{MP_j}$ are complex. In the Bragg geometry, $S(\gamma_h) = -1$, and, even in a non-absorbing crystal, X_{oj}, X_{hj}, and $\overline{MP_j}$ are complex for $|\eta| < 1$. In particular, the real and imaginary parts of the deviation parameter ($\eta = \eta_r + i\eta_i$) are given by:

$$\eta_r = \frac{\Delta\theta \sin 2\theta_B \cos\beta + (1-\gamma)(\chi_{ro}\cos\beta + \chi_{io}\sin\beta)/2}{\sqrt{|\gamma|}|C|\sqrt{|\chi_h\chi_{\bar{h}}|}}, \qquad (2.24)$$

$$\eta_i = A\eta_r + B = -\eta_r \tan\beta + \frac{(1-\gamma)\chi_{io}}{2\cos\beta\sqrt{|\gamma|}|C|\sqrt{|\chi_h\chi_{\bar{h}}|}}. \qquad (2.25)$$

2.2.6. *Amplitudes of the diffracted waves*

The ratio of reflected, D_{hj}, and refracted, D_{oj}, waves is deduced from Eqs. (2.12), (2.14) and (2.15):

$$\xi_j = \frac{D_{hj}}{D_{oj}} = \left|\frac{D_{hj}}{D_{oj}}\right| \exp i\psi = \frac{2X_{oj}}{kC\chi_{\bar{h}}}$$
$$= -\frac{S(C)S(\gamma_h)}{\sqrt{|\gamma|}}\sqrt{\left|\frac{\chi_h}{\chi_{\bar{h}}}\right|}\exp(i\beta')[\eta \pm \sqrt{\eta^2 + S(\gamma_h)}], \qquad (2.26)$$

where the phase angle β' was defined in Eq. (2.7).

The values of the amplitudes are determined by the boundary conditions.

2.2.6.1. *Bragg or reflection geometry*

In Bragg geometry (Figs. 2.4(b) and 2.5), the normal to the entrance surface drawn from the extremity, M, of the incident wave vector intersects one branch only of the zero-absorption, infinite medium dispersion surface, at P_1' and P_1'' for branch 1, P_2' and P_2'' for branch 2. Let I_{o1} and I_{o2} be the intersections with T_o' of the tangents to the two branches of the dispersion surface, respectively, drawn parallel to the normal to the entrance surface (Fig. 2.5). If M lies between these two points, the intersections of **Mz** with the dispersion surface are complex and total reflection takes place.

The propagation direction of the wave fields is along the normal to the zero-absorption, infinite medium dispersion surface. For P_1' and P_2', it lies toward the inside of the crystal ($\mathbf{S_1'}$ and $\mathbf{S_2'}$), while for P_1'' and P_2'' it lies from the bottom of the crystal up toward the entrance surface ($\mathbf{S_1''}$ and $\mathbf{S_2''}$). For a thick absorbing crystal the corresponding wave fields do not exist and only *one* wave field propagates in the crystal (left inset of Fig. 2.5); in a

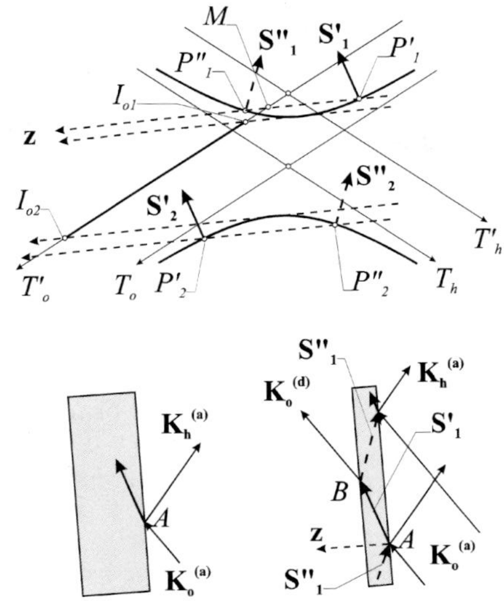

Fig. 2.5. Wave fields in the Bragg geometry.

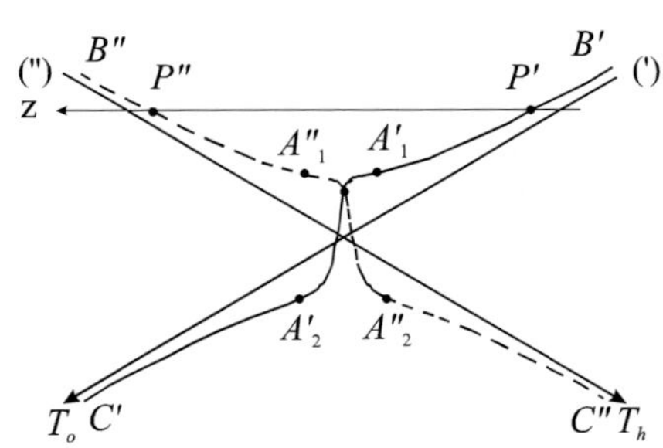

Fig. 2.6. Real part of the dispersion surface in the Bragg geometry for an absorbing crystal.

thin crystal, they are generated by partial reflection at the bottom surface (right inset) and two wave fields must therefore be taken into account.

In fact, when the boundary conditions have been introduced, the dispersion surface to be taken in account is the dispersion surface in the semi-infinite medium.[26] In the Bragg case, it is complex even in a non-absorbing crystal. Its real part is represented in Fig. 2.6 in the case of a GaAs crystal, a symmetric (111) reflection and Mo K_α. It consists of two branches, ($'$) corresponding to wave fields propagating towards the inside of the crystal (solid curve) and ($''$) corresponding to wave fields propagating towards the outside of the crystal (dashed curve). The coordinates of a tiepoint belonging to branch ($'$) are given by:

$$X^r_{oj} = -\mathcal{R}\left(\frac{\gamma_o}{2\Lambda_o}[\eta - S(\eta_r)\sqrt{\eta^2 - 1})]\right),$$

$$X^r_{hj} = \mathcal{R}\left(\frac{|\gamma_h|}{2\Lambda_o}[-\eta - S(\eta_r)\sqrt{\eta^2 - 1}]\right),$$

where $\mathcal{R}(\ )$ means real part of. Branch ($'$) tends asymptotically towards branch 1 of the infinite medium dispersion surface for $\eta_r \rightarrow -\infty$ and towards branch 2 for $\eta_r \rightarrow +\infty$.

The coordinates of a tiepoint belonging to branch ($''$) are given by:

$$X^r_{oj} = -\mathcal{R}\left(\frac{\gamma_o}{2\Lambda_o}[\eta + S(\eta_r)\sqrt{\eta^2 - 1})]\right),$$

$$X^r_{hj} = \mathcal{R}\left(\frac{|\gamma_h|}{2\Lambda_o}[-\eta + S(\eta_r)\sqrt{\eta^2 - 1}]\right).$$

Branch ($''$) tends asymptotically towards branch 2 of the infinite medium dispersion surface for $\eta_r \rightarrow -\infty$ and towards branch 2 for $\eta_r \rightarrow +\infty$. There is no jump from one branch to another.

When one considers the real part of the dispersion surface, the two tiepoints excited by the incident wave, P' and P'' lie on *different* branches. The propagation direction of the wave fields is along the normal to the real part of the dispersion surface when the imaginary part is small enough, namely along segments A'_1B' and A'_2C' of branch ($'$) and along segments $B''_1A''_1$ and A''_2C'' of branch ($''$).

(1) *Thick crystals.* The boundary conditions for the amplitudes are very simple here:

$$D_o = D_o^{(a)}; \quad D_h = D_h^{(a)}, \tag{2.27}$$

where $D_o^{(a)}$ and D_h^a are the amplitudes of the incident and reflected waves, respectively. The expression of the constituting waves of the wave fields are, therefore, using Eqs. (2.23) and (2.24), and noting that $\mathbf{MP_j}$ is along the normal to the crystal surface:

$$\mathcal{D}_o = D_o^{(a)} \exp[-2\pi i(\mathbf{K_o^{(a)}} \cdot \mathbf{r_s} + \overline{MP_j}z)], \tag{2.28}$$

$$\begin{aligned}\mathcal{D}_h = {} & S(C)(\sqrt{|(\chi_h/\chi_{\bar{h}})|}/\sqrt{|\gamma|}) \\ & \times [\eta \pm \sqrt{\eta^2 - 1}] \exp(i\beta') \exp[2\pi i\mathbf{h} \cdot \mathbf{r}]\mathcal{D}_o,\end{aligned} \tag{2.29}$$

where z is the depth inside the crystal. The sign to be chosen is that for which the amplitude $\mathcal{D}_h$ converges. It depends on which side of the rocking curve one is and the proper solution is:

$$\mathcal{D}_h = S(C)(\sqrt{|(\chi_h/\chi_{\bar{h}})|}/\sqrt{|\gamma|})[\eta - S(\eta_r)\sqrt{\eta^2-1}] \times \exp(i\beta')\exp[2\pi i\mathbf{h}\cdot\mathbf{r}]\mathcal{D}_o. \tag{2.30}$$

Setting $\eta - S(\eta_r)\sqrt{\eta^2-1} = Z\exp(i\psi')$, there comes:

$$\mathcal{D}_h = S(C)(\sqrt{|(\chi_h/\chi_{\bar{h}})|}/\sqrt{|\gamma|})Z\exp[(i(\beta'+\psi')] \times \exp[2\pi i\mathbf{h}\cdot\mathbf{r}]\,\mathcal{D}_o. \tag{2.31}$$

When the departure from Bragg's law $\Delta\theta$ and η_r tend towards $\pm\infty$,

$$\eta - S(\eta_r)\sqrt{\eta^2-1} = Z\exp(i\psi') \to \frac{S(\eta_r)}{2|\eta_r|\cos\beta}\exp(i\beta),$$

which shows that the limits of the phase, $\psi = \psi' + \beta'$, of the ratio $\mathcal{D}_h/\mathcal{D}_h$ when η_r tends towards $-\infty$ and $+\infty$ are, respectively:

$$\begin{aligned}\psi_{\eta_r\to-\infty} &= \pi + \beta + \beta' = \pi + \varphi_h,\\ \psi_{\eta_r\to+\infty} &= \beta + \beta' = \varphi_h,\end{aligned} \tag{2.32}$$

where φ_h is the phase of the structure factor.[27,28] The intensity of the two waves is:

$$\begin{aligned}|\mathcal{D}_o|^2 &= \exp[4\pi\mathcal{I}(\overline{MP_j}z)]|D_o^{(a)}|^2,\\ |\mathcal{D}_h|^2 &= (|(\chi_h/\chi_{\bar{h}})|/|\gamma|)Z^2\exp[4\pi\mathcal{I}(\overline{MP_j}z)]|D_o^{(a)}|^2,\end{aligned}$$

where $\mathcal{I}(Y)$ means imaginary part of Y. There comes, using (2.5) and (2.23):

$$\exp[4\pi\mathcal{I}(\overline{MP_j}z)] = \exp\left[-\frac{\mu_o}{\gamma_o} - \frac{2\pi k|C|\sqrt{|(\chi_h\chi_{\bar{h}})|}}{\sqrt{\gamma_o|\gamma_h|}}Z\sin\psi\right]z.$$

In the absence of absorption, this expression reduces to 1 for $|\eta| > 1$ and to $|\exp-[2\pi\sqrt{1-\eta^2}/2\Lambda_o]z$ for $|\eta| \leq 1$. This absorption of the waves as they progress towards the inside of the crystal within the domain of total reflection is called *extinction*.

Figure 2.7(a) represents the variations of the phase ψ of the ratio $\xi = D_h/D_o$ (full curve) and those of the reflectivity $I_h = \gamma|\mathcal{D}_h|^2$

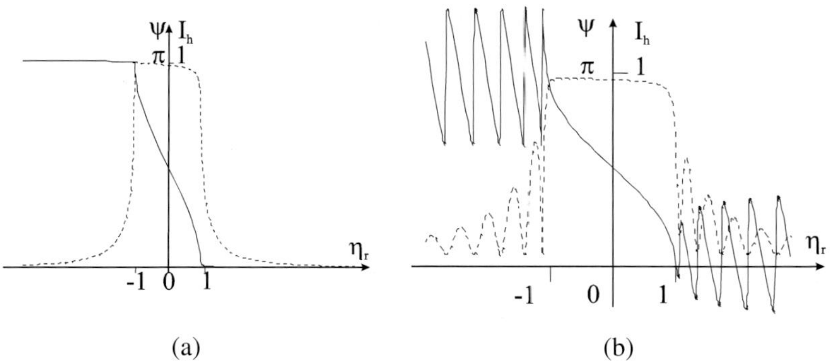

Fig. 2.7. Variations of the phase (full line) and the reflectivity (dashed line) of the reflected wave Bragg geometry. Si, 220, Mo K$_\alpha$ (a) thick crystal, (b) thin crystal $t = 2$; $\Lambda_o = 13.65\,\mu$m.

(dashed curve) for the 220 reflection of a thick silicon crystal with Mo K$_\alpha$ radiation.

(2) *Thin crystals.* One has here to take two wave fields into account, of tiepoints P_1' and P_1'', or P_2' and P_2'', respectively, (Fig. 2.5) and to write the boundary conditions both at the entrance and the back surface:

(a) *Entrance surface*:

$$\begin{aligned} D_{o'} + D_{o''} &= D_o^{(a)}, \\ D_{h'} + D_{h''} &= D_h^{(a)}. \end{aligned} \tag{2.33}$$

(b) *Back surface.* The wave field of tiepoint P_1' propagating inside the crystal generates at the back surface a back-reflected wave field of tiepoint P_1'' and to a partially transmitted wave, of amplitude $D_o^{(d)}$ and wave vector $\mathbf{K_o^{(d)}}$:

$$\begin{aligned} &D_{o'} \exp(-2\pi i \mathbf{K_{o'}} \cdot \mathbf{r}) + D_{o''} \exp(-2\pi i \mathbf{K_{o''}} \cdot \mathbf{r}) \\ &\quad = D_o^{(d)} \exp(-2\pi i \ \mathbf{K_o^{(d)}} \cdot \mathbf{r}), \\ &D_{h'} \exp(-2\pi i \mathbf{K_{h'}} \cdot \mathbf{r}) + D_{h''} \exp(-2\pi i \mathbf{K_{h''}} \cdot \mathbf{r}) = 0. \end{aligned}$$

The wave vectors are related by the condition of the continuity of their tangential components:

$$\begin{aligned} \mathbf{K_{o'}} &= \mathbf{K_o^{(a)}} + \mathbf{MP_j'}; \quad \mathbf{K_{o''}} = \mathbf{K_o^{(a)}} + \mathbf{MP_j''}, \\ \mathbf{K_{h'}} &= \mathbf{K_h^{(a)}} + \mathbf{MP_j'}; \quad \mathbf{K_{h''}} = \mathbf{K_h^{(a)}} + \mathbf{MP_j''}. \end{aligned}$$

The relations between the amplitudes at the back surface are, therefore:

$$\begin{aligned}D_{o'}\exp(-2\pi i\overline{MP'}t) + D_{o''}\exp(-2\pi i\overline{MP''}t) &= D_o^{(d)},\\ D_{h'}\exp(-2\pi i\overline{MP'}t) + D_{h''}\exp(-2\pi i\overline{MP''}t) &= 0,\end{aligned} \tag{2.34}$$

where t is the crystal thickness.

Using (2.23), (2.26), (2.33) and (2.34), and

$$D_{h'}/D_{o'} = \xi' = \frac{S(C)}{\sqrt{|\gamma|}}\sqrt{\left|\frac{\chi_h}{\chi_{\bar{h}}}\right|}\exp(i\beta')[\eta - \sqrt{\eta^2-1}],$$

$$D_{h''}/D_{o''} = \xi'' = \frac{S(C)}{\sqrt{|\gamma|}}\sqrt{\left|\frac{\chi_h}{\chi_{\bar{h}}}\right|}\exp(i\beta')[\eta + \sqrt{\eta^2-1}],$$

one finds for the amplitudes of the four waves:

$$D_{o'} = \frac{(\eta - \sqrt{\eta^2-1})\mathcal{E}'' D_o^{(a)}}{\eta(\mathcal{E}' - \mathcal{E}'') + \sqrt{\eta^2-1}(\mathcal{E}' + \mathcal{E}'')}, \tag{2.35}$$

$$D_{o''} = \frac{(\eta + \sqrt{\eta^2-1})\mathcal{E}' D_o^{(a)}}{\eta(\mathcal{E}' - \mathcal{E}'') + \sqrt{\eta^2-1}(\mathcal{E}' + \mathcal{E}'')}, \tag{2.36}$$

$$D_{h'} = -\frac{S(C)}{\sqrt{|\gamma|}}\sqrt{\left|\frac{\chi_h}{\chi_{\bar{h}}}\right|}\frac{\exp(i\beta')\mathcal{E}'' D_o^{(a)}}{\eta(\mathcal{E}_1 - \mathcal{E}'') + \sqrt{\eta^2-1}(\mathcal{E}' + \mathcal{E}'')}, \tag{2.37}$$

$$D_{h''} = \frac{S(C)}{\sqrt{|\gamma|}}\sqrt{\left|\frac{\chi_h}{\chi_{\bar{h}}}\right|}\frac{\exp(i\beta')\mathcal{E}' D_o^{(a)}}{\eta(\mathcal{E}' - \mathcal{E}'') + \sqrt{\eta^2-1}(\mathcal{E}' + \mathcal{E}'')}, \tag{2.38}$$

where $\mathcal{E}' = \exp(\pi\ i\sqrt{\eta^2-1}t/\Lambda_o)$, $\mathcal{E}'' = \exp-(\pi\ i\sqrt{\eta^2-1}t/\Lambda_o)$, and Λ_o is given by (2.17). The oscillations of the amplitudes are due to the interference between the two wave fields and their period is $\Lambda_o/\sqrt{\eta^2-1}$. It is to be noted that the *Pendellösung* is here between wave fields whose tiepoints belong to the same branch of the dispersion surface. Figure 2.7(b) represents the variations of the reflectivity $I_h = |\gamma|\,|D_{h'} + D_{h''}|^2$ and of the phase of the total reflected amplitude $D_h^{(a)} = D_{h'} + D_{h''}$ for a thin silicon crystal, 220 reflection, Mo K_α reflection and a crystal thickness $t = 2\Lambda_o = 13.65\,\mu$m

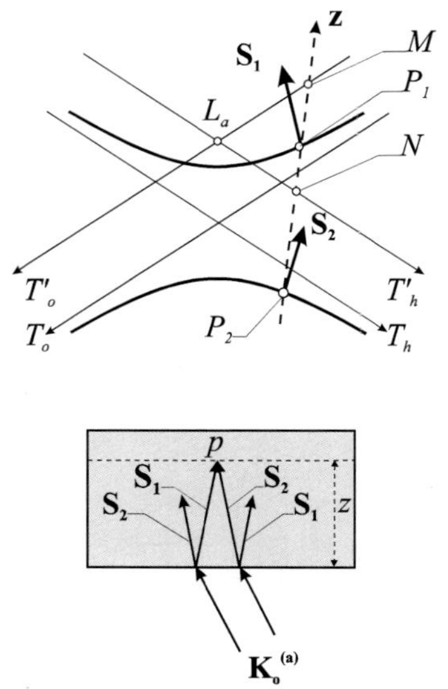

Fig. 2.8. Pendellösung–Laue geometry.

2.2.6.2. *Laue or transmission geometry*

In the Laue geometry (Fig. 2.8), the normal $\mathbf{Mz}$ to the entrance surface intersects both branches of the dispersion surface and the excited wave fields propagate both towards the inside of the crystal. For an incident plane wave, at any point p inside the crystal there will be interference between the two wave fields of tiepoints P_1 and P_2. The corresponding four waves are:

$$\mathcal{D}_{\mathbf{hj}} = D_{hj}\exp(-2\pi i[\mathbf{K}_{\mathbf{hj}}\cdot(\mathbf{r}-\mathbf{r}^{(\mathbf{a})}) + \mathbf{K}_{\mathbf{h}}^{(\mathbf{a})}\cdot\mathbf{r}^{(\mathbf{a})}]),$$
$$\mathcal{D}_{\mathbf{oj}} = D_{oj}\exp(-2\pi i[\mathbf{K}_{\mathbf{oj}}\cdot(\mathbf{r}-\mathbf{r}^{(\mathbf{a})}) + \mathbf{K}_{\mathbf{o}}^{(\mathbf{a})}\cdot\mathbf{r}^{(\mathbf{a})}]),$$

where $\mathbf{r}$ is the position vector of p, $\mathbf{r}^{(\mathbf{a})}$ the position vector of a point on the entrance surface, $\mathbf{K}_{\mathbf{hj}} = \mathbf{HP}_{\mathbf{j}}$, $\mathbf{K}_{\mathbf{oj}} = \mathbf{OP}_{\mathbf{j}}$ the wave vectors of the four waves, $\mathbf{K}_{\mathbf{o}}^{(\mathbf{a})} = \mathbf{OM}$ the incident wave vector and $\mathbf{K}_{\mathbf{h}}^{(\mathbf{a})} = \mathbf{HN}$ (N intersection of $\mathbf{Mz}$ with the sphere of center H and radius k, here approximated by its tangential plane, T'_h). Using (2.26), the four waves can be written:

$$\mathcal{D}_{\mathbf{hj}} = D_{hj}\exp(-2\pi i[\overline{MP_j}z + \mathbf{K}_{\mathbf{h}}^{(\mathbf{a})}\cdot\mathbf{r}]),$$
$$\mathcal{D}_{\mathbf{oj}} = D_{oj}\exp(-2\pi i[\overline{MP_j}z + \mathbf{K}_{\mathbf{o}}^{(\mathbf{a})}\cdot\mathbf{r}]).$$

The amplitudes D_{hj} and D_{oj} are determined by the boundary conditions at the entrance surface:

$$D_{o1} + D_{o2} = D_o^{(a)},$$
$$D_{h1} + D_{h2} = 0.$$

Using these relations and Eq. (2.26) where $S(\gamma_h) = 1$, one finds, for the four waves:

$$\mathcal{D}_{h1} = -\mathcal{A}\frac{S(C)}{\sqrt{\gamma}}\sqrt{\left|\frac{\chi_h}{\chi_{\bar{h}}}\right|}\exp(i\beta')\frac{\exp[-\mu(\gamma_o^{-1}+\gamma_o^{-1})z/4]\mathcal{E}_2}{2\sqrt{1+\eta^2}}\exp(2\pi i\mathbf{h}\cdot\mathbf{r}), \tag{2.39}$$

$$\mathcal{D}_{h2} = \mathcal{A}\frac{S(C)}{\sqrt{\gamma}}\sqrt{\left|\frac{\chi_h}{\chi_{\bar{h}}}\right|}\exp(i\beta')\frac{\exp[-\mu(\gamma_o^{-1}+\gamma_o^{-1})z/4]\mathcal{E}_1}{2\sqrt{1+\eta^2}}\exp(2\pi i\mathbf{h}\cdot\mathbf{r}), \tag{2.40}$$

$$\mathcal{D}_{o1} = \mathcal{A}\frac{\exp[-\mu(\gamma_o^{-1}+\gamma_o^{-1})z/4](\sqrt{1+\eta^2}-\eta)\mathcal{E}_2}{2\sqrt{1+\eta^2}}, \tag{2.41}$$

$$\mathcal{D}_{o2} = \mathcal{A}\frac{\exp[-\mu(\gamma_o^{-1}+\gamma_o^{-1})z/4](\sqrt{1+\eta^2}+\eta)\mathcal{E}_1}{2\sqrt{1+\eta^2}}, \tag{2.42}$$

where

$$\mathcal{E}_1 = \exp(\pi i\sqrt{1+\eta^2}\;t/\Lambda_o); \quad \mathcal{E}_2 = \exp-(\pi i\sqrt{1+\eta^2}t/\Lambda_o)$$

Λ_o is given by (2.17), t is the crystal thickness, and:

$$\mathcal{A} = \exp\left[-2\pi i\frac{k\Delta\theta\sin 2\theta_B}{2\gamma_h}\right]\exp\left[-2\pi ik\frac{\chi_{rh}}{4}\left(\frac{1}{\gamma_o}+\frac{1}{\gamma_o}\right)\right]$$
$$\times\exp[-2\pi i\mathbf{K_o^{(a)}}\cdot\mathbf{r}]D_o^{(a)}.$$

Figure 2.9 represents the variations of the reflectivities $I_h = |\gamma||D_{h1} + D_{h2}|^2$ and $I_o = |D_{o1} + D_{o2}|^2$ for a silicon crystal, 220 reflection, Mo K_α reflection and two crystal thicknesses.

2.3. Standing Wave Field in the Reflection (Bragg) Geometry

The intensity of the total wave field is $|\mathcal{D}|^2 = |\mathcal{D}_o+\mathcal{D}_h|^2$ and the normalized intensity is:

$$I_{SW}(\Delta\theta, z) = |\mathcal{D}|^2/|\mathcal{D}_o^{(a)}|^2 = 1 + |\xi|^2 + 2C|\xi|\cos(2\pi\mathbf{h}\cdot\mathbf{r}+\psi), \tag{2.43}$$

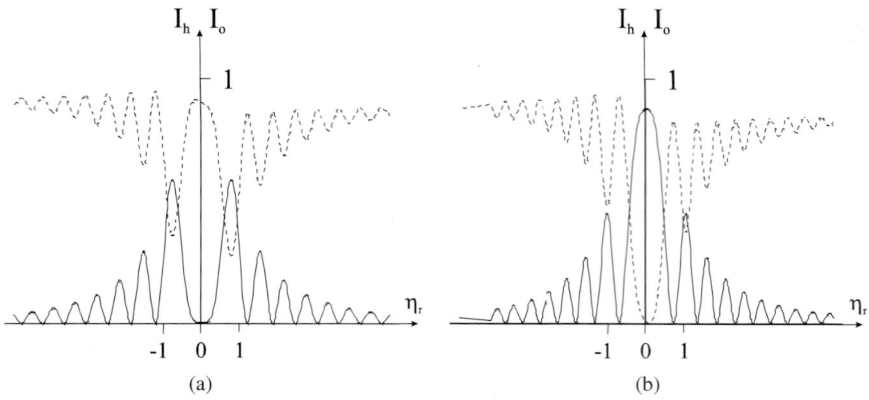

Fig. 2.9. Variation of the reflected (full line) and refracted (dashed line) reflectivities — Laue geometry. (a) $t = 2\Lambda_o = 72.63\,\mu$m. (b) $t = 2.5\Lambda_o = 90.80\,\mu$m.

where ξ is given by Eq. (2.26) with $S(\gamma_h) = -1$, ψ is its phase and $\mathbf{r}$ is the position vector of the observation point. This expression shows that the intensity of the wave field is spatially periodic. Its maxima and minima lie on families of planes $\mathbf{h} \cdot \mathbf{r} =$ constant in direct space whose spacing $d_{hk\ell}$ is the lattice spacing d divided by the order of the reflection. The product $\mathbf{h} \cdot \mathbf{r}$ can therefore be written as

$$\mathbf{h} \cdot \mathbf{r} = N + \Delta d / d_{hk\ell}, \tag{2.44}$$

where Δd is the distance of the extremity of the position vector from the origin of the unit cell along the normal to the reflecting planes (Fig. 2.10(a)) and N an integer. The nodes and antinodes of the standing waves extend over the volume where the incident and reflected beam overlap, both inside the crystal and above it (Fig. 2.10(b)).

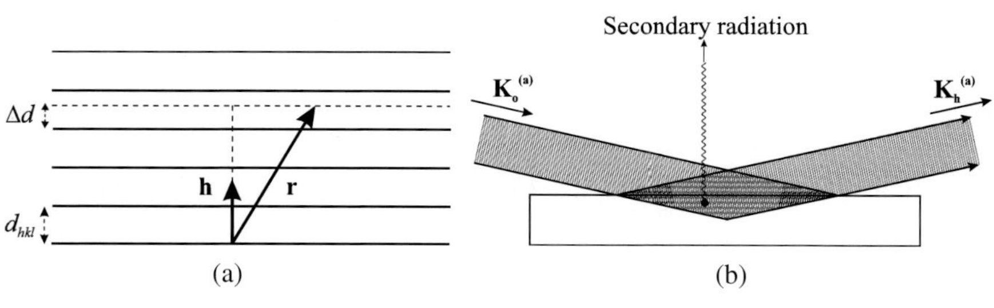

Fig. 2.10. Standing waves in the Bragg geometry. (a) Position of a plane. (b) Standing waves.

The position of the nodes and antinodes of standing waves are, respectively:

$$\text{Nodes:} \quad \Delta d/d_{hk\ell} = N + 0.5 - \psi/2\pi$$
$$\text{Antinodes:} \quad \Delta d/d_{hk\ell} = N - \psi/2\pi.$$

As the crystal is rocked through the reflection domain, the phase ψ varies continuously from $\pi + \varphi_h$ to φ_h and the position of the nodes and antinodes is shifted by half the spacing, $d_{hkl}/2$, of the reflecting planes. In order to better understand this continuous shift, it suffices to consider the real part of the dispersion surface in the Bragg geometry (Fig. 2.6). In the case of Bragg reflection on a thick crystal, the tiepoints of the excited wave fields remain continuously on branch ($'$), without any jump from one branch to another. For $\eta_r \rightarrow -\infty$, Eq. (2.32) shows that the position of the corresponding nodes is

$$\Delta d/d_{hk\ell} = N - \varphi_h/2\pi.$$

The position of the antinodes is

$$\Delta d/d_{hk\ell} = N + 0.5 - \varphi_h/2\pi.$$

When $\eta_r \rightarrow +\infty$, it is the opposite; the nodes lie on the planes of minimum electronic density and their antinodes on the planes of maximum electronic density. As examples, Fig. 2.11 shows in two cases the variations of the relative positions of the nodes and antinodes as the crystal is rocked through the reflection domain:

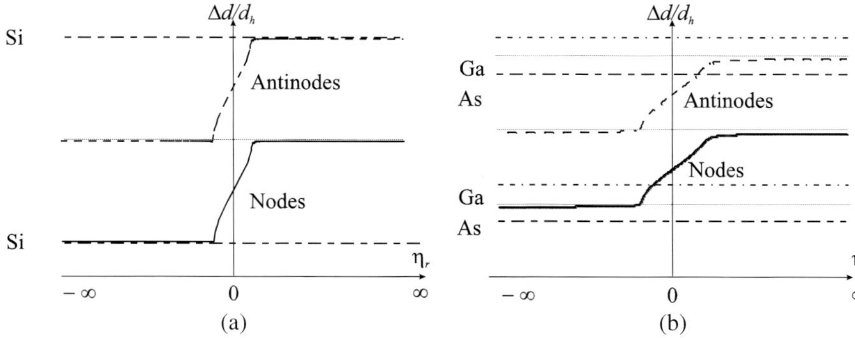

Fig. 2.11. Variations of the position of the nodes and antinodes. The ordinate axis is oriented towards the inside of the crystal from the surface. (a) Si, 220 reflection, Mo K_α radiation, (b) GaAs, 111, Mo K_α.

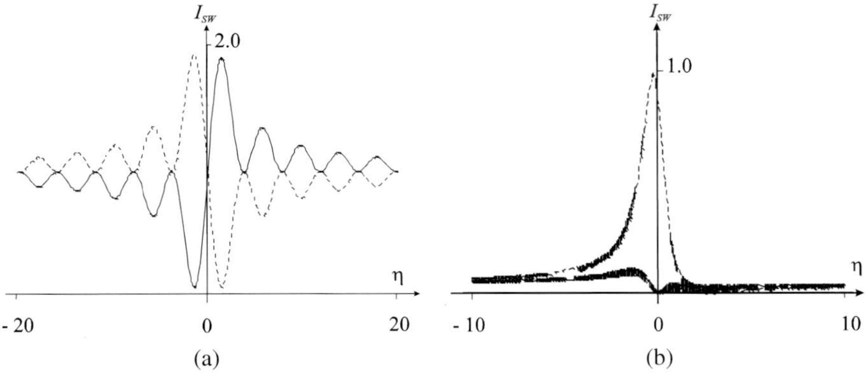

Fig. 2.14. Angular dependence of the standing wave field at the exit surface of a silicon crystal (Mo K_α, 220 symmetric reflection, Laue geometry) with deviation parameter η and position in unit cell, $\Delta d/d_{hk\ell}$; solid line: $\Delta d/d_{hk\ell} = 0$; dashed line: $\Delta d/d_{hk\ell} = 0.5$. (a) thickness 10 μm, (b) thickness 2.2 mm.

and half-way between them, are symmetrical for a thin crystal, but are very different for a thick crystal, due to the anomalous absorption effect. For a thick crystal, the total standing wave field comes essentially from type 1 wave fields. This can be confirmed on Fig. 2.15 which shows, for the same two positions in the unit cell, $\Delta d/d_{hk\ell} = 0$ and 0.5, the variations of the intensity of the total standing wave field, T, and of the types 1 and 2 standing wave fields across the reflection domain as a function of depth z in a 2.2-mm-thick silicon crystal (symmetric 220 reflection, Mo K_α). At a depth z close to the exit surface of the crystal, type 2 wave field is nearly absorbed out and the total standing wave field T is practically identical to type 1 standing wave field, except for small oscillations due to the Pendellösung effect.

It can also be noted that on the low-angle side of the rocking curve it is the type 1 wave fields that have the highest intensity, while it is the type 2 ones on the high-angle side.

In the applications of standing waves for the study of impurities or of local atomic displacements, what is important is the variations of the standing wave field as a function of the position in the unit cell and of the deviation parameter at a given depth in the crystal. As examples, they are represented on Fig. 2.16 at depth $z = 0.49$ mm and at the exit surface of a 2.2-mm-thick silicon crystal (symmetric 220 reflection, Mo K_α). At medium depths it shows a complicated pattern due to the interference between the type 1 and type 2 standing wave fields. At the exit surface of a thick crystal,

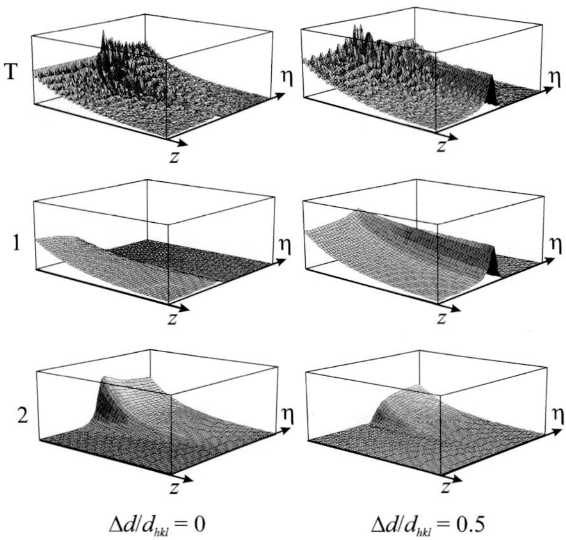

Fig. 2.15. Depth dependence of the intensity of the total (T) and types 1 and 2 standing wave fields across the reflection domain ($-10 \leq \eta \leq 10$) in a 2.2-mm-thick silicon crystal (symmetric 220 reflection, Mo K_α, Laue geometry), for two positions in the unit cell.

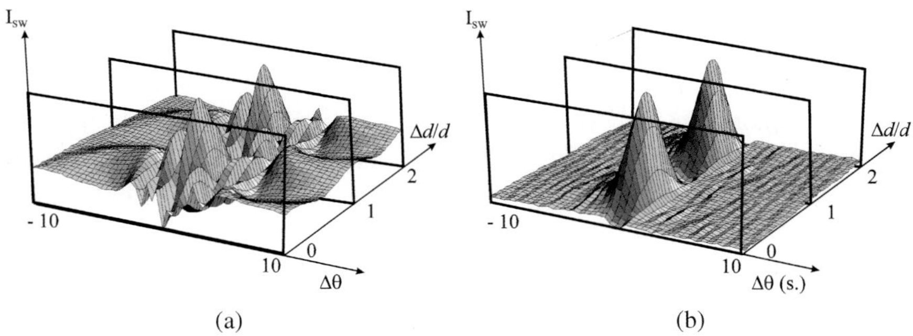

Fig. 2.16. Variations of the intensity of the total standing wave field with angle of incidence and position $\Delta d/d$ in the unit cell across two unit cells; the thick lines represent successive reflecting plane Laue geometry; silicon, symmetric 220 reflection, Mo K_α. (a) Thickness 0.49 mm, (b) thickness 2.2 mm.

the total standing wave field is nearly identical, except for some ripples, to that of type 1 wave field. These figures may be compared to Fig. 2.13 in the Bragg case, which illustrates the great difference in behavior of the wave field in the Bragg and Laue geometries.

2.5. Applications of X-ray Standing Waves in the Laue Geometry

2.5.1. *Introduction*

As in the case of Bragg geometry standing waves, local variations of the intensity of the standing wave field induce variations of the secondary emissions (photoelectrons, X-ray fluorescence, Compton scattering, thermal diffuse scattering, etc.), which are used to analyze the distribution and location of impurities or local atomic displacements. The first observations of secondary emissions associated with the presence of the standing wave field in the Laue geometry were:

- anomaly of the X-ray fluorescence scattering[31,32]
- anomaly of the neutron inelastic scattering[33]
- anomaly of the X-ray thermal diffuse scattering[32]
- photoemission[34,35]
- photoemission from a disturbed surface layer[36]
- photoelectric voltage excited by X-ray standing waves in a silicon crystal with a $p-n$ junction[37]
- Compton effect[38] and determination of the valence electron component of the atomic scattering factor[39]
- location of impurities.[40,41]

The obvious advantage of Bragg-case standing waves is that one can directly explore the unit cell by shifting the nodal and antinodal planes, which is simply achieved by rocking the crystal through the reflection domain. This cannot be done with Laue-case standing waves but it will be shown in Sec. 2.5.2. that the secondary emission signal is nevertheless very sensitive to the incidence angle and varies greatly when rocking the crystal, enabling the location of impurities[40,41] or the detection of small lattice distortions in multilayered systems (for a review, see Ref. 42).

The possibility of moving the exploring plane across the unit cell can, however, be achieved in the transmission geometry by using a L-L-L interferometer, as shown by Materlik[43] in a surface registration study. The Bonse–Hart L-L-L interferometer[44] consists of three crystalline slabs cut in a monolithic silicon crystal, all of them operating in the transmission geometry. These three crystals act as beam splitter, mirror and analyzer, respectively. The two beams generated by the first crystal are brought together by the mirror and generate a set of standing waves at the entrance surface of the third crystal, the analyzer. Because of the anomalous transmission in the first two crystals, type 2 wave fields have been practically absorbed out and the antinodal planes are situated halfway between the scattering planes of the analyzer. If one now puts a phase shifter between the splitter and the mirror, one can then effectively shift the position of the nodal and antinodal planes, as in a Bragg-case standing waves experiment. In the study by Materlik[43] the crystal surfaces were cut parallel to (111) and a 220 reflection was used. A parallel-sided Lucite phase shifter placed after the beam splitter could be rotated so as to change the phase step by step. This set-up was used to study the position of bromine atoms adsorbed on the entrance surface of the analyzer, the total coverage being of about 1/4 of a monolayer. The triangulation method introduced by Golovchenko[45] was used to determine the exact position of the adsorbed bromine atoms in the silicon unit cell, by combining the measurement using the 220 reflection in transmission with measurements using the 111 reflection in the Bragg geometry on the same surface.

2.5.2. *Integrated yield*

In a transmission experiment, the secondary emissions take place in the whole bulk of the crystal, or in the successive layers of a multilayered system. The expression of their intensity therefore requires an integration over the crystal thickness where they occur. The integrated yield of the secondary emission is of the form[46]:

$$Y_{SW}(\Delta\theta) = \int_0^t P(z) I_{SW}(\Delta\theta, z) dz,$$

where $P(z)$ is the probability density of the secondary emission, z the depth of the point where the secondary emission is emitted and t the crystal thickness and $I_{SW}(\Delta\theta, z)$ the intensity of the standing wave field.

(a) In the case of *photoemission*, the probability density of photoelectron emission, $P(z)$, has a complicated form[46,47] that can only be reconstructed empirically, which has been done in the case of a silicon crystal, a 422 symmetric reflection, Cu K_α radiation and an inclined Laue geometry.[48] It can be approximated by an exponential[49]:

$$P(z) = \exp -\mu_{PE} z,$$

where $1/\mu_{PE}$ is the absorption length of the photoelectrons, assuming here that the photoemission is measured on the entrance surface side of the crystal. In general, this absorption length is much shorter than the extinction distance (Bragg case) or the Pendellösung distance (Laue case). The experimental curves of the angular dependence of the yield are therefore related to the standing wave field exactly at the position of the atoms emitting the photoelectrons.[42]

(b) In the case of *fluorescence*, the probability density function is simply[31,50]:

$$P(z) = \exp -\mu_F(t - z)/\gamma_F$$

where μ_F is the absorption coefficient of the X-ray fluorescence radiation, γ_F the direction cosine of the direction along which the fluorescent radiation is measured, and assuming that it is measured on the exit surface side of the crystal. The absorption length $1/\mu_F$ is usually longer than the extinction distance (Bragg case) or the Pendellösung distance (Laue case). For instance, for a silicon crystal, germanium impurities, Mo K_α radiation and a symmetric 220 reflection in Laue geometry, $1/\mu_F = 128\,\mu$m and $\Lambda_o = 36.3\,\mu$m. Figure 2.17 shows for these conditions and a 2.2-mm-thick crystal, the variations with depth in the crystal of the angular dependence of the yield, $\exp - \mu_F(t-z)I_{SW}(\Delta\theta, z)$, in the case of germanium impurities located at mid-distance between the silicon reflecting planes. The integrated yield $Y_{SW}(\Delta\theta)$ results therefore in practice from integration over several Pendellösung distances.

The intensity of the standing wave field is given by (2.45) if the crystal is perfect but, in practice, corrections have to be introduced to take into account imperfections such as random displacements of atoms or strains. Random displacements can be taken into account by a static Debye–Waller factor $\exp - W_j(z)$ multiplying the relevant atomic scattering factor f_j in

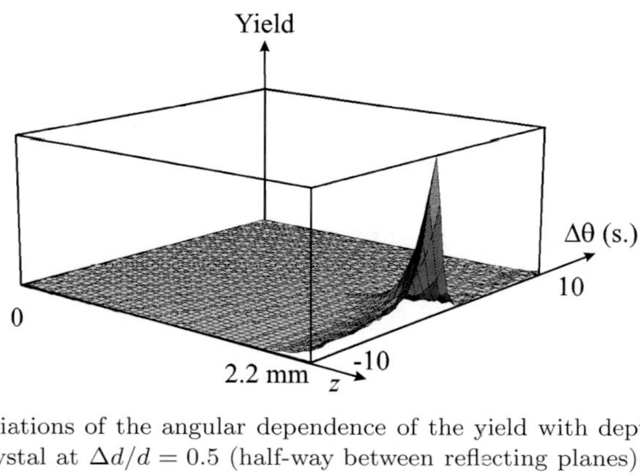

Fig. 2.17. Variations of the angular dependence of the yield with depth in a 2.2-mm-thick silicon crystal at $\Delta d/d = 0.5$ (half-way between reflecting planes); symmetric 220 reflection, Mo K_α.

Eq. (2.4)[51]; it is called the coherent fraction.[42] Strains in the bulk of the crystal or in epilayers are taken into account by solving Takagi equations. This was done both for the Bragg[46,47,52] and Laue geometries[37,41]; for a review see Ref. 42.

2.5.3. *Angular dependence of the X-ray fluorescence integrated yield*

The angular dependence of the integrated yield in the Laue geometry is very sensitive to both the crystal thickness and the position of the emitting atoms in the unit cell. To illustrate them, and in order to make a qualitative comparison with the experimental results by Kazimirov, Koval'chuk and Kohn,[41] the integrated yield has been calculated for a perfect silicon crystal slab parallel to (001), a 111 reflection, Mo K_α radiation and germanium impurities. This reflection is asymmetric, the angle between the [001] and [111] directions is 54.74 degrees and the asymmetry ratio is $\gamma = 0.851$. The (111) reflecting planes of a silicon crystal can be considered as a stacking of $AA'BB'CC'\ldots$ planes. Taking the origin at mid-distance between the A and A' layers, let u be the position of an emitting atom along the normal to the (111) planes. The silicon structure being centrosymmetric, it is enough to calculate the integrated yield for the following positions in the unit cell: $u = 0$ (half-way between A and A' planes); $u = 0.125$ (on A' reflecting planes); $u = 0.25$; $u = 0.375$; $u = 0.5$ (half-way between A' and B reflecting planes). Position $u = 0.125$ corresponds to substitutional impurities and $u = 0, 0.25, 0.375, 0.5$ to interstitial impurities.

(a) *Dependence on crystal thickness.* Figure 2.18 compares the angular dependence of the integrated yield for three crystal thicknesses, $t = 0.49, 1.03$ and 2.2 mm, respectively, and in the case of substitutional impurities of germanium in the silicon reflecting planes ($u = 0.125$). It can be observed that the shape of the curve varies significantly with crystal thickness. The curves for the 0.49 and 2.2-mm-thick crystals are in very good qualitative agreement with the corresponding experimental curves obtained for the same conditions (Figs. 2 and 3 of Ref. 41, respectively). The calculated results given in Ref. 41 take into account the lattice distortions.

(b) *Dependence on position in unit cell.* Figures 2.19(a) and 2.19(b) show the variations of the angular dependence of the integrated yield for the positions $u = 0, 0.125, 0.25, 0.375, 0.5$ (grey curves) and for uniform doping (black curve), for crystal thicknesses 0.49 mm and 2.2 mm,

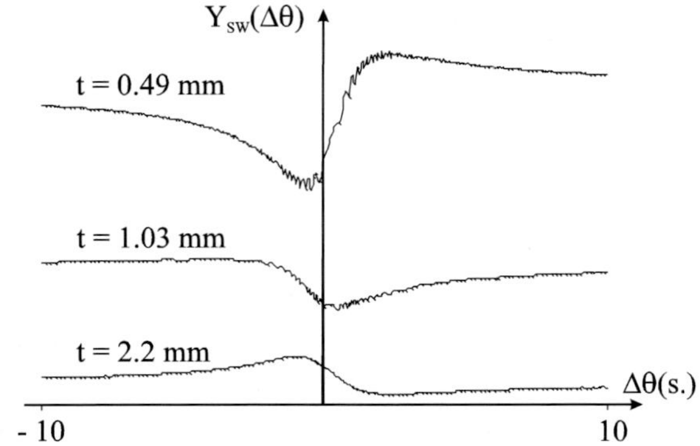

Fig. 2.18. Variations of the angular dependence of the integrated yield for substitutional germanium impurities ($u = 0.125$) and three crystal thicknesses (silicon, asymmetric 111 reflection, Mo K_α).

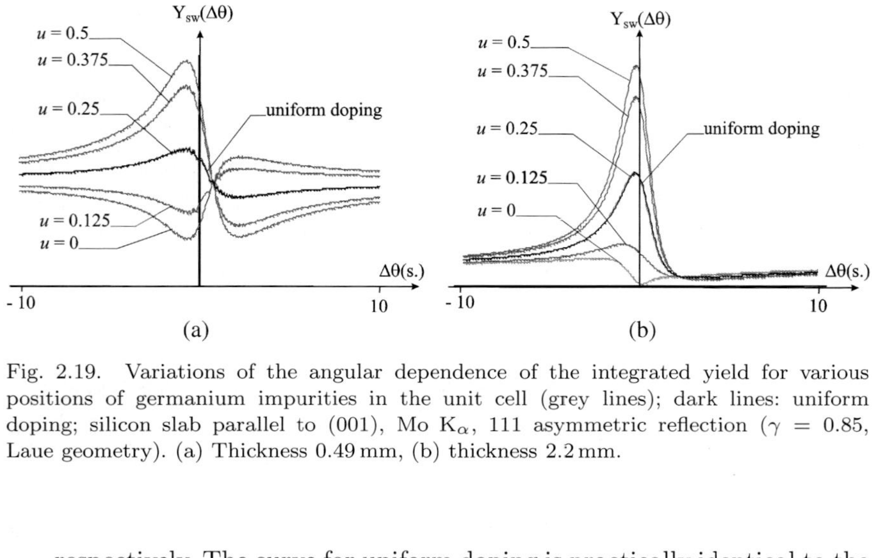

Fig. 2.19. Variations of the angular dependence of the integrated yield for various positions of germanium impurities in the unit cell (grey lines); dark lines: uniform doping; silicon slab parallel to (001), Mo K_α, 111 asymmetric reflection ($\gamma = 0.85$, Laue geometry). (a) Thickness 0.49 mm, (b) thickness 2.2 mm.

respectively. The curve for uniform doping is practically identical to the curves for $u = 0.25$. It can be seen that for small angles of incidence, the values of $Y_{SW}(\Delta\theta)$ increase for increasing values of u, while they decrease for high angles of incidence.

The high sensitivity of the curves to the position in the unit cell can be noticed, which shows that Laue geometry standing waves are well suited to determine the positions of impurities, as was proved by Kazimirov, Koval'chuk and Kohn.[41]

References

1. M. von Laue, Der optische Reziprozitätsatz in Anwendung auf die Röntgenstrahlinterferenzen, *Naturwiss* **23** (1935) 373, in German.
2. M. von Laue, Die Fluoresenzröntgenstrahlung von Einkristalle, *Ann. Phys.* **23** (1935) 705–746, in German.
3. M. von Laue, Helligkeitswechsel längs Kossellinien, *Ann. Phys.* **28** (1937) 528–532, in German.
4. M. von Laue, *Röntgenstrahl–Interferenzen*, Akademische Verlagsgesellschaft, Becker and Erler, Leipzig (1941), in German.
5. W. Kossel, H. Loeck and H. Voges, Die Richtungsverteilung der in einem Kristall entstandenen charakteristischen Röntgenstrahlung, *Z. Phys.* **94** (1935) 139–144, in German.
6. G. Borrmann, Über die Interferenzen aus Gitterquellen bei Anregung durch Röntgenstrahlen, *Ann. Phys. Lpz.* **27** (1936) 669–693, in German.
7. M. von Laue, *Röntgenstrahl-Interferenzen*, Akademische Verlagsgesellschaft, Frankfurt am Main (1960), in German.
8. P. P. Ewald, Zur Theorie der Interferenzen der Röntgentstrahlen in Kristallen, *Phys. Z.* **14** (1913) 465–472, in German.
9. P. P. Ewald, Zur Begründung der Kristalloptik. I. Theorie der Dispersion, *Ann. Phys.* **54** (1916) 1–38, in German.
10. P. P. Ewald, Zur Begründung der Kristalloptik. II. Theorie der Reflexion und Brechung, *Ann. Phys.* **54** (1916) 117–143, in German.
11. P. P. Ewald, Zur Begründung der Kristalloptik. III. Die Kristalloptik der Röntgenstrahlen, *Ann. Phys.* **54** (1917) 519–597, in German.
12. G. Borrmann, Über die Röntgeninterferenzen des selbstleuchtenden Eisens, *Z. Krist.* (*A*), **100** (1938) 228–233, in German.
13. G. Borrmann, Röntgenwellenfelder, *Beit. Phys. Chem. 20 Jahrhunderts, Vieweg und Sohn, Braunschweig* (1959), pp. 262–282, in German.
14. J. W. Knowles, Anomalous absorption of slow neutrons and X-rays in nearly perfect single crystals, *Acta Crystallogr.* **9** (1956) 61–69.
15. B. W. Batterman, Effect of dynamical diffraction in X-ray fluorescence scattering, *Phys. Rev. A* **133** (1964) 759–764.
16. S. Annaka, S. Kikuta and K. Kohra, Intensity anomaly of thermal and Compton scatterings accompanying of X-rays accompanying the Bragg reflection, *J. Phys. Soc. Jpn.* **20** (1965) 2093–2093.
17. S. Annaka, S. Kikuta and K. Kohra, Intensity anomaly of X-ray Compton and thermal scatterings accompanying the Bragg reflection from perfect Si and Ge crystals, *J. Phys. Soc. Jpn.* **21** (1966) 1559–1564.
18. B. W. Batterman, Detection of foreign atom sites by their X-ray fluorescence scattering, *Phys. Rev. Lett.* **22** (1969) 703–705.
19. M. V. Koval'chuk, A. Yu. Kazimirov and S. I. Zheludeva, Surface sensitive X-ray diffraction methods: Physics, applications and related X-ray and SR instrumentation, *Nucl. Instrum. Meth. Phys. Res. B* **101** (1995) 435–452.
20. J. R. Patel, *X-Ray Standing Waves.* Eds. A. Authier, S. Lagomarsino and B. K. Tanner, *X-Ray and Neutron Dynamical Diffraction: Theory and Applications, NATO ASI Series, B: Physics*, Plenum Press, New York (1996).

was used for both the Bragg[3] and for the Laue[5] cases to analyze specific experimental results and was later summarized in the most recent form in Ref. 6.

The chapter is organized as follows. In the Sec. 3.2., the analytical solution for the local reflection and transmission amplitudes is derived for a single crystalline layer. Then, the YPF is introduced and the integration over the thickness of the sample is performed. It will be followed by the description of a numerical algorithm. The set of parameters required to compute the XSW yield from a multilayer crystalline structure will be presented and discussed, followed by a computational example and summary.

3.2. Solution for One Crystal Layer

Let us consider a single crystal of a lamina-like shape and a case of the two-beam diffraction on a reciprocal lattice vector **h**. The solution of the Maxwell's equation can be sought in the form

$$\mathbf{E}(\mathbf{r},\omega) = \exp(i\mathbf{k}_0\mathbf{r})[\mathbf{e}_0 E_0(z) + \mathbf{e}_h E_h(z)\exp(i\mathbf{hr})], \tag{3.1}$$

where $\mathbf{e}_0$, $\mathbf{e}_h$ are the unit polarization vectors, $\mathbf{k}_0$ is the wave-vector of the incident plane wave in the air, $|\mathbf{k}_0| = K$ where $K = \omega/c = 2\pi/\lambda$, c is the speed of light, and λ is the wavelength of X-rays, z is the depth inside the crystal. The complex functions $E_{0,h}(z)$ are slowly varying in space compared to the exponential $\exp(i\mathbf{hr})$. We assume the incident wave to be a plane-polarized wave, which is a valid assumption for synchrotron radiation. In the case of a nonpolarized radiation, one has to consider two standard polarization states separately and average intensity over polarizations states.

The integration of the Maxwell's equation over unit cell allows us to write the set of two equations for $E_0(z)$, and $E_h(z)$:

$$\begin{aligned} 2\gamma_0\frac{dE_0}{dz} &= iK\{\chi_0 E_0 + C\chi_{\bar{h}}\exp(i\varphi - W)E_h\}, \\ 2\gamma_h\frac{dE_h}{dz} &= iK\{[\chi_0 - \alpha]E_h + C\chi_h\exp(-i\varphi - W)E_0\}, \end{aligned} \tag{3.2}$$

where $\gamma_0 = k_{0z}/K$, and $\gamma_h = k_{hz}/K$ are the geometrical parameters, $\alpha = [\mathbf{k}_h^2 - \mathbf{k}_0^2]/K^2$ is the parameter of deviation from the Bragg condition, $\mathbf{k}_h = \mathbf{k}_0 + \mathbf{h}$ is the wave vector of the diffracted wave, $C = (\mathbf{e}_0\mathbf{e}_h)$ is the polarization factor, and $\varphi(z) = \mathbf{hu}(z)$ is an additional phase due to a mean displacement of atoms from their equilibrium positions by a vector $\mathbf{u}$. The quantities χ_0, χ_h, and $\chi_{\bar{h}}$ are the Fourier coefficients of the crystal susceptibility with the reciprocal lattice vectors 0, $\mathbf{h}$, $-\mathbf{h}$. Finally, the factor $\exp[-W(z)]$ describes dephasing of the scattered wave due to random displacements of atoms from their mean value at depth z. This factor was introduced for the first time in Ref. 1 and called the static Debye–Waller factor on analogy with a well-known thermal Debye–Waller factor, which is incorporated into the crystal susceptibility.

The boundary conditions for Eq. (3.2) depend on a sign of the geometrical parameter γ_h. In the Laue case, $\gamma_h > 0$ and the diffracted beam is escaping the crystal through the back surface and absent at the entrance surface; therefore, we have $E_0(0) = 1$ and $E_h(0) = 0$. Here and later on we assume that the entrance surface is at $z = 0$ and the incident intensity is normalized to unity. In the Bragg case, $\gamma_h < 0$ and the diffracted beam is escaping from the entrance surface and absent at the back surface, so we have $E_0(0) = 1$ and $E_h(d) = 0$, where d is the thickness of the crystal plate.

3.2.1. *Local reflection amplitude*

We will consider the Bragg and the Laue cases simultaneously. The boundary conditions do not allow us to move from the entrance surface step by step. It is convenient to divide a problem into two parts and to introduce first a local reflection amplitude as the ratio

$$R(z) = \frac{E_h(z)}{E_0(z)} \frac{\exp[i\varphi(z)]}{Y\beta^{1/2}}. \tag{3.3}$$

This variable obeys the nonlinear equation which can be derived from the set of equation (3.2)

$$\frac{dR(z)}{dz} = -\frac{2is}{L_{ex}}[y - y_\varphi(z) + iy_0]R(z) + \frac{iC_1}{L_{ex}}[s + R^2(z)], \tag{3.4}$$

where the variables are introduced: $C_1 = C(1 - ip)\exp(-W)$,

$$L_{ex} = \frac{\lambda\gamma_0}{\pi\beta^{1/2}X'}, \quad X = (\chi_h\chi_{\bar{h}})^{1/2} = X' + iX'' = X'(1 - ip), \tag{3.5}$$

$$Y = \left(\frac{\chi_h}{\chi_{\bar{h}}}\right)^{1/2} = |Y|\exp(i\Phi_Y), \quad y = -\frac{[\alpha\beta - s\chi_0'(1 + s\beta)]}{2\beta^{1/2}X'}, \tag{3.6}$$

$$y_0 = \frac{s\chi_0''(1 + s\beta)}{2\beta^{1/2}X'}, \quad y_\varphi(z) = s\frac{L_{ex}}{2}\frac{d\varphi(z)}{dz}, \quad \beta = \frac{\gamma_0}{|\gamma_h|}. \tag{3.7}$$

Here we use notations a' and a'' for the real and imaginary parts of a complex value a. The parameter s is equal to 1 for the Bragg case and -1 for the Laue case. The boundary conditions for the local reflection amplitude are as follows: $R(0) = 0$ in the Laue case and $R(d) = 0$ in the Bragg case.

There are two ways of changing the parameter of deviation from the Bragg condition α. The first one is to change the angle of incidence θ of the X-ray beam by $\Delta\theta$ while keeping the energy constant. In this case we have

$$y = C_{y\theta}\Delta\theta, \quad y_\varphi(z) = C_{y\theta}\Delta\theta_B(z), \quad C_{y\theta} = \pi\frac{L_{ex}}{\lambda|\gamma_h|}\sin 2\theta_B, \tag{3.8}$$

where θ_B is the Bragg angle in a perfect crystal while $\Delta\theta_B(z)$ is a local shift of the Bragg angle at the depth z due to distortions of the crystal lattice. The angle $\Delta\theta$ is positive if $\theta > \theta_B$. The second way is to change the energy of X-ray photons, keeping constant the direction of the beam. In this case

$$y = C_{y\omega}\Delta(\hbar\omega), \quad y_\varphi(z) = C_{y\omega}\Delta(\hbar\omega_B(z)), \quad C_{y\omega} = \frac{L_{ex}}{\hbar c|\gamma_h|}\sin^2\theta_B \tag{3.9}$$

where $\hbar = h/2\pi$, h is the Planck constant, $\hbar\omega_B$ is the Bragg energy of X-ray photons. Parameters $\Delta\theta_B(z)$ and $\Delta(\hbar\omega_B(z))$ describe a local shift of the Bragg angle or energy at depth z due to distortions. The origin of the y-axis corresponds to the center of the diffraction peak for a perfect crystal. The $\Delta\theta$-dependence is common for experiments in a nondispersive arrangement, e.g., with laboratory X-ray sources. The $\Delta(\hbar\omega)$-dependence is often used in experiments with synchrotron radiation when the energy is scanned by the upstream monochromator, e.g., at near backscattering conditions with $\theta_B \approx \pi/2$. Note that the parameter y_0 can also be expressed through the dimensionless variables $y_0 = (s\mu_0/4\gamma_0)L_{ex}(1 + s\beta)$ where $\mu_0 = 2\pi\chi_0''/\lambda$ is a linear absorption coefficient.

The method based on a direct numerical solution of the differential equation (3.4) was proposed in Ref. 7 and discussed in Ref. 8. However, such an approach has a disadvantage for a crystal containing thick layers with approximately constant parameters and a large difference in parameters between the layers. Indeed, to have sufficient accuracy in numerical processing of Eq. (3.4), a very small step Δz is required over the total thickness of the layer even if parameters within this layer are almost constant. It is more convenient to consider a crystal as a set of layers with the parameters that are constant within each layer and can be changed only at the layers boundaries. Equation (3.4) for a layer with the constant parameters y_φ has an analytical solution. Here we present the general solution[6] that is valid for both the Bragg and the Laue cases. In addition, we assume that the boundary conditions for each layer does not contain zero amplitudes, i.e., the amplitudes $R(0)$ in the Laue case and $R(d)$ in the Bragg case are finite and known, where d is now the thickness of the layer, not the thickness of a sample. We omit the derivation and present the solution in the form

$$R(z) = \frac{x_1 - x_2 B \exp(-is\sigma z)}{F_d(z)}, \quad F_d(z) = 1 - B \exp(-is\sigma z), \tag{3.10}$$

where

$$x_{1,2} = -\frac{s}{C_1}[-a \pm \sqrt{a^2 - sC_1^2}], \quad \sigma = \frac{2}{L_{ex}}\sqrt{a^2 - sC_1^2}, \tag{3.11}$$

$$a = y - y_\varphi + iy_0, \quad B = \frac{(x_1 - R(z_b))}{(x_2 - R(z_b))} \exp(i\sigma z_b). \tag{3.12}$$

Here and later on it is assumed that square roots have positive imaginary parts. One can verify the solution by the direct substitution. Equations (3.10) to (3.12) allow one to derive the recurrent relation for the reflection amplitude at the exit surface $z = z_e$ from the known value at the entrance surface $z = z_b$.

$$R(z_e) = \frac{(x_1 - x_2)R(z_b) + x_2[x_1 - R(z_b)][\exp(i\sigma d) - 1]}{x_1 - x_2 + [x_1 - R(z_b)][\exp(i\sigma d) - 1]}. \tag{3.13}$$

The parameters z_e and z_b are $z_e = d$ and $z_b = 0$ in the Laue case and $z_e = 0$ and $z_b = d$ in the Bragg case.

The accurate solution presented above allows us to easily consider analytical kinematical approximations. If $d \to 0$, we obtain the same expression, which can be obtained directly from Eq. (3.4) if one takes the right-hand side of the equation at the boundary and replace a derivative by $[R(z_e) - R(z_b)]/(-sd)$. In a pure kinematical case, when $|R(z)| \ll 1$ and $|aR(z)| \ll 1$, we have a simple expression $R(z_e) = R(z_b) - idC_1/L_{ex}$ meaning that the reflection amplitude linearly increases with thickness independently of the parameter of deviation from the Bragg condition. Another kinematical approximation can be obtained for a large deviation from the Bragg condition, $|a| \gg |C_1|$. Under this condition in the Bragg case we have $\sqrt{a^2 - sC_1^2} \approx a$. Then $x_1 \approx 0$, $x_2 \approx 2a/C_1$, $|x_2| \gg 1$, $\sigma d = \Phi = 2ad/L_{ex}$, and we obtain from Eq. (3.13) that $R(z_e) = R(z_b) \exp(i\Phi)$. This means that the layer changes the phase of the reflection amplitude which may have a large modulus due to reflection at the substrate. The layer does not influence practically the modulus of the reflection amplitude. As for the phase, it can be measured by means of XSW.

3.2.2. *Local transmission amplitude*

Taking into account the definition (3.3), we can write a straightforward solution of the first Takagi equation as follows

$$E_0(z) = \exp\left\{ i\frac{\pi\chi_0}{\lambda\gamma_0}z - i\frac{C_1}{L_{ex}}\int_0^z dz' R(z') \right\} E_0(0) = T(z)E_0(0). \quad (3.14)$$

The solution may be used for a numerical calculation, however, again with a disadvantage owing to the integral. For a constant parameter y_φ the function $R(z)$ has the analytical expression (3.10) and the integral can be calculated analytically by means of a table integral. The result looks as follows

$$T(z) = \exp\left(\frac{i}{2}Gz\right)\frac{F_d(z)}{F_d(0)}, \quad (3.15)$$

where

$$G = G' + iM = 2\frac{\pi\chi_0}{\lambda\gamma_0} - 2\frac{C_1}{L_{ex}}x_1, \quad M = \frac{\mu_0(1-s\beta)}{2\gamma_0} + s\sigma''. \quad (3.16)$$

Here $M = G''$ and the values x_1 and $F_d(z)$ are defined above. As one can observe, the recurrent relation for the intensity of the transmitted wave has a simple form. However, to use this expression, one needs to know the value $R(z_b)$ for this layer. Therefore, this recurrent relation may be used only after the recurrent relation (3.13) is applied.

3.3. Secondary Radiation Yield

We consider the XSW techniques based on measuring the intensity of the secondary radiation scattered via photoelectron emission or fluorescence. This radiation involves many spherical waves originating from individual atoms. If a SRY detector counts all electrons or photons that reach the surface, the YPF must be averaged over the surface. Then, the averaged YPF $P_{yi}(z)$ depends only on the z-coordinate, which is the distance between atoms emitting radiation and the surface. As we discussed in Sec. 3.1, we consider the YPF in the form of an exponential function $P_{yi}(z) = \exp(-\mu_{yi}z)$.

Below we assume that the SRY detector collects radiation from the entrance surface. For the radiation collected from the exit surface we may formally consider a negative value of μ_{yi}. Both types of secondary radiation, fluorescence and photoelectrons, are generated via a resonant interaction of X-rays with atoms. Within a dipole approximation the intensity of the SRY emitted from atom is proportional to the intensity of the X-ray field at atomic position and the size of atom is assumed to be negligibly small. To take into account the size of atom we need to add a quadrupole term of the multipole expansion.[8] Thermal vibrations and static atomic displacements from equilibrium positions are accounted for by the thermal and static Debye–Waller factors.

Consider again a layered crystal. Each layer is uniform, i.e. the structural parameters are constant within the thickness of the layer, however, they may differ for different layers. Then, the total yield of secondary radiation I_{SR} is a sum over all layers

$$I_{SR} = \sum_{n=1}^{N} Z_{n-1} I_{SR}^{(n)}, \quad Z_n = |E_0(z_n)|^2 P_{yi}(z_n), \tag{3.17}$$

where $I_{SR}^{(n)}$ is a contribution of the layer with the back boundary at z_n ($z_0 = 0$). The thickness of the nth layer is $d_n = z_n - z_{n-1}$. Using solution (3.10), we write expression for $I_{SR}^{(n)}$ in terms of the local reflection amplitude

$$\begin{aligned} I_{SR}^{(n)} = \chi''_{0a} \int_0^d dz' P_{yi}(z') |T(z')|^2 \big[1 + |R(z')|^2 |Y|^2 \beta + 2\,\mathrm{Re}\big\{R(z') \\ \times Y \beta^{1/2} C(\chi''_{ha}/\chi''_{0a}) \exp[-i\varphi(z') + i\varphi_a(z')]\big\} \exp[-W_a]\big] \end{aligned} \tag{3.18}$$

where the index a indicates that the yield is calculated only for atoms contributing into the SRY, $\varphi_a(z) = \mathbf{h}\mathbf{u}_a(z)$. All the parameters must be taken for the nth layer.

To move further we accept a reasonable assumption that the difference $\varphi(z) - \varphi_a(z) = \Delta\varphi_a$ does not depend on z. If the emitting atoms occupy crystal lattice nodes, then $\Delta\varphi_a = 0$. In a general case of atoms occupying position defined by the vector $\mathbf{u}_a$ within the unit cell (e.g. impurity atoms in interstitial positions) this parameter is nonzero. Under this assumption, the integral can be calculated analytically. The result can be written as:

$$I_{SR}^{(n)} = \frac{d_n \chi''_{0a}}{|1 - B|^2} [A_1 \Psi_1 + A_2 \Psi_2 - \mathrm{Re}(A_3 \Psi_3)] \tag{3.19}$$

one the angular dependence of the SRY is measured by rotating sample through the Bragg reflection. The incident beam is conditioned by using a crystal collimator. The energy spread of the incident beam is determined by the natural width of a characteristic line when using laboratory sources or by the properties of the upstream monochromator when using synchrotron radiation. In a typical XSW set-up, a sample and monochromator (or, postmonochromator) crystals are arranged in a nondispersive $(n, -n)$ set-up. Then, the parameter X' of the monochromator is equal or close to X'_0 and a well-known formula can be used for convolution:

$$I_c(y) = \frac{1}{S}\sum_j \int dy_1 P_R^{(m)}(j, y_1) I(j, y + y_1[\beta_m \beta]^{1/2}), \tag{3.29}$$

$$S = \sum_j \int dy_1 P_R^{(m)}(j, y_1). \tag{3.30}$$

Here β_m is an asymmetry factor for a monochromator crystal and j is an index of polarization. A notation $I(j, y)$ is used for any of the functions $P_R(j, y)$, $P_T(j, y)$, and $I_{SR}(j, y)$.

The second technique is often utilized when using SR source and based on scanning the energy of the incident beam by rotating the monochromator while keeping the angle of incidence fixed. Therefore, the energy dependence of the SRY is measured. In particular, this technique is standard for the near backscattering geometry. The energy spread of the incident beam is determined by the properties of the monochromator set-up and the properties of a SR source. It is a common practice to approximate it by a Gaussian function. Then, in the y-scale we have

$$I_c(y) = (\sigma_y \sqrt{\pi})^{-1} \int dy_1 \exp(-y_1^2/\sigma_y^2) I(j, y + y_1) \tag{3.31}$$

where σ_y is used as a variable fitting parameter. Convolution with the Gaussian function can also be useful in the first case to account for some mosaicity.

Many of the 11 parameters describing each layer are usually very well known or can be calculated based on the knowledge of a preparation procedure, sample history, and the results of independent measurements by using complementary techniques. Other parameters have to be determined by fitting. The list of fitting parameters usually includes the thickness of the layer d, the shift of the Bragg angle $\Delta\theta_B$ or energy $\Delta(\hbar\omega_B)$, and the static Debye–Waller factor $\exp(-W)$. Fitting is performed by minimizing the function

$$\chi^2 = \sum_i [I_c(\theta_i) - K I_{ex}(\theta_i)]^2, \tag{3.32}$$

where $I_{ex}(\theta_i)$ is the experimental data and $I_c(\theta_i)$ is the theoretical value calculated for the same data points. If experimental data are normalized, the scaling factor $K = 1$, otherwise the K value is determined for each combination of fitting parameters by using a well-known solution $K = \left(\sum_i I_{ex}^2(\theta_i)\right)^{-1} \sum_i I_c(\theta_i) I_{ex}(\theta_i)$.

3.4.1. *Example: InGaP/GaAs(111)*

As an example, consider $In_{0.5}Ga_{0.5}P$ film grown by the liquid phase epitaxy on $GaAs(111)$ surface. The polarity of the $GaAs$ substrate and the film is known and fixed such that the Ga atoms in the substrate and the In and Ga atoms in the film occupy top half of the (111) double layer. We are interested in the fluorescence yield from the In and P atoms from the film excited by the (111) Bragg reflection.[10] Both the substrate and the layer materials belong to a zincblend structure with four Ga atoms in the substrate and two In atoms and two Ga atoms in the film occupy the $(0,0,0;0,1/2,1/2;1/2,0,1/2;1/2,1/2,0)$ fcc sublattice while four As atoms in the substrate and four P atoms in the film occupy the $fcc + (1/4,1/4,1/4)$ sites. Then, the geometrical structure factors for In atoms $S_{In} = 1$ and for P atoms $S_P = i$ and, accordingly, $\varphi_c^{In} = 0$ and $\varphi_c^P = -\pi/2$. The X-ray reflectivity curve from the sample is shown on the bottom panel of Fig. 3.1, and the In-L and the P-K fluorescence yields from the film are on the top panels. Striking difference in the shape of the XSW curves for the In and P atoms is due to the difference by $\pi/2$ in the phases of their structure factors. The reflectivity and the fluorescence data were fitted by using the layer thickness, the difference in the Bragg angles for the substrate and the layer and the static DW factor as fitting parameters. It was assumed that the static DW factor for both In and P sublattices are the same and equal to the static DW factor of the layer as a whole. The best fit within a single layer model is shown by a thin line. The quality of the fit can be improved if a thin layer with different lattice constant is introduced at the interface (thick solid line, see Ref. 6 for details). The remaining discrepancy in the fit of the P fluorescence data may be due to the secondary excitations not accounted for in this model. These results (i) proved that the XSW fluorescence bears direct information about the phases of the

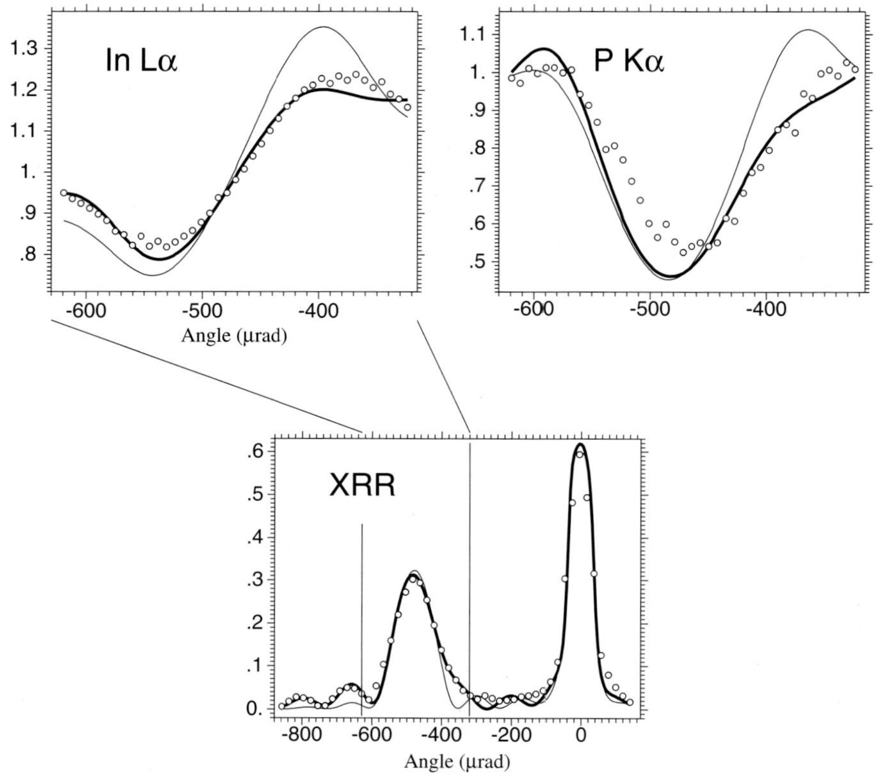

Fig. 3.1. X-ray reflectivity (bottom panel) and fluorescence yield from the In (top left panel) and P (top right panel) atoms from the $In_{0.5}Ga_{0.5}P$ films grown by liquid phase epitaxy on GaAs(111) substrate. The fluorescence data are in the angular range of the Bragg peak from the film. The best fit within a single layer model (thin line) and the model with a thin interface layer (thick line) are shown. The shape of the fluorescence curve is determined by the phase of the structure factor of the corresponding sublattice. (From Refs. 6 and 10.)

structure factors of the individual sublattices in multicomponent films and (ii) demonstrated that a more detailed information about the depth profile can be obtained when the X-ray reflectivity is assisted by a phase sensitive XSW data.

3.5. Brief Historical Overview and Summary

The first version of the computer program based on the theory presented in this section was developed by the author in 1980s stimulated by the pioneering experiments performed by Russian scientists in which photoelectron emission excited by the XSW from perfect crystals was studied.[11] Potential applications of this technique as a tool to study the structure of surface layers had been quickly realized and first experiments on crystals with amorphous layers,[12] ion implanted layers,[13] and epitaxial films[3] were performed. In the 1990s, the author continued developing his program motivated by the new experimental results and working in close collaboration with the researches from the laboratory of Michael Kovalchuk, Institute of Crystallography, Russian Academy of Science. In particular, the program was extended to calculate fluorescence yield from the crystals with multicomponent epitaxial layers[10,14] and from single crystals and crystals with epitaxial films in the Laue case[5] (see Ref. 15 for more references). In the last decade, the program was used to analyze experimental data from a variety of research projects in which the XSW method was applied to interesting physical and material science problems such as the isotopic effect on the lattice constants of Ge[16] and Si[17] (Chapter 17), structure of thin HT_c films,[18] polarity of thin GaN,[19] and ferroelectric[20] films (Chapter 16), and others.

In conclusion, the theory and the computer algorithm to calculate secondary radiation yield from the crystal consisting of several layers has been presented. The introduction of multilayer crystals into the XSW method significantly broadened the application areas of this technique. Indeed, in a modern world, a large variety of man-made structures can be considered as layered crystals. These are the homo- and hetero-epitaxial films grown on single crystal substrates (such as semiconductor lasers, photodiodes, and other optoelectronic devices), superlattices, bicrystals, etc. By using the theoretical approach presented in this section and the computer program available nowadays as a free software, the XSW technique in its different modifications can be effectively utilized to perform their structural characterization.

References

1. A. M. Afanasev and V. G. Kohn, *Sov. Phys. JETP* **47** (1978) 154.
2. M. V. Kovalchuk, D. Lil'equist and V. G. Kohn, *Sov. Phys. Solid State* **28** (1986) 1918.
3. M. V. Kovalchuk, V. G. Kohn and E. F. Lobanovich, *Sov. Phys. Solid State* **27** (1985) 2034.
4. J. Zegenhagen, *Surf. Sci. Rep.* **18** (1993) 199.

5. A. Yu. Kazimirov, M. V. Kovalchuk and V. G. Kohn, *Acta Crystallogr.* A **46** (1990) 649.
6. V. G. Kohn, *Phys. Status Solidi B* **231** (2002) 132.
7. V. G. Kohn and M. V. Kovalchuk, *Phys. Status Solidi A* **64** (1981) 359.
8. M. V. Kovalchuk and V. G. Kohn, *Sov. Phys. — Usp.* **29** (1986) 426.
9. V. G. Kohn (2007), url: http://kohnvict.narod.ru
10. A. Yu. Kazimirov, M. V. Kovalchuk and V. G. Kohn, *Sov. Tech. Phys. Lett.* **14** (1988) 587.
11. V. N. Schemelev, M. V. Kruglov and V. P. Pronin, *Sov. Phys. Solid State* **12** (1971) 2005–2006; V.N. Schemelev and M.V. Kruglov, *Sov. Phys. Solid State* **14** (1973) 2988–2991; *Sov. Phys. Solid State* **16** (1974) 942–944; *Sov. Phys. Solid State* **17** (1975) 253–256; *Sov. Phys. Crystallogr.* **20** (1975) 153–157.
12. M. V. Kruglov, E. A. Sozontov, V. N. Schemelev and B. G. Zaharov, *Sov. Phys. Crystallogr.* **22** (1977) 397–400.
13. M. V. Kruglov, V. N. Schemelev and G. G. Kareva, *Phys. Status Solidi A* **46** (1978) 343–350.
14. A. Yu. Kazimirov, M. V. Kovalchuk, A. N. Sosphenov, V. G. Kohn, J. Kub, P. Novak and M. Nerviva, *Acta. Crystallogr.* B **48** (1992) 577.
15. I. A. Vartanyants and M. V. Kovalchuk, *Rep. Prog. Phys.* **64** (2001) 1009.
16. A. Kazimirov, J. Zegenhagen and M. Cardona, *Science* **282** (1998) 930–932.
17. E. Sozontov, L. X. Cao, A. Kazimirov, V. G. Kohn, M. Cardona and J. Zegenhagen, *Phys. Rev. Lett.* **86** (2001) 5329.
18. A. Kazimirov, T. Haage, L. Ortega, A. Stierle, F. Comin and J. Zegenhagen, *Solid State Commun.* **104** (1997) 347–350.
19. A. Kazimirov, G. Scherb, J. Zegenhagen, T.-L. Lee, M. J. Bedzyk, M. K. Kelly, H. Angerer and O. Ambacher, *J. Appl. Phys.* **84** (1998) 1703; A. Kazimirov, N. Faleev, H. Temkin, M. J. Bedzyk, V. Dmitriev and Yu. Melnik, *J. Appl. Phys.* **89** (2001) 6092.
20. M. J. Bedzyk, A. Kazimirov, D. L. Marasco, T.-L. Lee, C. M. Foster, G.-R. Bai, P. F. Lyman and D. T. Keane, *Phys. Rev.* B **61** (2000) R7813; D. L. Marasco, A. Kazimirov, M. J. Bedzyk, T.-L. Lee, S. K. Streiffer, O. Auciello and G.-R. Bai, *Appl. Phys. Lett.* **79** (2001) 515.

Chapter 4

X-RAY STANDING WAVE IN A BACKSCATTERING GEOMETRY

D. P. WOODRUFF

Physics Department, University of Warwick, Coventry CV4 7AL, UK

The special advantages, but also associated constraints, of conducting XSW at or close to a 180° backscattering geometry are described. In this condition, one is relatively insensitive to crystal mosaicity, making a wider range of crystalline materials accessible to the technique, but this condition also constrains the photon energy to the specific values determined by the crystal lattice plane spacings. Typically, this leads to relatively low photon energies, favoring the use of photoelectron or Auger electron monitoring of the X-ray absorption. This combination of parameters is particularly appropriate for ultra-high vaccum (UHV) studies of surface adsorption structure.

The great majority of X-ray scattering wave (XSW) investigations, particularly during the early development of the technique, has been performed using "hard" X-rays with typical photon energies of 8–18 keV. As the associated wavelength of ~0.1 nm is significantly less than typical interplanar spacings of crystalline solids, at least those corresponding to low Miller indices, the associated XSW condition is achieved at relatively grazing incidence to the scatterer planes. Under these conditions the rocking curve width, i.e. the width of the Darwin reflectivity curve, is very narrow, often only a few seconds of arc. This places three key requirements on the XSW experiment, namely that the incident radiation must be highly collimated, the crystalline sample should be highly perfect, with a mosaicity smaller than this rocking curve width, and high stability of the experimental set-up. Typically, the most perfect crystals are those of covalently-bonded least one of the available low Miller-index reflections, the X-ray emission yield is substantial and X-ray fluorescence measurements are also perfectly possible.

One should perhaps also remark that if sub-surface penetration is explicitly required for investigations of structural effects within the bulk or in thin films, NIXSW is still applicable using higher Miller-index reflections, including higher-order reflections of widely-spaced scatterer planes. These higher energies do increase the demands on the spectral resolution of the incident radiation, however, as remarked above, and while the smaller associated interplanar spacings potentially lead to greater precision in atomic positions, they also lead to greater ambiguities. This is a general issue in all XSW measurements; the measured coherent positions lead to a determination of the absorber position relative to the extended bulk scatterer planes, and not the actual atomic planes. In particular, atomic positions differing by any integral number of scattering interplanar distances are indistinguishable in XSW. For atoms located within the solid, this ambiguity is generally irrelevant, because for the lowest-order reflections, these displacements correspond to symmetrically identical locations within the solid. At the surface, however, the fact that the solid is terminated means this is not the case; in this case, the same location relative to different extended scatterer planes leads to different distances from the surface. In practice, even a very approximate knowledge of the likely bonding distance of the adsorbed atoms relative to the underlying outermost surface layer removes this ambiguity, but solving this problem is clearly much easier if the interplanar spacings (and thus the positional ambiguity given by XSW) are large, rather than small.

Both Auger electron and photoelectron detection have been used in NIXSW studies of surface structure. Photoelectron detection offers both advantages and complications. An important advantage is the ability to achieve chemical-state specificity through "chemical shifts" to the photoelectron binding energy associated with atoms of the same element in inequivalent bonding configurations, and exploiting this potential is described more fully in Chapter 25. In principle, of course, such chemical shifts must also occur in Auger electron emission, but in practice, the most intense Auger electron signals are associated with transitions involving one or two valence states, leading to relative broad spectral lines and thus less ability to resolve small shifts. The complication associated with photoelectron emission arises from the intrinsic angular dependence of the core level photoemission. Photoelectron detection of the photoemission involves components of the signal arising from both the incident and reflected X-ray beams (and their coherent interference to give the standing wave), so if the angular dependence of these two components differs, and angle-resolved measurement of the photoemission yield will measure the two components differently and thus will not properly reflect the relative photo-absorption in the standing wave.

In this regard, we note that it has been usual to assume that photoemission is governed by dipole selection rules, and in this case the angular dependence is defined entirely relative to the polarization vector of the radiation. For this reason, if the dipole approximation is valid, an NIXSW experiment ensures that the angle-resolved photoemission signal is proportional to the photo-absorption cross-section, because the incident and reflected X-ray beams are collinear. This ensures that the polarization vectors of the incident and reflected waves are also collinear, so the angular dependence of the photoemission from the two components is identical. This is not true, of course, at a more general incidence angle, if the polarization vector of the incident radiation contains a component within the plane defined by the incident and reflected X-ray beams (π-polarization). However, it has been found that in NIXSW experiments the dipole approximation is not generally adequate to fully describe the effects of the angular dependence of the photoemission.[10] Instead, quadrupolar contributions that lead to an inequivalence in the angular dependence depending on the direction of photon propagation are significant, and these effects on the photoemission monitoring of XSW must be accounted for to correctly interpret the absorption profiles, even for the NIXSW geometry. However, an established procedure for applying this correction has been demonstrated, and the relevant parameters have been reported for emission from initial $1s$ states for all atoms from C to Cl.[11,12] This problem is described more fully in Chapter 11 and in Ref. 3.

While the low photon energies characteristic of NIXSW experiments are generally beneficial in photoemission monitoring of the X-ray absorption profile in UHV surface science experiments, they are also especially relevant in extending the applicability of the XSW method to low atomic number species. Particularly challenging are the important elements of the first full row of the Periodic Table, notably C, N and O. These are key ingredients of many organic molecules of interest in surface chemistry and organic semiconductor, and, of course, are key components of carbides, nitrides and oxides, the latter group being a particularly and generally important class of solids. For these elements the deepest core levels are only in the binding energy range of ~280–530 eV, so the photoionization (and hence photoabsorption) cross-sections for these atoms are extremely small even at typical NIXSW energies of ~3 keV. Nevertheless, these lower energies do make the experiments possible, and a number of NIXSW studies of submonolayer coverages of molecules containing these atoms have proved successful, significantly expanding the range of systems to which the general XSW methodology can be applied.

One final issue concerning the NIXSW technique is the problem of establishing accurately the instrumental parameters of the experiment, and particularly the absolute energy scale, required in the fitting of the experimental absorption profiles to extract the structural parameters. In XSW, generally this is typically achieved by a measurement of the reflectivity profile, R, as one scans through the standing wave condition in angle or photon energy. For normal incidence to the scatterer planes the incident and reflected beams are collinear, so a separate measurement of the reflectivity is not possible. A number of solutions to this problem are possible. One is to simply insert a detector into the incident beam and thus monitor the sum of the incident beam intensity, I_0, and the reflected intensity, $I_0(1 + R)$. Another is to work at an incidence angle slightly (~1–2°) removed from true normal incidence, a condition which is generally advisable in any case to avoid problems with multiple-beam excitation, as mentioned above. Under this condition the insensitivity to the exact incidence angle (and thus to beam collimation and crystal mosaicity) is retained, but the incident and reflected beam are sufficiently separated a few centimeters from the sample to allow separate measurements of the incident and reflected beam intensities. A third option is to dispense with the reflectivity measurement completely and use the XSW absorption profile of the substrate (if the structure is known) as the reference spectrum for extracting the non-structural parameter values. In practice, all three of these methods have been used in different studies.

References

1. G. Materlik, J. Zegenhagen and W. Uelhoff W, *Phys. Rev. B* **32** (1985) 5502.
2. H. D. Abruna, T. Gog T, G. Materlik and W. Uelhoff, *J. Electroanal. Chem.* **360** (1993) 315.
3. D. P. Woodruff, *Rep. Prog. Phys.* **68** (2005) 743.
4. J. P. Sutter, E. E. Alp, M. Y. Hu, P. L. Lee, H. Sinn, W. Sturhahn, T. T. Toellner, G. Bortel and R. Colella, *Phys. Rev. B* **63** (2001) 094111.

5. D. P. Woodruff, D. L. Seymour, C. F. McConville, C. E. Riley, M. D. Crapper, N. P. Prince and R. G. Jones, *Surf. Sci.* **195** (1988) 237.
6. R. Jaeger, J. Feldhaus, J. Haase, J. Stöhr, Z. Hussain, D. Menzel and D. Norman, *Phys. Rev. Lett.* **45** (1980) 1870.
7. D. P. Woodruff, D. L. Seymour, C. F. McConville, C. E. Riley, M. D. Crapper, N. P. Prince and R. G. Jones, *Phys. Rev. Lett.* **58** (1987) 1460.
8. T. Ohta, H. Sekiyama, Y. Kitajima, H. Kuroda, T. Takahashi and S. Kikuta, *Jpn. J. Appl. Phys.* **24** (1985) L485.
9. T. Ohta, Y. Kitajima, H. Kuroda, T. Takahashi and S. Kikuta, *Nucl. Instrum. Methods A* **246** (1986) 760.
10. C. J Fisher, R. Ithin, R. G. Jones, G. J. Jackson, D. P. Woodruff, B. C. C. Cowie and M. F. Kadodwala, *J. Phys.: Condens. Matter* **10** (1998) L623.
11. G. J. Jackson, B. C. C. Cowie, D. P. Woodruff, R. G. Jones, M. S. Kariapper, C. J. Fisher, A. S. Y. Chan and M. Butterfield, *Phys. Rev Lett.* **84** (2000) 2346.
12. J. Lee, C. Fisher, D. P. Woodruff, M. G. Roper, R. G. Jones and B. C. C. Cowie, *Surf. Sci.* **494** (2001) 166.

Chapter 5

X-RAY STANDING WAVE AT THE TOTAL REFLECTION CONDITION

MICHAEL J. BEDZYK

Department of Materials Science and Engineering, Northwestern University, Cook Hall Evanston, IL 60208, USA

Argonne National Laboratory, 9700 South Cass Avenue Argonne, IL 60439, USA

Fresnel theory is used to derive the complex electric fields above and below an X-ray reflecting interface that separates two materials with differing refraction indices. The interference between the incident and reflected waves produces an X-ray standing wave (XSW) above the reflecting interface. The XSW intensity modulation is strongly enhanced by the total external reflection condition, which occurs at incident angles less than the critical angle. At these small milliradian incident angles, the XSW period ($\lambda/2\theta$) becomes very large, which makes the TR-XSW an ideal probe for studying low-density structures that extend 1 to 1000 nm above the reflecting interface.

5.1. Introduction

The original (and most widely used) method for generating an X-ray standing wave (XSW) has been to use dynamical diffraction from a perfect single crystal in a Bragg reflection geometry.[1–3] However, as with any standing wave phenomena, the minimum requirement is the superposition of two coherently-coupled plane waves. Therefore, one can imagine several alternative geometries for generating an XSW. This chapter discusses the case of generating an XSW by total external reflection from an X-ray mirror surface.[4]

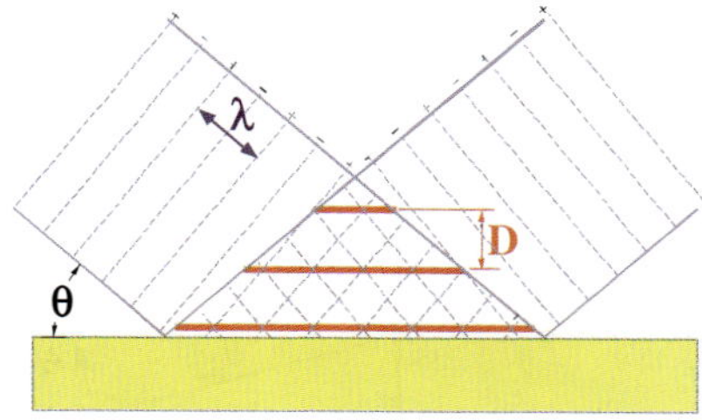

Fig. 5.1. Illustration of XSW generated by interference between incident and specular reflected plane waves.

Referring to Fig. 5.1, the primary distinguishing feature for the total reflection (TR) case is that the length of the XSW period above the mirror surface,

$$D = \frac{\lambda}{2\sin\theta} = \frac{2\pi}{Q}, \tag{5.1}$$

is much longer, since TR occurs at very small incident angles, θ. Also, the length of the XSW period, D, will continuously decrease as θ increases through the range of TR. This long-period XSW is ideally suited for measuring surfaces, interfaces, and supported nanostructures with structural features that range from 50 to 2000 Å. Examples include studies of Langmuir–Blodgett (LB) multilayers,[4–7] layer-by-layer self-assembly of metal–organic films,[8,9] the diffuse double-layer formation at the electrified water/solid interface,[10,11] biofilm ion adsorption,[12] and metal nanoparticle dispersion in polymer films.[13,14]

5.2. X-Ray Transmission and Reflection at a Single Interface

Based on Maxwell's equations, an electromagnetic traveling plane wave impinging on a boundary separating two different refractive media, splits into a reflected and transmitted (or refracted) plane wave.[15] At X-ray frequencies, the index of refraction,

$$n_j = 1 - \delta_j - i\beta_j \tag{5.2}$$

is less than unity and therefore (as illustrated in Fig. 5.2) the angle of refraction, θ_2, is less than the incident angle, θ_1.[16] Parameters δ_j and β_j, which account, respectively, for refraction and absorption effects by the

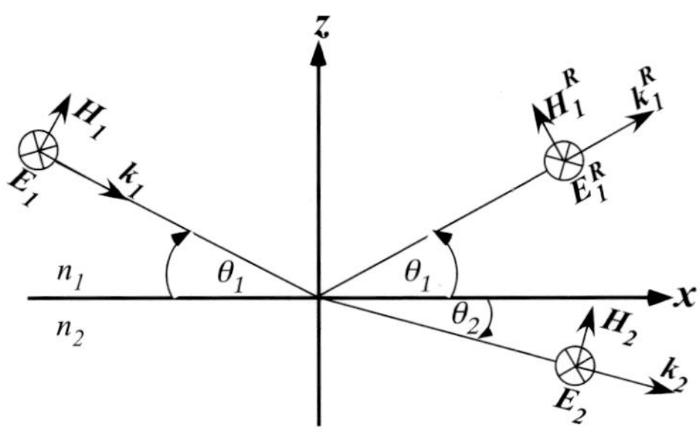

Fig. 5.2. The σ-polarization case for the reflection and refraction of X-rays at a boundary separating two media with indices of refraction $n_1 > n_2$.

j^{th} medium, can be expressed as:

$$\delta = -\frac{1}{2}\chi_0' = \frac{r_e\lambda^2}{2\pi}N_e', \tag{5.3}$$

$$\beta = -\frac{1}{2}\chi_0'' = \frac{\lambda\mu_0}{4\pi}, \tag{5.4}$$

where N_e' is the real part of the effective electron density.

The E-fields associated with the incident, reflected, and transmitted plane waves are expressed respectively as:

$$\bar{\varepsilon}_1(\mathbf{r},t) = \mathbf{E}_1 \exp(-i(\mathbf{k}_1\bullet\mathbf{r} - \omega t)) \tag{5.5a}$$

$$\bar{\varepsilon}_1^R(\mathbf{r},t) = \mathbf{E}_1^R \exp(-i(\mathbf{k}_1^R\bullet\mathbf{r} - \omega t)) \tag{5.5b}$$

$$\bar{\varepsilon}_2(\mathbf{r},t) = \mathbf{E}_2 \exp(-i(\mathbf{k}_2\bullet\mathbf{r} - \omega t)). \tag{5.5c}$$

At $z = 0$, the space and time variations of all three fields must be equivalent. This produces the "law of co-planarity," which requires the transmitted and reflected wave vectors, $\mathbf{k}_2$ and $\mathbf{k}_1^R$, to be confined to the same plane as the incident wave vector, $\mathbf{k}_1$ (the xz-plane in Fig. 5.2). The continuity of the tangential components of the three wave vectors at the boundary dictates the kinematical properties corresponding to the "law of reflection" $\theta_1^R = \theta_1$ and the "law of refraction" (Snell's Law) $n_2 \cos\theta_2 = n_1 \cos\theta_1$. Using these relationships the spatial components in Eq. (5.5) can be expressed as:

$$\mathbf{k}_1\bullet\mathbf{r} = k_1(x\cos\theta_1 - z\sin\theta_1) \tag{5.6a}$$

$$\mathbf{k}_1^R\bullet\mathbf{r} = k_1(x\cos\theta_1 + z\sin\theta_1) \tag{5.6b}$$

$$\mathbf{k}_2\bullet\mathbf{r} = \frac{n_2}{n_1}k_1\left[x\frac{n_1}{n_2}\cos\theta_1 - z\sqrt{1-\left(\frac{n_1}{n_2}\cos\theta_1\right)^2}\,\right]. \tag{5.6c}$$

Total reflection occurs when the transmitted plane wave $\bar{\varepsilon}_2(\mathbf{r}, t)$ propagates strictly in the x-direction and is attenuated in the inward negative z-direction. From Eqs. (5.5c) and (5.6c), TR occurs when $\theta_1 < \theta_C$. For $n_1 = 1$ (e.g., vacuum or air) and $n_2 = 1 - \delta - i\beta$, the critical angle[16] is

$$\theta_C = \sqrt{2\delta}. \tag{5.7}$$

The scattering vector at the critical angle is

$$Q_c = 4\pi \sin\theta_c/\lambda \cong 4\pi\theta_c/\lambda = 4\sqrt{\pi r_e N_e'}. \tag{5.8}$$

If dispersion corrections are ignored $N_e' = N_e$ and Q_c becomes a wavelength-independent property, in which case, $Q_c = 0.0315$ Å^{-1} for Si and $Q_c = 0.0812$ Å^{-1} for Au.

The continuity of the tangential components of the E-fields and magnetic-fields at the $z = 0$ boundary dictates the dynamical properties of the fields, corresponding to the Fresnel equations, which for the σ-polarization case and for small angles θ_1 can be expressed as

$$F_{1,2}^R = \frac{E_1^R}{E_1} = \left|\frac{E_1^R}{E_1}\right| e^{iv} = \frac{q - \sqrt{q^2 - 1 - ib}}{q + \sqrt{q^2 - 1 - ib}} \tag{5.9}$$

$$F_{1,2}^T = \frac{E_2}{E_1} = \frac{2q}{q + \sqrt{q^2 - 1 - ib}}, \tag{5.10}$$

where the normalized angle $q = \theta_1/\theta_C = Q/Q_c$ and $b = \beta/\delta$ for the case of $n_1 = 1$ and $n_2 = 1 - \delta - i\beta$. At $\lambda = 0.71$ Å, $b = 0.005$ for Si and $b = 0.1$ for Au.

Figure 5.3 shows the q dependence of the reflectivity, $R = |F_{1,2}^R|^2$, the normalized E-field intensity at the surface, $I_{z=0} = |F_{1,2}^T|^2$, the phase of the reflected plane wave relative to the phase of the incident at $z = 0$, $v = \mathrm{Arg}(F_{1,2}^R)$, and the penetration depth

$$\Lambda = 1/\mu_e = \left[Q_c \,\mathrm{Re}\left[\sqrt{1 - q^2 + ib}\right]\right]^{-1}. \tag{5.11}$$

As can be seen, TR occurs for $q < 1$, where the reflectivity approaches unity, the phase shifts by π radians, and E-field intensity below the surface forms an evanescent wave[17] with a penetration depth approaching Q_c^{-1}, which is 32 Å for Si and 12 Å for Au. For q increasing above unity, the reflectivity quickly reduces (approaching zero) and the transmitted wave propagates into the medium with a penetration depth quickly approaching the normal absorption process, where $\Lambda = \sin\theta/\mu_0$.

5.4. X-Ray Fluorescence Yield from an Atomic Layer within a Thin Film

The q dependence for the normalized E-field intensity at $z = 0$ is shown in Fig. 5.3(b). Figure 5.5 shows the Eq. (5.12) calculation for the two additional heights above the surface. These three curves illustrate the basis for the TR-XSW technique as a positional probe, since (in the dipole approximation for the photo-effect) the XRF yield, $Y(q)$, from an atomic-layer at a discrete height z will follow such a curve. Note that in the TR range, $0 < q < 1$, the number of modulations in the E-field intensity is equivalent to $z/D_c + (1/2)$. The extra 1/2 modulation is due to the π phase shift shown in Fig. 5.3(b). Therefore, for an XRF-marker atom layer within a low-density film on a high-density mirror, the atomic layer height can be quickly approximated by counting the number of modulations in the XRF yield that occur between the film critical angle and the mirror critical angle. Referring to Fig. 5.6, this effect can be seen in the experimental results and analysis for the case of a Zn atomic layer trapped at the topmost bilayer of a 1000-Å-thick LB multilayer that was deposited on a gold mirror. There are $11\frac{1}{2}$ Zn K$_\alpha$ XRF modulations as the incident angle is advanced over this range, indicating that the Zn layer is at a height of 11 critical periods (or 900 Å) above the gold surface. From the simultaneously collected reflectivity shown in Fig. 5.6, the critical angles for the LB film and Au mirror are at 2.15 and 7.52 mrad, respectively. A more rigorously determined Zn atomic distribution profile, $\rho(z)$, is determined by a fit of

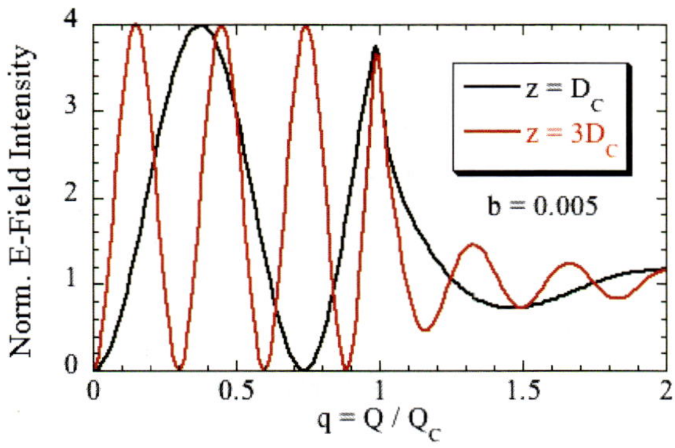

Fig. 5.5. The normalized-angle dependence of the normalized E-field intensity for two different heights above the mirror surface. The calculations are for the case of $b = 0.005$ in Eq. (5.12). The critical period $D_C = 199$ Å for Si and 77 Å for Au.

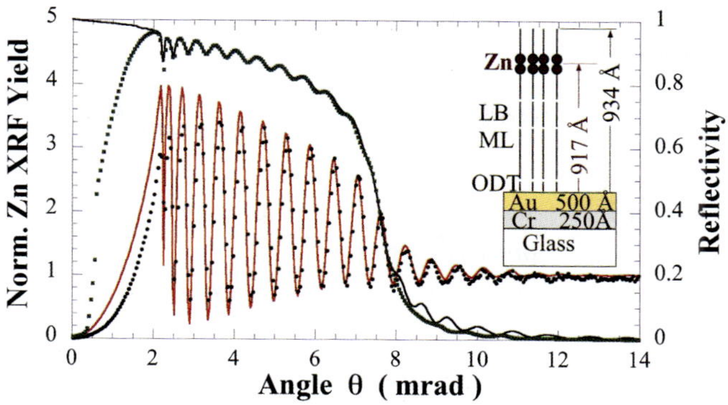

Fig. 5.6. The experimental and theoretical reflectivity and Zn K_α XRF yield versus incident angle at $\lambda = 1.215$ Å from a LB multilayer-coated gold mirror depicted in the inset. From the reflectivity fit, the film thickness is measured to be $t_F = 934$ Å and the interface roughness $\sigma_1 = 3$ Å. From the XSW Zn yield fit with a modeled Zn Gaussian distribution, $\langle z \rangle = 917$ Å and $\sigma_{\text{Zn}} = 15$ Å (FWHM = 35.3 Å). The data deviation from theory for $\theta < 2$ mrad is due to X-ray foot-print geometrical effects. See Ref. 5, for details from a similar measurement on a similar sample.

the modeled XRF yield $Y(\theta) = \int \rho(z) I(\theta, z) dz$ to the data in Fig. 5.6, where the E-field intensity $I(\theta, z)$ within the refracting (and absorbing) film was calculated by an extension of Parratt's recursion formulation[18] described in the section entitled "XSW in Multilayers" (Chapter 7). This same model described in the inset was also used to generate a fit to the reflectivity data (Fig. 5.6), which is independently sensitive to the density and thickness of the film and the widths of the interfaces. The very sharp drop in the reflectivity at the film critical angle (2.15 mrad) is due to the excitation of the first mode of a resonant cavity that was observed to produce a 20-fold enhancement in the E-field intensity at the center of the film.[19]

5.5. Fourier Inversion for a Direct Determination of $\rho(z)$

Similar to the Bragg diffraction XSW case (see the section entitled "XSW Imaging", Chapter 14), the TR-XSW XRF yield is also directly linked to the Fourier transform of the atomic distribution. In this case, however, the Fourier transform is measured at low-Q, over a continuous range in Q and only in the Q_z direction. This section describes how the Fourier transform can be extracted from the TR-XSW data to produce a model-independent measure of the atomic distribution profile, $\rho(z)$.

To account for the refraction and absorption effects that will influence the observed reflectivity and XRF yields from a film-coated mirror, the earlier described two-layer model (Fig. 5.2) needs to be replaced by a three-layer model (or double interface model) formed by vacuum (the $j = 1$ layer), a thin low-density film ($j = 2$), and a higher-density mirror ($j = 3$). The $j = 2$ film/mirror interface is at $z = 0$ and the $j = 1$ vacuum/film interface is at $z = t_F$. For the present case study, $\delta_1 = \beta_1 = 0$, $\delta_2 \ll \delta_3$, and $\beta_2 \ll \beta_3 \ll \delta_3$. For the Fig. 5.6 example of 1.215 Å X-rays reflecting from a gold mirror coated with a LB multilayer, $\delta_2 = 2.31 \times 10^{-6}$, $\delta_3 = 2.83 \times 10^{-5}$, $\beta_2 = 1.90 \times 10^{-9}$, $\beta_3 = 1.96 \times 10^{-6}$, $\theta_{c,2} = 2.15$ mrad, $\theta_{c,3} = 7.52$ mrad. At $\theta_1 = \theta_{c,3}$, the XSW period is 80.8 Å in the vacuum and 84.2 Å inside the LB film.

TR exists at the interface above the jth layer when $\theta_1 < \theta_{c,j} = (2\delta_j)^{1/2}$. When $\theta_1 > \theta_{c,2}$, a refracted (or transmitted) traveling wave penetrates through the film and is reflected from the mirror surface. The total E-field intensity within the film is then described as

$$I_2^T(q_2, Z) = I_2(q_2, Z) + I_2^R(q_2, Z) + 2\sqrt{I_2(q_2, Z) I_2^R(q_2, Z)} \times \cos(v_2(q_2) - 2\pi q_2 Z), \tag{5.13}$$

where I_2 and I_2^R are the respective intensities of the incident (transmitted) and reflected plane waves that form an interference field within the film. The refraction-corrected normalized angle (or normalized scattering vector) within the film is defined as

$$q_2 = Q_2/Q_{c,2} = \left(\theta_1^2 - \theta_{c,2}^2\right)^{1/2} \left(\theta_{c,3}^2 - \theta_{c,2}^2\right)^{-1/2}. \tag{5.14}$$

Here Z is the normalized height above the mirror surface in units of the refraction corrected critical period. Namely, $Z = z/D_{c,2}$, where $D_{c,2} = \lambda_1/(2\theta_{c,3})/(1 - (1/2)\,\delta_2/\delta_3)$. The use of generalized coordinates q_2 and Z makes the description independent of wavelength and index of refraction. The phase of the reflected plane wave relative to the incident at $z = Z = 0$ is expressed as v_2. Based on the dipole approximation for the photoelectric effect, the fluorescence yield from a normalized atomic distribution $\rho(Z)$ within the film is

$$Y(q_2) = \int_0^{t'_F} I_2^T(q_2, Z)\rho(Z)dZ, \tag{5.15}$$

where $t'_F = t_F/D_{C,2}$ is the normalized film thickness.

Weakly Absorbing Film Approximation: If the attenuation depth of I_2 and I_2^R within the film is large in comparison to the spread $\langle Z^2 \rangle^{1/2}$ of $\rho(Z)$, then Eq. (5.15) can be simplified, so that the yield for a normalized incident intensity and normalized distribution is expressed as

$$\begin{aligned} Y(q_2) &= I_2(q_2, \langle Z \rangle) + I_2^R(q_2, \langle Z \rangle) \\ &\quad + 2\sqrt{I_2(q_2, \langle Z \rangle) I_2^R(q_2, \langle Z \rangle)}\, y(q_2), \end{aligned} \tag{5.16}$$

where the modulation in the yield due to the interference fringe field is

$$y(q_2) = \int_0^{t'_F} \rho(Z) \cos(v_2(q_2) - 2\pi q_2 Z) dZ. \tag{5.17}$$

Since $I_2(q_2, \langle Z \rangle)$ and $I_2^R(q_2, \langle Z \rangle)$ can be calculated from Parratt's recursion formulation, this reduced yield, $y(q_2)$, can be extracted from the measured yield $Y(q)$. Figure 5.7 shows this for the yield data shown in Fig. 5.6. The inverse Fourier transform of this reduced yield can be directly used to generate the fluorescence selected atom distribution $\rho(z)$ to within a resolving limit defined by the range of Q over which the visibility of the interference fringes is significant.

Linear Phase Approximation: If the phase v_2 were zero in Eq. (5.17), $y(q_2)$ would simply be the real part of the Fourier transform of $\rho(Z)$. Based

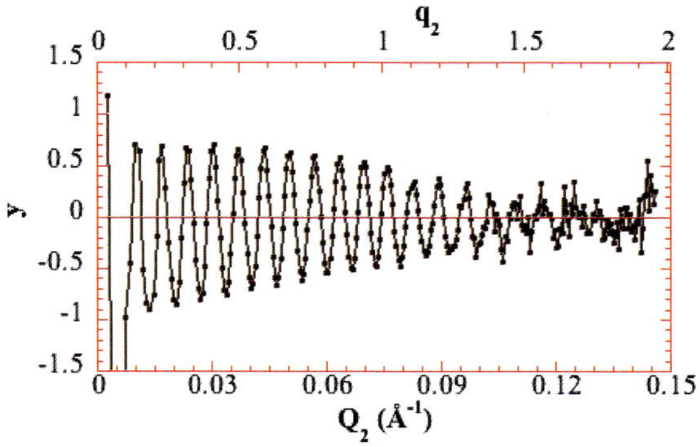

Fig. 5.7. The reduced yield that was extracted from the Zn XRF yield data in Fig. 5.6. See Eqs. (5.16) and (5.17). The line connecting data points is drawn to guide the eye. The oscillation period and envelop width are inversely related to the mean height and intrinsic width, respectively, of the Zn distribution profile.

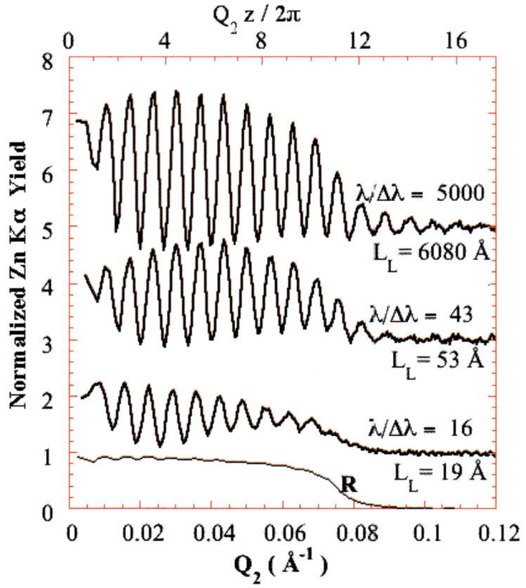

Fig. 5.9. Experimental demonstration of TR-XSW sensitivity to longitudinal coherence (L_L). The bottom and top curves are the Q_2 dependence of the reflectivity (R) and Zn XRF yield for the data from Fig. 5.6 that was taken with a Si(111) monochromator. The mean Zn height is at $z = 917$ Å. The lower two XRF yield curves are from the same sample, but taken with Si/Mo and C/Rh multilayer monochromators with reduced monochromaticity ($\lambda/\Delta\lambda$) and therefore reduced longitudinal coherence. The top two curves are vertically offset by 2 and 4 units, respectively.

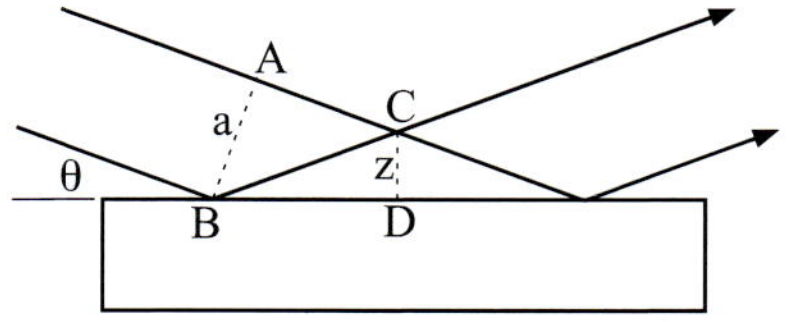

Fig. 5.10. Schematic ray diagram used for illustrating coherence effects between the incident and specular reflected X-ray beams at height z above the mirror surface.

Acknowledgments

Colleagues who inspired and assisted in this work include Donald Bilderback, Boris Batterman, Martin Caffrey, Hector Abruna, Mark Bommarito, Jin Wang, Thomas Penner, Jay Schildkraut, Paul Fenter, and Gordon Knapp. The data in this chapter were collected at the D-Line station of the Cornell High Energy Synchrotron Source (CHESS), which is supported by the US National Science Foundation. The work was also partially supported by the US Department of Energy.

References

1. B. W. Batterman, *Phys. Rev. Lett.* **22** (1969) 703.
2. P. L. Cowan, J. A. Golovchenko and M. F. Robbins, *Phys. Rev. Lett.* **44** (1980) 1680.
3. J. A. Golovchenko, J. R. Patel, D. R. Kaplan, P. L. Cowan and M. J. Bedzyk, *Phys. Rev. Lett.* **49** (1982) 560.
4. M. J. Bedzyk, G. M. Bommarito and J. S. Schildkraut, *Phys. Rev. Lett.* **62** (1989) 1376.
5. J. Wang, M. J. Bedzyk, T. L. Penner and M. Caffrey, *Nature* **354** (1991) 377.
6. M. J. Bedzyk, D. H. Bilderback, G. M. Bommarito, M. Caffrey and J. S. Schildkraut, *Science* **241** (1988) 1788.
7. S. I. Zheludeva, M. V. Kovalchuk, N. N. Novikova, A. N. Sosphenov, V. E. Erochin and L. A. Feigin, *J. Phys. D* **26** (1993) A202.
8. W. B. Lin, T. L. Lee, P. F. Lyman, J. J. Lee, M. J. Bedzyk and T. J. Marks, *J. Am. Chem. Soc.* **119** (1997) 2205.
9. J. A. Libera, R. W. Gurney, C. Schwartz, H. Jin, T. L. Lee, S. T. Nguyen, J. T. Hupp and M. J. Bedzyk, *J. Phys. Chem. B* **109** (2005) 1441.
10. M. J. Bedzyk, M. G. Bommarito, M. Caffrey and T. L. Penner, *Science* **248** (1990) 52.
11. J. Wang, M. Caffrey, M. J. Bedzyk and T. L. Penner, *Langmuir* **17** (2001) 3671.
12. A. S. Templeton, T. P. Trainor, S. J. Traina, A. M. Spormann and G. E. Brown, *Proc. Nat. Acad. Sci.* **98** (2001) 11897.
13. R. S. Guico, S. Narayanan, J. Wang and K. R. Shull, *Macromolecules* **37** (2004) 8357.
14. B. H. Lin, T. L. Morkved, M. Meron, Z. Q. Huang, P. J. Viccaro, H. M. Jaeger, S. M. Williams and M. L. Schlossman, *J. Appl. Phys.* **85** (1999) 3180.
15. M. Born and E. Wolf, *Principles of Optics*, 6th edn. (Pergamon Press, Oxford, 1980).
16. A. H. Compton and S. K. Allison, *X-Rays in Theory and Experiment* (van Nostrand, New York, 1935).
17. R. S. Becker, J. A. Golovchenko and J. R. Patel, *Phys. Rev. Lett.* **50** (1983) 153.
18. L. G. Parratt, *Phys. Rev.* **95** (1954) 359.
19. J. Wang, M. J. Bedzyk and M. Caffrey, *Science* **258** (1992) 775.

Chapter 6

X-RAY STANDING WAVE AT GRAZING INCIDENCE AND EXIT

OSAMI SAKATA

National Institute for Materials Science,
Synchrotron X-ray Laboratory at SPring-8,
1-1-1 Kouto, Sayo, Hyogo 679-5148, Japan
SAKATA.Osami@nims.go.jp

TERRENCE JACH

Material Measurement Laboratory, MS8371,
National Institute of Standards and Technology,
100 Bureau Drive, Gaithersburg, MD 20899-8371, USA
terrence.jach@nist.gov

Grazing-angle X-ray standing waves (GAXSWs) generated dynamically are discussed under a geometry where incident, specularly reflected, and specularly diffracted waves make grazing angles to a crystal surface. The grazing angles are close to the critical angle for total external reflection. A Bragg condition is satisfied on lattice planes perpendicular to the surface. We explain all waves excited inside and outside the crystal using a three-dimensional dispersion surface. The perpendicular components of the waves are derived from the geometrical consideration on the basis of phase continuity on the boundary surface. Electric fields are formulated by solving simultaneous equations of boundary conditions for the tangential components of electric field and the normal components of the magnetic fields. The GAXSW field is formed by the interference of incident, specularly reflected, and specularly diffracted beams above the surface. The external field intensity, modulated along the surface, is expressed, while the position-dependent behaviors of the intensity are discussed. The output, proportional to the local field intensity at an adsorbed atom, is normally parametrized by the coherent position and the coherent fraction. We introduce two examples of in-plane structural analyses of iodine adsorbed on Ge(111) and 10-nm-thick $Ca_{0.39}Sr_{0.61}F_2$ epilayer film on GaAs(111).

6.1. Introduction

One of the prominent applications of the X-ray standing wave (XSW) technique is to determine the positions of adsorbate atoms on a surface. Since the interference of the X-ray field continues to produce a standing wave above the surface of a crystal at the Bragg condition, the XSW technique can be applied to atoms above the crystal, which may constitute an adsorbate layer on the surface, an epitaxial layer, or an interface. Although triangulation (or imaging, cf. Chapter 14), using various reflections is often possible, one may ask how XSW measurements can yield directly structural information parallel to the crystal surface. With the availability of well-collimated synchrotron radiation beams, X-ray diffraction at grazing angles with the crystal surface became feasible in order to study surface structures.[1] In this technique, the X-ray beam arrives nearly parallel to the surface, is diffracted by crystal planes normal to the surface by a substantial angle, and also departs nearly parallel to the surface. The technique is commonly called surface X-ray diffraction (SXRD) or grazing incidence diffraction (GIXRD or GID) and has been widely used to analyze surface reconstructions.

It was a logical consequence to consider dynamical diffraction at a crystal surface at grazing incidence. This process is much more complex theoretically than the common two-beam dynamical diffraction problem. The geometry involves incident, specularly reflected, and reflected–diffracted beams outside the crystal, in addition to beams that propagate in the initial and diffracted directions within the crystal,[2–8] and it requires a three-dimensional dispersion surface.[5,6,8,9] As an example of the novel effects of diffraction in this geometry, the reflected–diffracted wave becomes evanescent at a certain incidence condition. Thus its X-ray field is confined to a thin layer right above the surface.

GID is also experimentally more demanding, since the incident radiation has to be collimated both normal to the surface (the angle of incidence) and parallel to the surface (the diffraction angle). Good collimation along both directions requires a synchrotron light source with a small source size and a beamline with a long tangent point. Only a few of the second-generation light sources originally met these criteria.

Effects due to the formation of XSWs in the GID geometry were initially observed inside perfect crystals.[6,10–12] For this variant of the XSW technique we shall refer here to grazing-angle X-ray standing waves (GAXSW), since the reflected and diffracted beams are also always detected at grazing angles. The reciprocal-lattice vector is directed nearly parallel to the crystal surface, unlike the conventional Bragg standing wave geometry field (Refs. 2–6).

$$E_s = \{(\phi - \Phi_2)(\phi_h + \Phi_1)\xi_1 - (\phi - \Phi_1)(\phi_h + \Phi_2)\xi_2\}E_o/\Lambda, \quad (6.12)$$

$$E_h = 2\phi(\Phi_1 - \Phi_2)\xi_1\xi_2 E_o/\Lambda, \quad (6.13)$$

$$D_{o1} = -2\phi(\phi_h + \Phi_2)\xi_2 E_o/\Lambda, \quad (6.14)$$

$$D_{o2} = 2\phi(\phi_h + \Phi_1)\xi_1 E_o/\Lambda, \quad (6.15)$$

$$\Lambda = (\phi + \Phi_2)(\phi_h + \Phi_1)\xi_1 - (\phi_h + \Phi_2)(\phi + \Phi_1)\xi_2. \quad (6.16)$$

The amplitude ratio ξ_j is defined from the fundamental equation as $\xi_j \equiv D_{hj}/D_{oj} = (\Phi_{oj}^2 - \phi^2 - \chi_o)/C^2\chi_{\bar{h}}$. The above angles except ϕ are complex numbers; Φ_j and ϕ_h are written approximately when ϕ is small: $\Phi_j \equiv \Phi_{oj} \approx K_{oj\perp}/k_o = K_{hj\perp}/k_o \approx \Phi_{hj}$ and $\phi_h \approx -k_{h\perp}/k_o$; ϕ_h is positive or imaginary as Eq. (6.2) suggests. Then we obtain

$$\Phi_{oj} = \Phi_{hj} \approx (\alpha + i\beta), \quad (6.17)$$

$$\phi_h \approx \begin{cases} -i(-y - \phi^2)^{1/2} & \text{for } y + \phi^2 < 0, \\ (\phi^2 + y)^{1/2} & \text{for } y + \phi^2 \geq 0. \end{cases} \quad (6.18)$$

The internal beams in the crystal have actually been observed as beams emerging from the backside of thin crystals[7].

6.3. The Standing Wave Field Above a Surface

The external GAXSW field intensity I above the crystal surface is given by expressions derived in Ref. 13. Here, z is pointing toward the interior of the crystal, $z = 0$ is defined as the surface position, and we consider only σ-polarization (X-ray beams with E vectors normal to the diffraction plane).

$$\begin{aligned} I &= |E_o \exp(-2\pi i\mathbf{k}_o \cdot \mathbf{r}) + E_s \exp(-2\pi i\mathbf{k}_s \cdot \mathbf{r}) + E_h \exp(-2\pi i\mathbf{k}_h \cdot \mathbf{r})|^2 \\ &= E_o^2[1 + R_s^2 + R_h^2 \exp\{-4\pi k_o \operatorname{Im}(\phi_h)z\} + 2R_s \cos[-\nu_s - 4\pi k_o\phi z] \\ &\quad + 2R_h \exp\{-2\pi k_o \operatorname{Im}(\phi_h)z\} \cos[\nu_h + 2\pi k_o\{\operatorname{Re}(\phi_h) + \phi\}z - 2\pi\mathbf{H}\cdot\mathbf{r}_{\|}] \\ &\quad + 2R_s R_h \exp\{-2\pi k_o \operatorname{Im}(\phi_h)z\} \\ &\quad \times \cos[-\nu_h + \nu_s - 2\pi k_o\{\operatorname{Re}(\phi_h) - \phi\}z + 2\pi\mathbf{H}\cdot\mathbf{r}_{\|}]], \end{aligned} \quad (6.19)$$

where R_s and R_h are proportional to the real amplitude of E_s and E_h while ν_s and nu_h are the phases of E_s and E_h, respectively; $E_s/E_o = R_s \exp(i\nu_s)$ and $E_h/E_o = R_h \exp(i\nu_h)$.

The output (fluorescence, photoelectrons, etc.) Y expressed in Eq. (6.20), proportional to the local field intensity at the atoms, is normally parametrized by the coherent position P_H and the coherent fraction F_H. The coherent position P_H ($\equiv \mathbf{H}\cdot\mathbf{r}_{\|}$) is defined as the phase of the hth Fourier component of the normalized atom distribution. The coherent fraction F_H is defined as the amplitude of the Fourier component for the reciprocal-lattice vector $\mathbf{H}$ of the normalized atom distribution.[19] It is noted that the sign of P_H is here different from that of P_H defined in Refs. 13 and 14.

$$\begin{aligned} Y &= 1 + R_s^2 + R_h^2 \exp\{-4\pi k_o \operatorname{Im}(\phi_h)z\} + 2R_s \cos[-\nu_s - 4\pi k_o \phi z] \\ &\quad + 2R_h F_H \exp\{-2\pi k_o \operatorname{Im}(\phi_h)z\}[\cos[\nu_h + 2\pi k_o\{\operatorname{Re}(\phi_h) + \phi\}z - 2\pi P_H] \\ &\quad + R_s \cos[-\nu_h + \nu_s - 2\pi k_o\{\operatorname{Re}(\phi_h) - \phi\}z + 2\pi P_H]]. \end{aligned} \tag{6.20}$$

Figure 6.3 shows the X-ray intensity versus $\Delta\theta$ calculated from Eq. (6.20) at four possible atomic positions using F_H =1 on the Ge unit cell for ϕ = 5.4 mrad at 8 keV. The calculation is made for atoms at a height of 0.25 nm

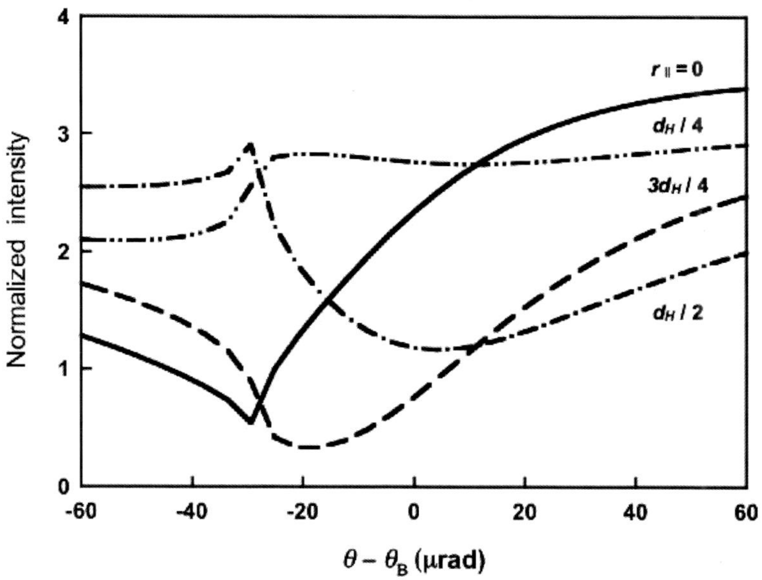

Fig. 6.3. Intensity of the electric field near the surface of a crystal versus $\Delta\theta$ for ϕ = 5.4 mrad. Curves are shown for atoms located on the diffraction planes ($\mathbf{r}_{\|} = 0$) and at three interplanar sites along the $\mathbf{H}$ direction. The intensity is calculated for a Ge(2$\bar{2}$0) reflection at 8 keV. Reprinted with permission from Ref. 6. Copyright (1989) by the American Physical Society. http://link.aps.org/doi/10.1103/PhysRevB.39.5739

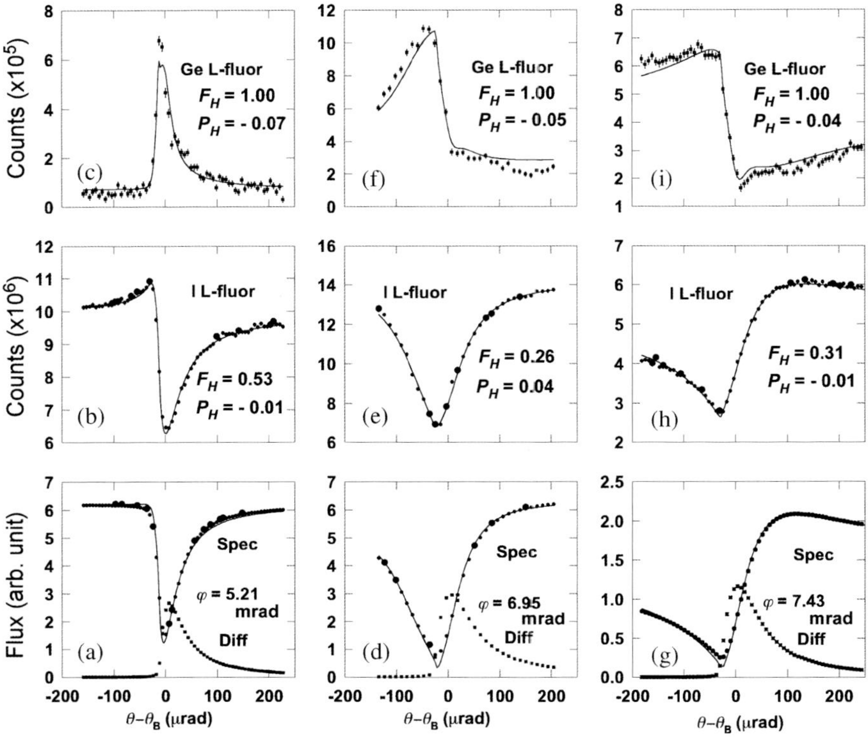

Fig. 6.5. (a), (d), (g) Specular and diffracted beam flux; (b), (e), (h) I fluorescence; (c), (f), (i) Ge fluorescence; for (a), (b), (c) $\phi = 5.21$; (d), (e), (f) $\phi = 6.95$; (g), (h), (i) $\phi = 7.43$ mrad. Reproduced with permission from the International Union of Crystallography. Reproduced with permission from Ref. 13. Copyright (1993) by the International Union of Crystallography. http://dx.doi.org/10.1107/S0108767392010687.

consisted of monitoring the specularly reflected beam flux, the reflected–diffracted beam flux, the Ge L fluorescence at 1.18–1.22 keV and the I L fluorescence at 3.93–4.80 keV.

Figures 6.5(a)–6.5(i) summarizes the results for angles of incidence ϕ that are significantly below the critical angle (a–c), near the critical angle (d–f), and well above the critical angle (g–i). A considerable difference is observed in the fluorescence yield between the I fluorescence and the Ge fluorescence. The fitting procedure followed was similar to Bragg diffraction XSW. In this case, the specular beam was fitted in Figs. 6.5(a), 6.5(d), and 6.5(g) to obtain first-order corrections to the diffraction-angle scale and the angle of incidence. Using the corrected values, the I fluorescence yield was fitted according to Eq. (6.20) and the Ge fluorescence yield according to Eqs. (6.18)–(6.23) in Ref. 13 to obtain the coherent fraction F_H and the coherent position P_H. A value of $z = -0.286$ nm, (above the surface), obtained from Bragg XSW measurements,[22] was used for the height above the (111) diffraction plane of the I chemisorbed atoms. The coherent position was consistent with $P_H = 0.0$ (within an error limit of ± 0.04). When combined with the vertical height from the additional measurement, this specified an atop site for the I atom. The coherent fraction decreased from $F_H = 0.53 \pm 0.05$ to $F_H = 0.26 \pm 0.05$ over the course of several hours due to the gradual disorder of the surface layer as the measurements at different incidence angles were completed.

The GAXSW method was also applied to a 10-nm-thick $Ca_{0.39}Sr_{0.61}F_2$ epilayer film on a GaAs(111) substrate.[18] The film is crystalline, with a lattice constant well matched with that of the substrate. In addition, the X-ray penetration depth varies from few nm to 10 nm at the grazing angles around the critical angle $\phi_c^{epi} = 2.1$ mrad for the film and $\phi_c^{sub} = 2.5$ mrad for the substrate used in the experiment mentioned below. Thus it is thought that the exciting X-ray field was modified by a dynamical diffraction in the film. Experimental GAXSW profiles were successfully explained by calculations taking into account four Bloch waves generated in the film. Fifteen waves were considered for the epilayer-substrate system in grazing-angle diffraction geometry; 14 amplitudes to be solved for: 8 for the epilayer waves, 4 for the substrate crystal waves, and 2 for the vacuum waves. All waves were expressed by solving eight boundary conditions and using six amplitude ratios.

The sample was epitaxially grown on a GaAs(111) substrate using molecular-beam-epitaxy facility.[23] By observation using Bragg XSW, the (111) lattice spacing, measured normal to the interface, is 0.73% larger in the film than in the substrate and that Sr atoms are ordered with a coherent fraction of 0.99 along the interface normal.[23] The set-up was described in Ref. 18. Expeimental data were collected around the $(\bar{2}20)$ Bragg planes normal to the (111) surface.

In Fig. 6.6, the Sr K emissions (14.2 keV) recorded were fit at various ϕ angles on the basis of a structural model having a lattice-matched interface using Eq. (6.15) in Ref. 18. The results of the fits indicates that the epilayer Sr planes are located on top of the $(\bar{2}20)$ planes of the substrate GaAs, on which all Ga and As atoms are situated. The coherent fraction F_H of 0.66 obtained for the epilayer suggests a disordered Sr-atom distribution in the in-plane direction, which is ascribed to combined effects of thermal

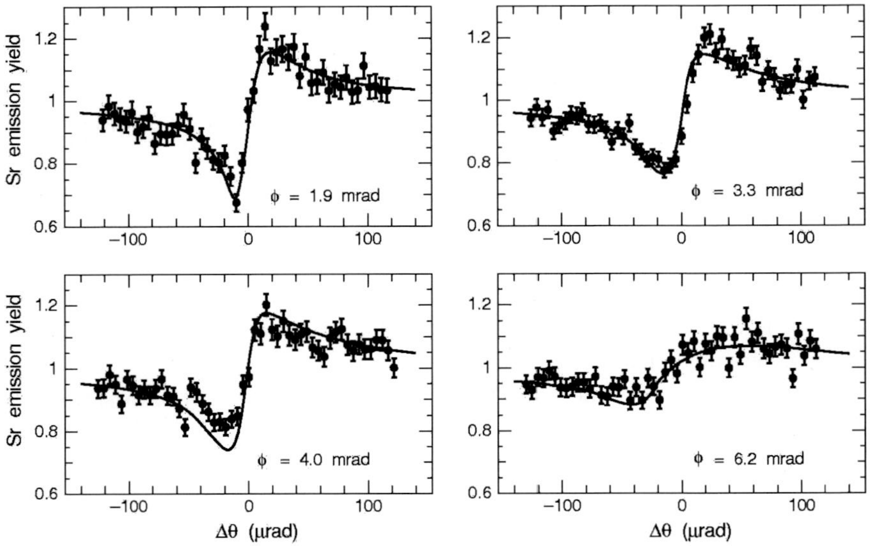

Fig. 6.6. Sr fluorescence signals (closed circles) from a 10-nm-thick $Ca_{0.39}Sr_{0.61}F_2$ epilayer film on a GaAs(111). Lines fitted on the basis of a structural model having a lattice-matched interface. Reproduced with permission from Ref. 18. Copyright (1997) by the International Union of Crystallography. http://dx.doi.org/10.1107/S0108767397008659.

vibration, interstitial atoms and interfacial dislocations as discussed in detail.[18]

Using the above two example studies, we showed that the GAXSW method is useful in identifying the in-plane location of adsorbed atoms at a surface and determining the in-plane structure of a thin film. Furthermore, the shallow penetration depth of the X-rays in case of GAXSW gives us a much higher surface sensitivity than the conventional XSW, which makes the technique particularly useful for surface science applications.

References

1. W. C. Marra, P. Eisenberger and A. Y. Cho, *J. Appl. Phys.* **50** (1979) 6927–6933.
2. A. M. Afanas'ev and M. K. Melkonyan, *Acta Crystallogr. A* **39** (1983) 207–210.
3. P. L. Cowan, *Phys. Rev. B* **32** (1985) 5437–5439.
4. N. Bernhard, E. Burkel, G. Gompper, H. Metzger, J. Peisl, H. Wagner and G. Wallner, *Z. Phys. B: Conden. Matter* **69** (1987) 303–311.

5. O. Sakata and H. Hashizume, *Rep. Res. Lab. Eng. Mat. Tokyo Inst. Tech.* **12** (1987) 45–57.
6. T. Jach, P. L. Cowan, Q. Shen and M. J. Bedzyk, *Phys. Rev. B* **39** (1989) 5739–5747.
7. S. M. Durbin and T. Gog, *Acta Crystallogr. A* **45** (1989) 132–141.
8. A. Authier, *IUCr Monographs on Crystallography*, Vol. 11 (Oxford University Press, Oxford, 2001), pp. 213–224.
9. H. R. Höche, O. Brümmer and J. Nieber, *Acta Crystallogr. A* **42** (1986) 585–587.
10. A. M. Afanas'ev, R. M. Imamov, A. V. Maslov and E. M. Pashaev, *Phys. Status Solidi A* **84** (1984) 73–78.
11. P. L. Cowan, S. Brennan, T. Jach, M. J. Bedzyk and G. Materlik, *Phys. Rev. Lett.* **57** (1986) 2399–2402.
12. H. Hashizume and O. Sakata, *Rev. Sci. Instrum.* **60** (1989) 2373–2375.
13. T. Jach and M. J. Bedzyk, *Acta Crystallogr. A* **49** (1993) 346–350.
14. T. Jach and M. J. Bedzyk, *Phys. Rev. B* **42** (1990) 5399–5402.
15. O. Sakata, H. Hashizume and H. Kurashina, *Phys. Rev. B* **48** (1993) 11408–11411.
16. O. Sakata and H. Hashizume, *Acta Crystallogr. A* **51** (1995) 375–384.
17. O. Sakata, N. Matsuki and H. Hashizume, *Phys. Rev. B* **60** (1999) 15546–15549.
18. O. Sakata and H. Hashizume, *Acta Crystallogr. A* **53** (1997) 781–788.
19. N. Hertel, G. Materlik and J. Zegenhagen, *Z. Phys. B* **58** (1985) 199–204.
20. O. Sakata and H. Hashizume, *Jpn. J. Appl. Phys.* **27** (1988) L1976–L1979.
21. O. Sakata, Ph.D. thesis, Tokyo Institute of Technology (1994).
22. M. J. Bedzyk, Q. Shen, M. E. O'Keeffe, G. Navrotski and L. E. Berman, *Surf. Sci.* **220** (1989) 419–427.
23. T. Niwa, M. Sugiyama, T. Nakahata, O. Sakata and H. Hashizume, *Surf. Sci.* **282** (1993) 342–356.

Chapter 7

X-RAY STANDING WAVE IN MULTILAYERS

MICHAEL J. BEDZYK

Department of Materials Science and Engineering, Northwestern University, Cook Hall, Evanston, IL 60208, USA
and
Argonne National Laboratory, Argonne, IL 60439, USA

JOSEPH A. LIBERA

Energy Systems Division, Argonne National Laboratory, 9700 S. Cass Ave, Argonne, IL 60439, USA

An extension of Fresnel theory is used to describe how reflectivity from a periodic multilayer mirror generates an X-ray standing wave (XSW) above the mirror surface. This long-period XSW is used to study distribution profiles of atoms within deposited ultra-thin organic films and ions at the liquid–solid interface.

7.1. Introduction

In Chapter 5, the case of generating a long-period X-ray standing wave (XSW) by total reflection from a mirror surface was discussed. The TR-XSW method is used to determine the X-ray fluorescence (XRF)-selected atomic density profile $\rho(z)$ in a surface overlayer structure by measuring its Fourier transform in the low-Q range that extends up to $Q_{max} \sim 1.6 \bullet Q_C$. Beyond this nominal limit, which is about 0.1 Å^{-1} for a gold mirror, the reflectivity from a simple mirror and the associated XSW fringe visibility become too weak for producing measurable XRF modulations.

In this chapter, we discuss how to extend the Q-range and thereby improve the π/Q_{max} intrinsic resolution of the long-period XSW method

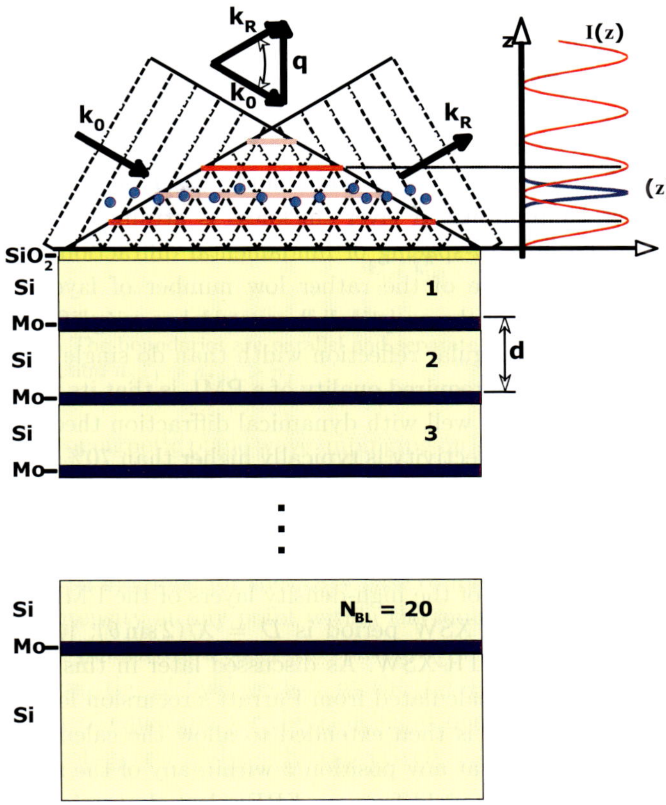

Fig. 7.1. Illustration of XSW generated by first-order Bragg diffraction from a Si/Mo periodic multilayer (PML). For $\lambda = 1$ Å and $d = 200$ Å, the incident angle is much smaller than depicted. Refraction effects at this very small incident angle cause the XSW period above the PML surface, $D = \lambda/(2\sin\theta)$, to be smaller than d.

by replacing the simple mirror with a periodic multilayer (PML) as depicted in Fig. 7.1. The XSW effect can then be observed over the $m = 0$, $m = 1$, and higher-order Bragg peaks and used for determining $\rho(z)$ for an atomic distribution contained within an overlayer that resides above the PML surface. Case studies include overlayers of sputter-deposited metal atoms,[1] electrochemically deposited metal atoms,[2] Langmuir–Blodgett multilayers,[3] organo-metallic multilayers,[4] and biomolecular adsorption at charged liquid–solid interfaces.[5] The XSW internal to the PML can also be used for characterizing the internal micro- or nanostructure of the PML,[6–10] or even the magnetic structure when coupled with circular-magnetic dichroism.[11]

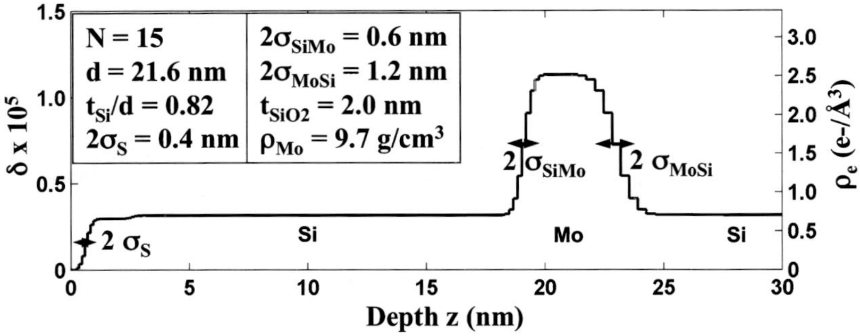

Fig. 7.4. The model used for calculating the reflectivity, E-field intensity, and XRF yield curves shown in Figs. 7.3 and 7.5. This is the electron density profile for the topmost period of the Si/Mo multilayer. The model also includes a 2-nm SiO_x layer at the air interface. From the model δ_j, β_j, and t_j are calculated for each layer in a graded interface model. The parameters listed in the inset were determined from the reflectivity fit in Fig. 7.3(a). See Refs. 21 and 22 for details.

The E-field intensity at depth z_j within the jth layer is:

$$\begin{aligned} I_j(Q, z_j) &= |E_j(z_j) + E_j^R(z_j)|^2 \\ &= |E_j(0)|^2 e^{-Q_j'' z_j} \left\{ 1 + \left| \frac{E_j^R(z_j)}{E_j(z_j)} \right|^2 + 2 \left| \frac{E_j^R(z_j)}{E_j(z_j)} \right| \cos(v_j - Q_j' z_j) \right\}. \end{aligned} \tag{7.10}$$

From Eq. (7.6), the modulus and relative phase of the E-field amplitude ratio are respectively defined as:

$$\left| \frac{E_j^R(z_j)}{E_j(z_j)} \right| = \left| \frac{E_j^R(0)}{E_j(0)} \right| e^{2Q_j'' z_j} = |A_{j,j+1}| e^{2Q_j'' z_j} \tag{7.11}$$

$$v_j = \arg(A_{j,j+1}). \tag{7.12}$$

If we normalize the incident intensity to unity, i.e., set $|E_1(0)|^2 = 1$, the intensity in the transmitted E-field at the top of the jth layer is:

$$|E_j(0)|^2 = \prod_{m=1}^{j-1} |T_{m,m+1}|^2 e^{-Q_m'' t_m} \tag{7.13}$$

where

$$T_{j,j+1} = \frac{E_{j+1}(0)}{E_j(t_j)} = \frac{1}{F_{j,j+1}^T} \left[1 - e^{iQ_j t_j} F_{j,j+1}^R A_{j,j+1} \right]. \tag{7.14}$$

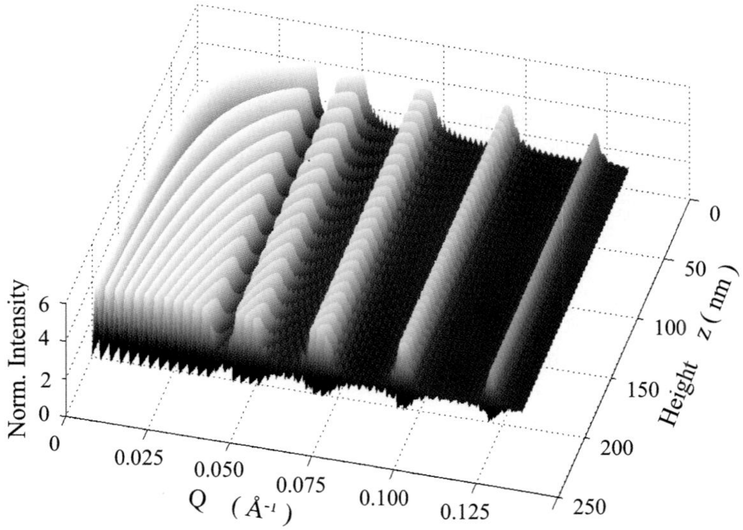

Fig. 7.5. The E-field intensity, $I(Q, z)$, in layer $j = 1$ (air) calculated from Eq. (7.10) for the multilayer described in Fig. 7.4 with reflectivity shown in Fig. 7.3(a). Notice how the XSW period decreases at successively higher values of "Q".

Figure 7.5 shows the calculated E-field intensity for the $d = 21.6\,\text{nm}$ Si/Mo multilayer described in Fig. 7.4. The height coordinate z in Fig. 7.5 and z_0 in Figs. 7.3(b) and 7.3(c) use the same origin as used for the depth coordinate in Fig. 7.4.

7.3. Analysis of the XRF Yield

The XRF yield from a distribution of atoms, $\rho(z)$, within the multilayer is

$$Y(Q) = \int \rho(z) I(Q, z) e^{-\mu_F z / \sin \alpha} dz, \qquad (7.15)$$

where μ_F and α are, respectively, the linear absorption coefficient and take-off angle for the emitted fluorescent X-rays.

The example XRF yield shown in Fig. 7.3(b) and 7.3(c) is for Hg-labeled RNA molecules adsorbed to an amine-terminated self-assembled monolayer that was grown on the top silica layer of the multilayer described in Fig. 7.4, Ref. 21. The modeled distribution, $\rho(z)$, for Hg atoms in this case study was partitioned into two parts. A fraction, C, of Hg atoms were assigned to occupy a Gaussian distribution with mean-height z_0 above the silica surface and width σ. The $(1 - C)$ remaining fraction of Hg atoms were set to occupy a uniform distribution extending from $z = 0$ to $z = 200\,\text{nm}$. The best fit value of $z_0 = 1.0\,\text{nm}$ is in reasonable agreement with the expected thickness of the primer layer and the RNA molecular radius. The sensitivity of the measurement of this parameter at the $m = 3$ and $m = 4$ Bragg peaks is demonstrated in Fig. 7.3(c). The fact that the Hg distribution profile width was only $\sigma = 0.3\,\text{nm}$ is consistent with the electrostatic attraction between the RNA polyanions and the positively charged amine groups at the surface.

In the above case study, a simple overlayer configuration that had no measurable effect on the X-ray reflectivity was used to demonstrate the XSW technique using periodic multilayers. However, many overlayer structures do have a significant effect on the reflectivity and must therefore be included in the reflectivity modeling step.[4,5,22] In this case, the unknown overlayer structure is determined using an iterative process of reflectivity-fitting followed by determination of the element distribution using XSW. In such cases the E-field intensity, $I(Q, z)$ is calculated inside the layer representing the unknown overlayer. This method is also able to successfully treat cases in which a resonant cavity occurs.[22,23]

Acknowledgments

Colleagues who inspired and assisted in this work include G. Mark Bommarito, Donald Bilderback, Boris Batterman, Martin Caffrey, Hector Abruna, Jin Wang, Thomas Penner, Jay Schildkraut, Chian Liu, Ray Conley, Hao Cheng, Kai Zhang, Monica Olvera de la Cruz, Alfonso Mondragon, and Zhong Zhong. The data in this chapter were collected at the X15A station of the National Synchrotron Light Source (NSLS), which is supported by the US Department of Energy. The work was also partially supported by the US National Science Foundation and National Institute of Health.

References

1. T. W. Barbee and W. K. Warburton, *Mater. Lett.* **3** (1984) 17.
2. M. J. Bedzyk, D. Bilderback, J. White, H. D. Abruna and M. G. Bommarito, *J. Phys. Chem.* **90** (1986) 4926.
3. M. J. Bedzyk, D. H. Bilderback, G. M. Bommarito, M. Caffrey and J. S. Schildkraut, *Science* **241** (1988) 1788.
4. J. A. Libera, R. W. Gurney, C. Schwartz, H. Jin, T. L. Lee, S. T. Nguyen, J. T. Hupp and M. J. Bedzyk, *J. Phys. Chem. B* **109** (2005) 1441.

5. J. A. Libera, H. Cheng, M. Olvera de la Cruz and M. J. Bedzyk, *J. Phys. Chem. B* **109** (2005) 23001.
6. S. K. Ghose and B. N. Dev, *Phys. Rev. B* (2001) 6324.
7. A. Iida, T. Matsushita and T. Ishikawa, *Jpn. J. Appl. Phys.* **24** (1985) L675.
8. J. B. Kortright and A. Fischer-Colbrie, *J. Appl. Phys.* **61** (1987) 1130.
9. S. I. Zheludeva, M. V. Kovalchuk, N. N. Novikova, A. N. Sosphenov, N. E. Malysheva, N. N. Salashenko, A. D. Akhsakhalyan, Y. Y. Platonov, R. I. Cernik and S. P. Collins, *Thin Solid Films* **259** (1995) 131.
10. D. K. G. de Boer, A. J. G. Leenaers and W. W. van den Hoogenhof, *X-Ray Spectrom.* **24** (1995) 91.
11. S. K. Kim and J. B. Kortright, *Phys. Rev. Lett.* **86** (2001) 1347.
12. T. W. Barbee, *Proc. Low Energy X-Ray Diagnostics* (AIP, New York, 1981), pp. 131–145.
13. B. L. Henke, *Proc. Low Energy X-Ray Diagnostics* (AIP, New York, 1981), pp. 146–155.
14. J. H. Underwood and T. W. Barbee, *Proc. Low Energy X-Ray Diagnostics* (AIP, New York, 1981), p. 170.
15. D. H. Bilderback, B. M. Lairson, T. W. Barbee, G. E. Ice and C. J. Sparks, *Nucl. Instrum. Meth. Phys. Res.* **208** (1983) 251.
16. E. Spiller and A. E. Rosenbluth, *Opt. Eng.* **25** (1986) 954.
17. L. G. Parratt, *Phys. Rev.* **95** (1954) 359.
18. L. Nevot and P. Croce, *Rev. Phys. Appl.* **15** (1980) 761.
19. M. Born and E. Wolf, *Principles of Optics*, 5th edn. (Pergamon Press, Oxford, 1975).
20. G. M. Bommarito, Ph.D. thesis, Cornell University (1992).
21. J. A. Libera, Ph.D. thesis, Northwestern University (2005).
22. J. A. Libera, R. W. Gurney, S. T. Nguyen, J. T. Hupp, C. Liu, R. Conley and M. J. Bedzyk, *Langmuir* **20** (2004) 8022.
23. J. Wang, M. J. Bedzyk and M. Caffrey, *Science* **258** (1992) 775.

Chapter 8

KINEMATICAL X-RAY STANDING WAVES

MARTIN TOLKIEHN*,† and DMITRI V. NOVIKOV†

**Diamond Light Source Ltd., Chilton, Didcot, Oxfordshire, OX11 0DE, UK*

†Hamburger Synchrotronstrahlungslabor HASYLAB at Deutsches, Elektronen-Synchrotron DESY, Notkestraße 85, D-22603 Hamburg, Germany

We present a theoretical investigation and experimental proof of X-ray standing wave formation in mosaic crystals under weak diffracted beam conditions. The obtained standing wave phase dependence, described using the kinematical approximation, is valid at large deviations from the Bragg angle for non-perfect crystals as well as for ideal crystals. The presented theory represents the basis for a novel variation of the XSW method, called kinematical X-ray standing waves.

8.1. Introduction

The X-ray standing wave (XSW) technique was first introduced as a tool for the investigation of single crystals under dynamical diffraction conditions.[1] It is based on the analysis of the shape of the secondary radiation yield curve in the vicinity of a Bragg reflection and requires a precise knowledge of the phase relation between the incident and the diffracted waves.[2] Initially applied to perfect single crystals, it could be later extended to a much wider range of objects by a transition to back-scattering conditions (NIXSW), which allows a significant mosaicity of the crystal.[3,4] Further variants of the XSW approach make use of X-ray wave fields formed by total external reflection of X-rays[5] or by Bragg scattering in artificial multilayers.[6,7]

A standing wave emerges in crystals also without dynamical diffraction. Any Bragg diffracted beam, even much weaker than that formed by dynamical diffraction, which interferes with the incident beam provides a well-pronounced periodic field distribution. Kazimirov *et al.*[8,9] used this effect for the application of XSW to thin films, where the reflectivity can be described in the frame of the first Born approximation. This approximation is also valid for bulk crystals at large deviations from the Bragg condition. In such a case, the phase of the standing wave field depends only on the sign of the deviation from the Bragg angle. This knowledge can be used as a basis for the new method of kinematical X-ray standing waves (KXSW).

8.2. Theory

Let us first consider a plane monochromatic incident X-ray wave $\mathbf{E}(\mathbf{r},t) = \mathbf{E}_0 e^{i\omega t - i\mathbf{K}_0\cdot\mathbf{r}}$. In the frame of the first Born approximation this plane wave causes a polarization $\mathbf{P}(\mathbf{r},t) = \epsilon_0\chi(\mathbf{r})\mathbf{E}(\mathbf{r},t)$ of the sample, where $\chi(\mathbf{r}) = -\frac{e_0^2}{\epsilon_0 m_e \omega^2}\rho(\mathbf{r})$ is the polarizability, ϵ_0 is the vacuum dielectric constant, m_e is the electron mass, e_0 is the elementary charge, and ρ is the electron density of the sample. This oscillating polarization is connected to a current density $\mathbf{J}(\mathbf{r},t) = \frac{\partial \mathbf{P}}{\partial t} = i\omega\epsilon_0\chi(\mathbf{r})\mathbf{E}(\mathbf{r},t)$. Using Biot–Savart's[10] law, one can now calculate the vector potential of the wave field emitted by the oscillating polarization $\mathbf{P}$:

$$\mathbf{A}(\mathbf{r},t) = e^{i\omega t}\frac{\mu_0}{4\pi}\int_{\mathbb{R}^3}\frac{\mathbf{J}(\mathbf{r}')}{\|\mathbf{r}-\mathbf{r}'\|}e^{-ik\|\mathbf{r}-\mathbf{r}'\|}\,d^3r' \tag{8.1}$$

$$= -e^{i\omega t}\frac{ie_0^2}{4\pi\epsilon_0 m_e c^2\omega}\mathbf{E}_0\int_{\mathbb{R}^3}\rho(\mathbf{r}')\frac{e^{-i\mathbf{K}_0\cdot\mathbf{r}'-ik\|\mathbf{r}-\mathbf{r}'\|}}{\|\mathbf{r}-\mathbf{r}'\|}\,d^3r'. \tag{8.2}$$

Introducing the Thomson scattering length[a] $r_e = \frac{e_0^2}{4\pi\epsilon_0 m_e c^2}$ and assuming a crystalline sample with electron density $\rho = \frac{S(\mathbf{r})}{V}\sum_{\mathbf{H}\in G^*}F_{\mathbf{H}}e^{-i\mathbf{H}\cdot\mathbf{r}}$, where V is the volume of the unit cell, G^* is the reciprocal-lattice, $F_{\mathbf{H}}$ is the structure factor, and $S(\mathbf{r})$ is the *shape function*, which is 1 inside the sample and 0 elsewhere, this can be written as

$$\mathbf{A}(\mathbf{r},t) = -e^{i\omega t}\frac{ir_e}{\omega V}\mathbf{E}_0\int_{\mathbb{R}^3}S(\mathbf{r}')\sum_{\mathbf{H}\in G^*}F_{\mathbf{H}}\frac{e^{-i\mathbf{K}_{\mathbf{H}}\cdot\mathbf{r}'-ik\|\mathbf{r}-\mathbf{r}'\|}}{\|\mathbf{r}-\mathbf{r}'\|}\,d^3r'. \tag{8.3}$$

Provided that k has a small imaginary part, the spherical wave in (8.3) can be written as Fourier transform

$$\frac{e^{-ik\|\mathbf{r}-\mathbf{r}'\|}}{\|\mathbf{r}-\mathbf{r}'\|} = \frac{1}{2\pi^2}\int_{\mathbb{R}^3}\frac{e^{-i\mathbf{q}\cdot(\mathbf{r}-\mathbf{r}')}}{\|\mathbf{q}\|^2-k^2}\,d^3q \tag{8.4}$$

[a] Also known as the classical electron radius.

and one obtains

$$\mathbf{A}(\mathbf{r},t) = \frac{r_e \mathbf{E}_0 e^{i\omega t}}{2\pi^2 i\omega V} \sum_{\mathbf{H}\in G^*} F_{\mathbf{H}} \int_{\mathbb{R}^3} \underbrace{\int_{\mathbb{R}^3} S(\mathbf{r}')e^{-i(\mathbf{K}_{\mathbf{H}}-\mathbf{q})\cdot\mathbf{r}'} d^3r'}_{=\widehat{S}(\mathbf{q}-\mathbf{K}_{\mathbf{H}})} \frac{e^{-i\mathbf{q}\cdot\mathbf{r}}}{\|\mathbf{q}\|^2 - k^2} d^3q. \tag{8.5}$$

The inner integral in (8.5) can be interpreted as the Fourier transform of $S(\mathbf{r})$, which is denoted by $\widehat{S}$. For centrosymmetric shape functions, this leads to

$$\mathbf{A}(\mathbf{r},t) = \frac{r_e \mathbf{E}_0 e^{i\omega t}}{2\pi^2 i\omega V} \sum_{\mathbf{H}\in G^*} F_{\mathbf{H}} \int_{\mathbb{R}^3} \widehat{S}(\mathbf{q}-\mathbf{K}_{\mathbf{H}}) \frac{e^{-i\mathbf{q}\cdot\mathbf{r}}}{\|\mathbf{q}\|^2 - k^2} d^3q \tag{8.6}$$

$$= \frac{r_e \mathbf{E}_0 e^{i\omega t}}{2\pi^2 i\omega V} \sum_{\mathbf{H}\in G^*} F_{\mathbf{H}} [\widehat{S} * \kappa_{\mathbf{r}}](\mathbf{K}_{\mathbf{H}}), \tag{8.7}$$

where $\kappa_{\mathbf{r}}(\mathbf{q}) = \frac{e^{-i\mathbf{q}\cdot\mathbf{r}}}{\|\mathbf{q}\|^2 - k^2}$ and $*$ stands for the convolution. For a mosaic crystal consisting of many crystallites with shape functions S_j, the charge density is given by

$$\rho(\mathbf{r}) = \sum_{j=1}^{N} S_j(\mathbf{r}) \frac{1}{V} \sum_{\mathbf{H}\in G^*} F_{\mathbf{H}} e^{-i\mathbf{H}\cdot R_j^{-1}(\mathbf{r}-\mathbf{d}_j)}, \tag{8.8}$$

where R_j is a rotation matrix describing the misalignment of the jth crystallite and $\mathbf{d}_j$ is an offset of the lattice of the crystallite with respect to a global coordinate system. The vector potential is then given by

$$\mathbf{A}(\mathbf{r},t) = \frac{r_e \mathbf{E}_0 e^{i\omega t}}{2\pi^2 i\omega V} \sum_{j=1}^{N} \sum_{\mathbf{H}\in G^*} F_{\mathbf{H}} e^{iR_j\mathbf{H}\cdot\mathbf{d}_j} \times \int_{\mathbb{R}^3} \widehat{S}_j(\mathbf{q}-\mathbf{K}_0 - R_j\mathbf{H}) \frac{e^{-i\mathbf{q}\cdot\mathbf{r}}}{\|\mathbf{q}\|^2 - k^2} d^3q.$$

Assuming that all crystallites have the same shape $\widehat{S}_j(\mathbf{q}) = e^{i\mathbf{q}\cdot\mathbf{d}_j}\widehat{S}(\mathbf{q})$, this leads to

$$\mathbf{A}(\mathbf{r},t) = \frac{r_e \mathbf{E}_0 e^{i\omega t}}{2\pi^2 i\omega V} \sum_{\mathbf{H}\in G^*} F_{\mathbf{H}} \sum_{j=1}^{N} e^{-i\mathbf{K}_0\cdot\mathbf{d}_j} [\widehat{S} * \kappa_{\mathbf{r}-\mathbf{d}_j}](\mathbf{K}_0 + R_j\mathbf{H}). \tag{8.9}$$

For a large number of crystallites N, the sum over j has a similar effect as a convolution with the mosaic spread function. The only difference comes from the phase factor $e^{-i\mathbf{K}_0\cdot\mathbf{d}_j}$. In a practical case, the mosaic spread in non-perfect single crystals becomes much larger than the broadening due to the mosaic block size effects.[11] In this case, it is possible to use the approximation

$$\widehat{S}(\mathbf{q}) = \frac{1}{N}(2\pi)^3\delta(\mathbf{q}), \tag{8.10}$$

which is in fact the normalized shape function of an infinite crystal. Substituting (8.10) into (8.9) gives

$$\mathbf{A}(\mathbf{r},t) = \frac{1}{N}\frac{4\pi r_e}{i\omega V}\mathbf{E}_0 \sum_{\mathbf{H}\in G^*} F_{\mathbf{H}} \sum_{j=1}^{N} e^{iR_j\mathbf{H}\cdot\mathbf{d}_j} \frac{e^{i\omega t - i(\mathbf{K}_0 + R_j\mathbf{H})\cdot\mathbf{r}}}{\|\mathbf{K}_0 + R_j\mathbf{H}\|^2 - k^2}. \tag{8.11}$$

Using Maxwell's equations, the electrical field can be determined:

$$\mathbf{E}_P(\mathbf{r},t) = -\frac{i}{\omega}e^{i\omega t}\left(c^2\nabla\times\nabla\times\mathbf{A}(\mathbf{r}) - \frac{1}{\epsilon_0}\mathbf{J}(\mathbf{r})\right). \tag{8.12}$$

Substituting $\mathbf{J} = i\omega\epsilon_0\chi\mathbf{E}$ and $\mathbf{A}$ from (8.11) and introducing $\phi_j = e^{iR_j\mathbf{H}\cdot\mathbf{d}_j}$, $\mathbf{q}_j = \mathbf{K}_0 + R_j\mathbf{H}$ and $\Gamma = \frac{4\pi r_e}{k^2 V}$, one obtains

$$\begin{aligned}\mathbf{E}_P(\mathbf{r},t) &= \sum_{\mathbf{H}\in G^*}\Gamma F_{\mathbf{H}}\frac{1}{N}\sum_{j=1}^{N}\Bigg(\phi_j(\mathbf{q}_j(\mathbf{q}_j\cdot\mathbf{E}_0) - \mathbf{E}_0\|\mathbf{q}_j\|^2)\frac{e^{i\omega t - i\mathbf{q}_j\cdot\mathbf{r}}}{\|\mathbf{q}_j\|^2 - k^2} \\ &\quad + e^{-i\mathbf{H}\cdot(R_j^{-1}\mathbf{r} - \mathbf{d}_j)}\mathbf{E}_0 e^{i\omega t - i\mathbf{K}_0\cdot\mathbf{r}}\Bigg) \\ &= \sum_{\mathbf{H}\in G^*}\Gamma F_{\mathbf{H}}\sum_{j=1}^{N}\frac{\phi_j}{N}\left(\underbrace{\mathbf{E}_0 - \mathbf{E}_{\perp j}}_{\approx 0} - \frac{k^2\mathbf{E}_{\perp j}}{\|\mathbf{q}_j\|^2 - k^2}\right)e^{i\omega t - i\mathbf{q}_j\cdot\mathbf{r}},\end{aligned}$$

where $\mathbf{E}_{\perp j} = \mathbf{E}_0 - \mathbf{q}_j(\mathbf{q}_j\cdot\mathbf{E}_0)/\|\mathbf{q}_j\|^2$. The term $\mathbf{E}_0 - \mathbf{E}_{\perp j}$ is zero for σ-polarization and for π-polarization it is the component of the electrical field parallel to the scattered wave vector, which can be usually neglected.[1,12]

In the vicinity of a single Bragg reflection with reciprocal-lattice vector $\mathbf{H}'$ the sum is dominated by the terms $\mathbf{H} = 0$ and $\mathbf{H} = \mathbf{H}'$. Now one can approximate $\|\mathbf{q}_j\|^2 - k^2 \approx -2k^2\Delta\theta_j \sin 2\theta_B$, where $\Delta\theta_j$ is the deviation from the corresponding Bragg angle θ_B in the jth crystallite. This leads to

$$\mathbf{E}_P(\mathbf{r},t) = \mathbf{E}_0 e^{i\omega t - i\mathbf{K}_0\cdot\mathbf{r}} + \Gamma F_{\mathbf{H}'}\frac{1}{N}\sum_{j=1}^{N}\phi_j e^{i\omega t - i\mathbf{q}_j\cdot\mathbf{r}}\frac{\mathbf{E}_{\perp j}}{2\Delta\theta_j \sin 2\theta_B}. \tag{8.13}$$

For σ-polarization one obtains the following standing wave intensity distribution inside the sample

$$\mathbf{I}(\mathbf{r}) \propto \|\mathbf{E}_P\|^2 \approx \|\mathbf{E}_0\|^2 \left(1 + \mathrm{Re}\left(\Gamma F_{\mathbf{H}'} \frac{1}{N} \sum_{j=1}^{N} \frac{\phi_j e^{-iR_j \mathbf{H}' \cdot \mathbf{r}}}{\Delta\theta_j \sin 2\theta_B}\right)\right), \tag{8.14}$$

where quadratic terms in $\Gamma/\Delta\theta$ have been neglected. For negligible absorption effects, the secondary signal of impurity atoms with a coherent position $\Phi_c = \mathbf{H}' \cdot \mathbf{r}_0 / 2\pi$ is then given by the average of the intensity at the corresponding positions in every crystallite:

$$\begin{aligned} Y(\Delta\theta) &\propto \sum_{m=1}^{N} I(R_m \mathbf{r}_0 + \mathbf{d}_m) \\ &\propto 1 + \frac{\Gamma}{\sin 2\theta_B} \mathrm{Re}\left(F_{\mathbf{H}'} \frac{e^{-i2\pi\Phi_c}}{N^2} \sum_{j,m=1}^{N} \frac{e^{iR_j \mathbf{H}' \cdot (\mathbf{d}_j - \mathbf{d}_m + R_j \mathbf{r}_0 - R_m \mathbf{r}_0)}}{\Delta\theta_j}\right). \end{aligned}$$

Neglecting the influence of the rotation matrices R_j in the exponent, this can be approximated by

$$\approx 1 + \frac{\Gamma}{\sin 2\theta_B} \left[w * f\right](\Delta\theta)\, \mathrm{Re}(g F_{\mathbf{H}'} e^{-i2\pi\Phi_c}),$$

where w denotes the mosaic spread function, $f(x) = 1/x$ and $g = \langle e^{i\mathbf{H}' \cdot (\mathbf{d}_j - \mathbf{d}_m)} \rangle$ is the ensemble average of the additional phase factor arising from the displacements of the crystallites. For each crystallite, one can now find a lattice vector $\mathbf{D}_j$ so that $|\mathbf{H}' \cdot (d_j - D_j)| < 2\pi$. For normal distributed difference vectors $\widetilde{\mathbf{d}}_j := d_j - D_j$ with an average $\langle \widetilde{\mathbf{d}}_j \rangle = 0$, one has according to the Baker–Hausdorff theorem

$$g = \langle e^{i\mathbf{H}' \cdot (\mathbf{d}_j - \mathbf{d}_m)} \rangle = \langle e^{i\mathbf{H}' \cdot (\widetilde{\mathbf{d}}_j - \widetilde{\mathbf{d}}_m)} \rangle = e^{-\langle (\mathbf{H}' \cdot \widetilde{\mathbf{d}}_j)^2 \rangle}. \tag{8.15}$$

The new generalized parameter g describes the additional disorder emerging from the mosaic structure of the object.

For fluorescing atoms at different non-equivalent positions, one can introduce the *coherent position* Φ_c and *coherent fraction* f_c in the usual way. Assuming for simplicity a Lorentz-shaped spread function w with width σ_m one finally gets

$$Y(\Delta\theta) \propto 1 + \frac{\Gamma}{\sin 2\theta_B} \frac{\Delta\theta}{\Delta\theta^2 + \sigma_m^2} \underbrace{|F_{\mathbf{H}'}| g f_c \cos(2\pi\Phi_c - \arg F_{\mathbf{H}'})}_{=:\Upsilon_c}. \tag{8.16}$$

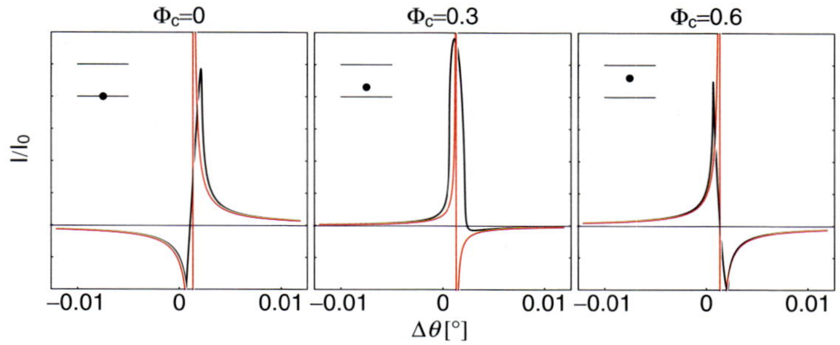

Fig. 8.1. Fluorescence yield curves calculated for different coherent positions of the emitting atom in the unit cell. The red lines show the fluorescence yield $Y(\Delta\theta)$ obtained in the kinematical approximation using Eq. (8.16) with $\sigma_m = 0$, while the black lines correspond to the full dynamical theory.

Equation (8.16) is only valid at the tails of the Bragg reflection, since it was derived using the kinematical theory. The same result can be obtained also from an approximation of the dynamical theory for large deviations $\Delta\theta$ from the Bragg angle.[13]

Figure 8.1 shows the fluorescence yield for a simple model of a single emitting atom in a unit cell. The yield $Y(\Delta\theta)$, shown in red, is calculated with Eq. (8.16). For comparison, the curves obtained by the full dynamical theory are given in black. Obviously, at sufficiently high deviations from the exact Bragg condition, there is no difference between both approaches. The tails of the fluorescence yield curves display a pronounced dependence on the coherent position: their amplitude and sign are directly dependent on Φ_c.

To facilitate the comparison of the kinematic and dynamic curves in Fig. 8.1, a perfect sample without mosaicity was assumed. For a real sample, the parameters σ_m (mosaic spread) and $\Delta\theta$ can be easily obtained from the reflectivity curve: σ_m can be determined from its full width at half maximum and $\Delta\theta$ is the deviation from its center of mass. For large deviations $\Delta\theta \gg \sigma_m$, the influence of the convolution with w becomes negligible and $\sigma_m = 0$ can be assumed. This is also true for arbitrary, non-Lorentz-shaped spread functions w.

The amplitude and sign of the fluorescence yield depends further on the dimensionless parameter

$$\Upsilon_c = |F_{\mathbf{H}'}| g f_c \cos(2\pi\Phi_c - \arg\ F_{\mathbf{H}'}), \tag{8.17}$$

which can be determined by fitting the experimental fluorescence yield data. This parameter itself depends on the coherent fraction f_c, the coherent position Φ_c and $g = e^{-\langle(\mathbf{H}'\cdot d_j)^2\rangle}$. The parameter g effectively reduces the coherent fraction f_c and depends only on the disorder of the crystallites in the object. It has the same value for all species of atoms. In a real experiment, g can be evaluated from the Υ_c value of any atom of the host lattice. It should be noted, that g is similar to a static Debye–Waller factor, however instead of atomic displacements, it depends on the displacements of the crystallites $\mathbf{d}_j$. Furthermore, Υ_c depends on the structure factor $F_{\mathbf{H}'}$. This means that a KXSW measurement can only be done by scanning the angle and not by scanning the energy. For the larger values of $\Delta\theta$, which are not necessary for standard XSW, the assumption that $F_{\mathbf{H}'}$ is constant for the whole energy range of the scan is no longer valid.

In order to determine the coherent fraction and the coherent position one can do measurements at two or more different energies. Since changing the energy of the incident beam does not influence these parameters but changes the phase of the structure factor $\arg F_{\mathbf{H}'}$, Eq. (8.17) leads to a simple set of equations from which both f_c and Φ_c can be calculated. However in many cases the knowledge of Υ_c is sufficient, for instance to distinguish between several existing models of atom arrangement.[13]

For fluorescing atoms in the bulk of the sample, angle-dependent absorption effects must be taken into account. This is usually done by defining the *effective thickness*

$$z_{\text{eff}} = \int_0^{\infty} \frac{e^{-\mu(\theta)\frac{z}{\sin\theta} - \mu_2\frac{z}{\sin\alpha}}}{\sin\theta} dz = \left(\mu(\theta) + \mu_2\frac{\sin\theta}{\sin\alpha}\right)^{-1}, \tag{8.18}$$

whereby the fluorescence yield of atoms in the bulk is given by

$$Y_{\text{bulk}}(\Delta\theta) = z_{\text{eff}} Y(\Delta\theta), \tag{8.19}$$

with $Y(\Delta\theta)$ from Eq. (8.16). The effective thickness depends on the angle of incidence θ, the exit angle of the secondary radiation α, and the absorption coefficients of incident and secondary radiation $\mu(\theta)$ and μ_2. The absorption of the incident wave depends again on θ and is given by[14]:

$$\mu(\theta) = \int_{\mathbb{R}^3} I(\mathbf{r}, \Delta\theta)\sigma_d(\mathbf{r}) d^3r = \mu_0 \left(1 + \frac{\Gamma}{\sin 2\theta_B} \underbrace{\frac{\operatorname{Im}(F_{\mathbf{H}'}F_{-\mathbf{H}'})}{2\operatorname{Im}F_0}}_{=:\Upsilon_\mu} \frac{1}{\Delta\theta}\right),$$

where $\sigma_d(\mathbf{r})$ is the dipole absorption[b] cross-section at $\mathbf{r}$.

---

[b]Quadrupole and higher order terms are neglected.

Taking into account absorption effects, the secondary signal for large $\Delta\theta$ is finally given by:

$$Y_{\text{bulk}}(\Delta\theta) \propto \frac{1}{1+M}\left(1+\frac{\Gamma}{\sin 2\theta_B}\left(\Upsilon_c - \frac{\Upsilon_\mu}{1+M}\right)\frac{1}{\Delta\theta}\right) \tag{8.20}$$

with $M = (\mu_2 \sin\theta)/(\mu_0 \sin\alpha)$. For large values of μ_2, which occur, e.g., if Auger or photoelectrons are used as secondary signal, the parameter M will be large and the structure-dependent absorption correction term $\frac{\Upsilon_\mu}{1+M}$ can be neglected. This can be also achieved by grazing exit conditions for the secondary radiation $\alpha \approx 0$. However, in a real experiment, this can become rather cumbersome, as the precision of the results depends strongly on the experimental error of α. For KXSW measurements it is therefore preferable to position detectors at exit angles $\alpha \approx \pi/2$. This is in contrast to the conventional XSW method, where extinction effects will complicate the analysis for large α.

If impurities in a substrate with known structure are considered, the parameter Υ_μ can be calculated from the structure factors of the substrate. For unknown structures Υ_μ can be determined by measuring the angle dependent absorption $\mu(\theta)$ or by measuring Y_{bulk} for different values of α. Such data could be also used for structure determination, since Υ_μ depends on the phase of the structure factor. For instance, for centrosymmetric crystals one can use

$$\Upsilon_\mu = \frac{|F_{\mathbf{H}'}|^2 \sin(2\arg F_{\mathbf{H}'})}{2\,\mathrm{Im}\,F_0}. \tag{8.21}$$

8.3. Application of KXSW to Mosaic Cu_3Au

The applicability of the KXSW method can be verified by experiments on samples with known structure.[13] A suitable test object is mosaic Cu_3Au, which has a fcc structure with Au atoms in the corners and Cu atoms in the face centers of the unit cell. Figure 8.2 shows experimental data of a Cu_3Au sample, for which the reflectivity curve was drastically broadened so that the dynamical theory cannot be used for data evaluation. A mosaic spread of $\sigma_m = 0.09°$ was determined from the half-width of the Bragg reflection. However, the broadening effects on the slowly varying tails of the secondary yield curves are much smaller, so that the KXSW approach is still feasible. This is shown for the Cu K_α fluorescence measured at the (002) reflection at two different energies of the incident beam (Fig. 8.2). The fluorescence curves were fitted using a modified version of Eq. (8.20). Since the curves

be performed typically by scanning angle or energy, the KXSW method should be performed by scanning the angle to assure that the energy dependence of the structure factor is not influencing the result. The KXSW theoretical approach can also be used for evaluation of holographic data from crystalline objects.[18,19]

Acknowledgments

We would like to thank B. Walz and P. Korecki for helpful discussions.

References

1. B. W. Batterman and H. Cole, *Rev. Mod. Phys.* **36** (1964) 681.
2. M. von Laue, *Röntgenstrahlinterferenzen*, 3rd edn. (Akademische Verlagsgesellschaft, Frankfurt am Main, 1960), in German.
3. A. Caticha and S. Caticha-Ellis, *Phys. Rev. B.* **25** (1982) 971.
4. J. Zegenhagen, *Surf. Sci. Rep.* **18** (1993) 202.
5. M. Bedzyk, D. Bilderback, G. Bommarito, M. Caffrey and J. Schildkraut, *Science.* **241** (1988) 1788.
6. J. Barbee, W. Troy and W. K. Warburton, *Matter Lett.* **3** (1984) 17.
7. S. Ghose and B. Dev, *Phys. Rev. B.* **63** (2001) 245409.
8. A. Kazimirov, T. Haage, L. Ortega, A. Stierle, F. Comin and J. Zegenhagen, *Solid State Commun.* **104** (1997) 347.
9. A. Kazimirov, G. Scherb, J. Zegenhagen, T.-L. Lee, M. J. Bedzyk, M. K. Kelly, H. Angerer and O. Ambacher, *J. Appl. Phys.* **84** (1998) 1703.
10. J. D. Jackson, *Classical Electrodynamics*, 2nd edn. (Wiley, New York, 1975).
11. P. Scherrer, *Gött. Nachr.* **2** (1918) 98, in German.
12. A. Authier, *Dynamical Theory of X-Ray Diffraction*, 2nd edn. (Oxford University Press, 2001).
13. M. Tolkiehn, D. Novikov and S. Fanchenko, *Phys. Rev. B.* **71** (2005) 165404.
14. M. Tolkiehn, *Untersuchung von nicht perfekten Einkristallen mit kinematischen stehenden Röntgenwellen, Ph.D. thesis*, Universität Hamburg (2005), in German.
15. D. R. Chipman, *J. Appl. Phys.* (1956), p. 739.
16. J. Zegenhagen, *Surf. Sci.* **554** (2004) 77.
17. M. J. Bedzyk and L. Cheng, *Rev. Mineral Geochem.* **49** (2002) 221.
18. B. Adams, D. V. Novikov, T. Hiort, G. Materlik and E. Kossel, *Phys. Rev. B.* **57**, (1998) 7526.
19. S. Marchesini, N. Mannella, C. S. Fadley, M. A. Van Hove, J. J. Bucher, D. K. Shuh, L. Fabris, M. J. Press, M. W. West, W. C. Stolte and Z. Hussain, *Phys. Rev. B.* **66** (2002) 94111.

Chapter 9

X-RAY WAVEGUIDES

IANNA BUKREEV*, ALESSIA CEDOLA*, DANIELE PELLICIA†, WERNER JARK‡ and STEFANO LAGOMARSINO§

Istituto di Fotonica e Nanotecnologie – CNR, V. Cineto Romano, 42 00156, Rome, Italy

†*School of Physics, Monash University, ARC Centre of Excellence for Coherent X-Ray Science, Victoria 3800, Australia*

‡*Sincrotrone Trieste S.c.p.A., S.S. 14 km 163.5 in Area Science Park, 34012 Basovizza, Italy*

Istituto Processi Chimico Fisici–CNR c/o Phys. Dept. Universita' Sapienza P.leA. Moro 2, 00185, Rome, Italy

This chapter deals with the fundamental properties of X-ray waveguides (WGs), whose development is a logical consequence of the theoretical and experimental work on X-ray standing waves. The different coupling modes and the formation of the wavefield inside the WG are reviewed. Some fabrication procedures and relevant applications are also briefly described.

9.1. Introduction

An important consequence of the intense research activity on X-ray standing waves (XSW) carried out in the 1980s and 1990s has been the development of X-ray waveguides (WGs), which can provide a nanosized intense X-ray beam utilizing a resonant enhanced XSW field inside a layer of lower electronic density sandwiched between two layers of higher electronic density. More than 10 years after the first demonstration of principle, WGs have been considerably improved and are now recognized as important optical elements for characterization of materials at the nanometer scale.

As we will see in the following sections, WGs have promising applications in several fields, namely in one of the most advanced X-ray imaging methods, i.e., in coherent X-ray diffraction imaging (CXDI). They are also proposed as possible optical element for the next generation synchrotron radiation sources, the Free Electron Lasers (FELs).

In this chapter, we will provide some basic information about the principle of WGs, and the main features of the beam they provide; we will then treat very briefly fabrication procedures and mention some relevant applications.

9.2. X-Ray WG Basic Principles

As shown in Fig. 9.1, an incident X-ray beam can be coupled into the guiding layer of WGs in two different ways: resonant beam coupling (RBC) and front coupling (FC).

RBC[1–3] into a WG takes place with the X-ray beam incident on the upper very thin layer of high Z at a grazing angle corresponding to one of the resonance angles. The beam is then transmitted by the intermediate guiding layer of low Z and reflected by the thick lower layer of high Z (Fig. 9.1(a)). With this scheme, the incoming beam of several tens of microns is trapped in the guiding layer, where it is compressed down to nanometer dimensions. A significant intensity gain (exit flux

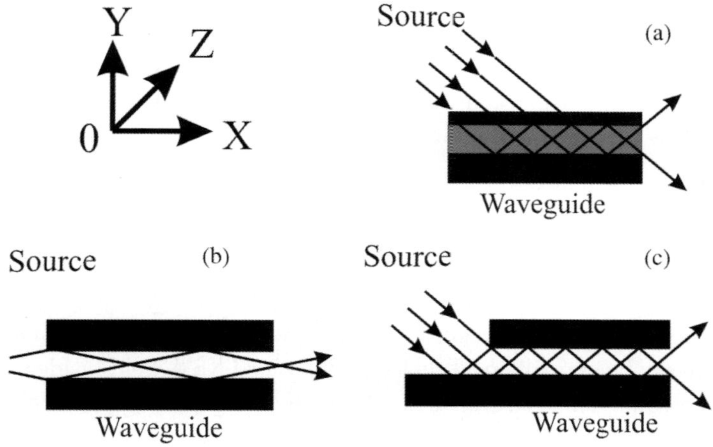

Fig. 9.1. Different coupling modes: (a) resonant beam coupling through the cover layer. (b) Front coupling directly into the guiding layer. (c) Front coupling with pre-reflection and generation of a standing wave pattern.

density/incident flux density >100) can then be achieved at the exit of the WG.[4] In the FC scheme,[5] the incoming radiation is directly side-coupled with the WG aperture, and the spatial acceptance is therefore of the order of the WG gap (Fig. 9.1(b)). However, in common cases, all modes are excited simultaneously in the FC WG. A variant of the FC scheme can be adopted with a preliminary reflection of the incoming beam just in front of the WG entrance at one of the resonance angles (Fig. 9.1(c)).[6] In this way, an XSW pattern is formed with a given spatial periodicity, and one particular resonance mode can thus be selected. This is a distinct advantage when mode mixing has to be avoided (e.g., when the structure of the guided beam at the exit has to be known). On the other hand, WGs with pre-selection mode are more complicated to fabricate with respect to the FC ones. The three WG types will be discussed at the end of this chapter with a short comparison of their important features.

9.2.1. *Resonant beam coupling*

The development of WGs working in RBC coupling mode is directly related to the intense research activity in XSW. Of particular importance has been the early work of Bedzyk *et al.*,[7] who showed that a long-period XSW is created above the surface of a reflecting material if the incidence angle θ_0 is less than the critical angle θ_c for total reflection. Later, Wang *et al.*[8] demonstrated that an enhancement of the electromagnetic field intensity takes place in a layer with a refractive index n_2 deposited on the surface of a material with a smaller refractive index n_1.

Efficient confinement of the electromagnetic radiation is obtained in a three-layer system if the refractive indices as presented in Fig. 9.2, fulfill the condition[9] $n_2 > n_1 \geq n_3$, where the index 2 refers to the guiding

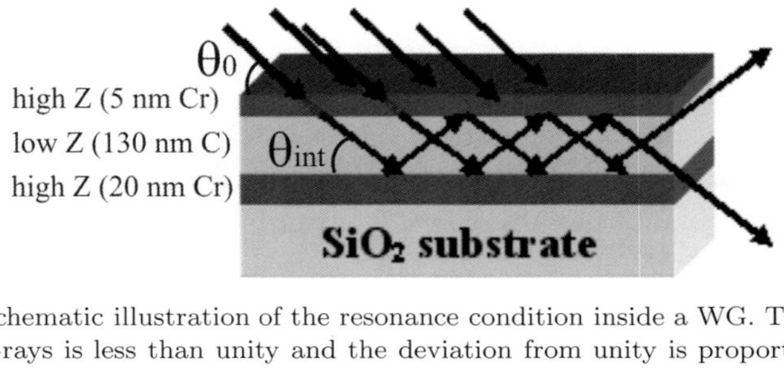

Fig. 9.2. Schematic illustration of the resonance condition inside a WG. The refractive index for X-rays is less than unity and the deviation from unity is proportional to the electron density. Thus, a layer with high Z has a smaller refractive index than the layer with low Z.

(i.e., the median) layer, and 1 and 3, respectively, to the upper and lower cladding layers. In the X-ray range, the refractive index is usually written as $n = 1 - \delta - i\beta$, where δ and β are small compared to unity. For X-rays, δ is a positive number, which varies with the electron density. Generally, lighter materials have smaller δ values compared with heavier materials. They usually also have a smaller absorption coefficient. Hence, X-ray mode excitation takes place in a lighter low absorbing layer sandwiched between heavier materials with higher absorption. The coupling and propagation of the beam into the core is more efficient for vacuum or air gap, where $\delta_0 \approx 0$, $\beta_0 \approx 0$. As shown in Fig. 9.1(a), the incident intensity coupled through a sufficiently thin cover layer directly excites permitted modes in the core layer.[1] In fact, the constructive interference, in the entire illuminated beam footprint, between the wave field coupled into the WG and the two internally reflected fields, produces a standing wave, stationary in the direction of the normal to the surface/interfaces. As a consequence, the intensity, i.e., photon flux density, in the guiding layer and at the exit of the WG can exceed the intensity of the incident wave field by far.[10] For the resonance condition to be met, the angle of grazing incidence onto the WG surface θ_0 and the grazing angle onto the internal interfaces θ_{int} must be properly related to the thickness of the guiding layer and to the wavelength λ.

We consider here symmetric WGs ($n_1 = n_3$). The thickness of the absorbing cover layer will always be assumed to be adjusted for maximum internal field enhancement. The guided mode has intensity maxima in the guiding layer and nodes close to the two interfaces between the guiding and the cladding layers. For proper resonance conditions, we must consider an effective resonator thickness defined as $D_{\text{eff}} = d_1 + d_2$, where d_2 is the WG aperture, and d_1 is the penetration depth of the electromagnetic field into the cladding layers that can be expressed as:

$$d_1 = (\lambda/2\pi)(\theta_c^2 - \theta_{\text{int}}^2)^{-1/2}.$$

The mode excitation condition, i.e., the standing wave condition for the X-ray beam of wavelength λ and an angle of grazing incidence θ_{int} (see Fig. 9.2) onto the internal interfaces, is given by

$$2D_{\text{eff}} \sin(\theta_{\text{int}}) = \lambda. \tag{9.1}$$

The relation of θ_{int} with the incident angle on the WG surface θ_{o} can be easily found considering the expression for critical angle for total reflection

$\theta_{c,i}$ (where the index i refers to the different layers),

$$\theta_{c,i} = \sqrt{2\delta_i} \tag{9.2}$$

and the refraction conditions at the interfaces (the Snell law). It results:

$$\vartheta_0 = \sqrt{\vartheta_{\text{int}}^2 + \theta_{c,2}^2}. \tag{9.3}$$

The electromagnetic modes excited into the guiding layer propagate along the WG length, and eventually exit at its terminal part. Thus the size of the produced X-ray beam is comparable with the thickness of the guiding layer. For the minimum beam size D_{min} that a WG can provide we obtain from Eqs. (9.1) and (9.3), in the approximation $\delta_1 \gg \delta_2$:

$$D_{\text{min}} > \frac{\lambda}{2\sqrt{2\delta_1}}. \tag{9.4}$$

In the hard X-ray range ($E > 4\,\text{keV}$) and sufficiently far from absorption edges, δ varies approximately as $\sim\lambda^2$. Thus D_{min} is nearly independent of the wavelength and depends only on the cladding layer material chosen. The latter is usually a material of high density as a metal. It is found that for all useful cladding materials D_{min} varies very little and is about $D_{\text{min}} = 10\,\text{nm}$. This number and Eq. (9.4) are also derived by Bergemann *et al.*[11] in the rigorous treatment of the minimum spot size obtainable in a tapered double plate X-ray WG.[12] The number $D_{\text{min}} = 10\,\text{nm}$ is in agreement with the experimental results of Pfeiffer,[13,14] who could still identify a weak exiting beam in a single-mode Ni/C WG with $d_1 + d_2 = (4.1\,\text{nm}) + (10.1\,\text{nm}) = 14.2\,\text{nm}$. Due to the resonant beam coupling, the WG provided still six times more output flux[13,14] than a hypothetical aperture with $d = 10.1\,\text{nm}$.

The other important aspect is the efficiency of the WG. This is strictly related to the angular and spatial acceptances. In fact, an X-ray WG is an optical resonator and thus it acts as a coherence filter. Therefore, the output flux cannot exceed the coherent fraction of the incoming radiation. According to Attwood,[15] the phase space extension for spatially coherent radiation is given by

$$A_{\text{FWHM}}\Delta\Phi_{\text{FWHM}} \approx 0.44\lambda \tag{9.5}$$

where A_{FWHM} and $\Delta\Phi_{\text{FWHM}}$ are, respectively, the spatial extension and the angular beam divergence in the y-direction (see Fig. 9.1). Jark and Di Fonzo[16,17] carried out a detailed analysis of the spatial and angular acceptances of a RBC-WG starting from Eq. (9.5). In Figs. 9.3 and 9.4, we

9.2.2. Front coupling with pre-reflection

This scheme (Fig. 9.1(c)) was first adopted by Zwanenburg *et al.*,[6] who used a configuration with two parallel plates accurately positioned one with respect to the other at a distance of few hundred nanometers. The maximum spatial acceptance is the double of the WG gap, and the correspondent angular acceptance for coherent radiation can be found from Eq. (9.5). We will analyze in the following the mode transmission in a WG with a vacuum gap and cladding layers of heavy materials, using both an analytical and a numerical solution approach. The computer simulations are described in detail elsewhere.[18,19] We based our analysis of the mode structure inside and outside the planar WG on the numerical solution of the parabolic wave equations ($|\partial^2 U/\partial^2 x| \ll k|\partial U/\partial x|$)

$$\begin{aligned} 2ik\,\partial U/\partial x + \partial^2 U/\partial y^2 &= 0, \quad \text{in vacuum} \\ 2ik\,\partial U/\partial x + \partial^2 U/\partial y^2 + k^2(\varepsilon - 1)U &= 0, \quad \text{inside WG} \end{aligned} \tag{9.6}$$

$$E(x, y) = U(x, y)\exp(ikx), \tag{9.7}$$

where $E(x, y)$ and $U(x, y)$ are the complex amplitudes of the electric field with, respectively, fast and slow change of the phase along the optical axis 0X. The dielectric constant of the materials ε is related to the refractive index n via $n = \varepsilon^{1/2} = (1 - \delta - i\beta)$, and $k = 2\pi/\lambda$ is the wave vector. As before, x is parallel to the interfaces in the plane of incidence, y is perpendicular to the guiding layer, and z is parallel to it (normal to the plane of incidence, see Fig. 9.1). It is assumed that the length along z is much larger than along y, therefore the field can be considered constant along z (planar WG).

Due to the high brilliance and the large distance between source and WG, synchrotron radiation is in general well approximated by a plane wave. In the pre-reflection geometry, the plane wave with grazing incidence angle θ_m is totally reflected just in front of the WG entrance, giving rise to a standing wave above the reflecting surface with the periodicity corresponding to the mth resonance mode. θ_m corresponds to the resonance mode condition

$$\tan\left[\frac{k\sin\theta_m\,d - (m-1)\pi}{2}\right] = \frac{\sqrt{\cos^2\theta_m - \cos^2\theta_c}}{\sin\theta_m}, \tag{9.8}$$

where d is the width of the vacuum gap between the two cladding layers, m is the mode number, and θ_c is the critical angle of total reflection. Equation (9.8) is equivalent to Eq. (9.1) but takes into account in a

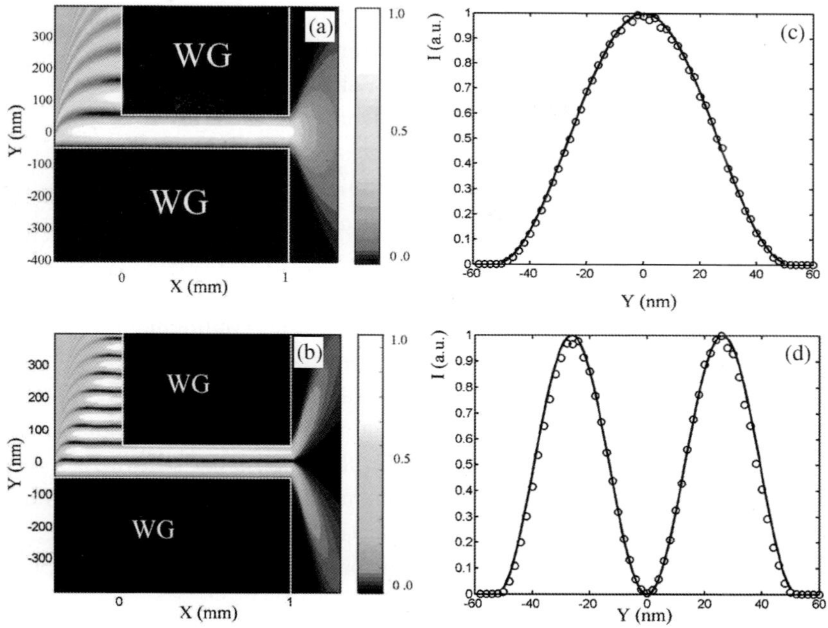

Fig. 9.5. Intensity distribution in front of and along a front coupling WG in mono-modal regime for: (a) first mode; (b) second mode; (c) and (d) present the vertical intensity distribution in the guiding layer of an X-ray WG for first (c) and second (d) modes calculated with computer code (full line) and with Eq. (9.9) (open circles).

quantitative way the penetration of the electromagnetic field into the cladding layers.

Simulations presented in Figs. 9.5(a) and 9.5(b) refer, respectively, to the first and second resonance modes from a WG with Pt cladding layers and gap of 100 nm at $\lambda \approx 0.154$ nm, where the spatial intensity distribution in front, inside, and at the exit of the WG is shown.

Figures 9.5(c) and 9.5(d) show in more detail the comparison, for the same WG, of the calculations of the intensity distribution based on the computer simulation (solid lines) and the analytical solution (empty circles) of Eq. (9.6)[20] for the mode m, given by:

$$U_m(y) = \begin{cases} \exp(\delta_m y) & y < 0 \\ \cos(\kappa_m y) + (\delta_m/\kappa_m)\sin(\kappa_m y) & 0 \le y \le d \\ [\cos(\kappa_m d) + (\delta_m/\kappa_m)\sin(\kappa_m d)]\exp[-\delta_m(y-d)] & y > d \end{cases} \tag{9.9}$$

where $\kappa_m = k\sin\theta_m$ and $\delta_m = [(1-\varepsilon)k^2 - \kappa_m^2]^{1/2}$.

The intensity distribution for the first mode in free space close to the WG exit coincides with the diffraction pattern of a wave with the amplitude given by Eq. (9.9) confined by a slit with size d corresponding to the WG gap. In the near field zone for distances from the WG end of $\Delta x < x_{\text{dif}} \sim d^2/4\lambda = 1.6 \times 10^4$ nm the beam cross-section is almost constant ($\sim d$). At $\Delta x = x_{\text{dif}}$, a focusing effect with a local minimum in the beam cross-section can be observed. In the far field zone $x > x_{\text{dif}}$ the beam diverges with an opening angle $\alpha \approx \lambda/d$.

A considerable difference of the FC compared to the RBC resides in the attenuation along the beam trajectory in the confinement region. In fact, a FC-WG can be fabricated with an air or vacuum gap, therefore absorption losses are due only to the penetration of the tails of the intensity distribution into the cladding material. Rigorous calculations should involve physical optical considerations, but a simple ray-tracing can provide very good approximate values for the estimation of absorption losses, as far as the resonance angle is smaller than the critical angle for the cladding.

In the case of independent mode propagation, the variation of power $W(X)$ along the WG length X ($\theta_m \ll \theta_c$) is given by[19]:

$$W(X) = W_0 \cdot R_{\text{Fr}}^N(\theta_m) \approx W_0 \exp\left(-\frac{\beta}{\delta^{3/2}} \frac{X}{d\sqrt{2}} \theta_m^2\right). \tag{9.10}$$

Here, W_0 is the energy of the radiation coupled into the WG, R_{Fr} is the Fresnel coefficient of reflection for the vacuum-cladding layer boundary, $N = X\theta_m/d$ is the number of reflections which the ray undergoes, d is the effective size of the mode confined by the cladding layers, θ_m is the resonance angle for the mode m; note the θ_m^2 dependence. The same function $W(X)$ can be calculated using a computer code.[20] Figure 9.6, shows very good agreement of the calculation using Eq. (9.10) and the result of a computer simulation based on the solution of the parabolic wave equation.

Front coupling with pre-reflection has the important advantage to select the resonance mode, and therefore to avoid mode mixing, as long as the divergence of the incoming radiation is smaller than the angular separation between adjacent modes, or, in other words, when the coherence length of the incoming beam is larger than the WG spatial acceptance. We have analyzed in detail this aspect, simulating an incoherent source very close to the WG entrance,[19] a situation that can be encountered for example with laboratory sources. However, a FC scheme with pre-reflection is not always possible or convenient and therefore we studied in detail also the direct FC mode.

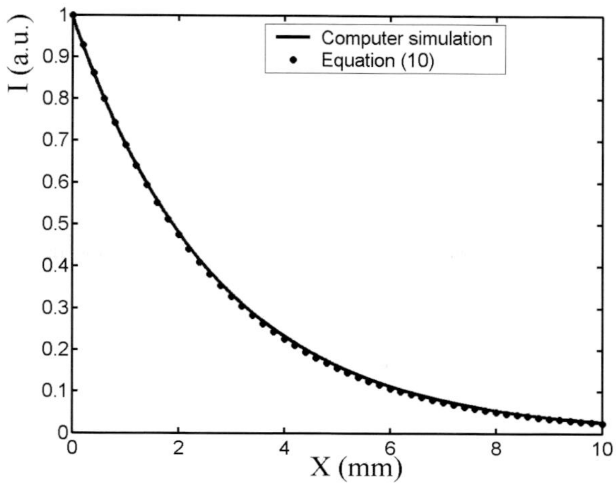

Fig. 9.6. Variation of the intensity for the second mode at $\lambda = 0.1$ nm along a WG with Pt cladding layer and vacuum gap 100 nm calculated with a computer code based on the parabolic wave Eq. (9.6) (black line) and with Eq. (9.10) (closed circle).

9.2.3. *Direct front coupling*

When the incoming beam is directly coupled to the WG (Fig. 9.1(b)), the detailed interaction of the beam with the cladding layers must be considered, especially if the absorption in the material of the cladding layers is weak. This analysis, which reveals several diffraction and refraction phenomena, gives a substantially different description of the wave field in the WG with respect to those generally found in the literature (see Ref. 20 for a complete analysis).

In the following, we consider a σ-polarized plane wave, i.e., with the vector of the E-field normal to the scattering plane, of wavelength $\lambda = 0.1$ nm incident at a right angle onto the side of a planar hollow X-ray WG (Fig. 9.1(b)). The gap d is limited by two cladding walls with refractive index $n = \varepsilon^{1/2} = (1 - \delta - i\beta)$, the material of the walls is Si, and $\beta \ll \delta$ for the energies considered in this chapter. We start by analyzing the interaction of the incident beam with a single dielectric corner (half of the WG in our case). It was shown by Kopylov and Popov[21] that the diffracted field $U(x, y)$ can be expressed, in the parabolic approximation, as the sum of three terms: (1) the standard Fresnel edge diffraction term, (2) a term, describing a correction to the Fresnel term due to the limited absorption of the wall material, and (3) a term, which represents a lateral plane wave propagating in the wall material along the material–vacuum interface $0X$

with the enhanced phase velocity $V_p = c/\varepsilon^{1/2}$ and entering into the vacuum at the critical angle θ_c.

The superposition of the direct and diffracted beams with the lateral wave gives rise to an interference pattern of successive maxima and minima (see Fig. 9.2 in Ref. 20). Similar considerations hold for the analysis of the field at the entrance aperture of the WG. In the following, the origin of the y-coordinate is in the center of the gap d, and the cladding walls are at $y = \pm d/2$. Also in this case the total field U can be expressed as the sum of the field

$$\Phi(x, y) \approx \tilde{\Phi}(k\theta) \exp(-i\pi/4 + ik\theta^2 x/2)/\sqrt{\lambda x},$$
$$\text{where} \quad \tilde{\Phi}(k\theta) \approx d\left[\frac{\sin(k\theta d/2)}{k\theta d/2} + \frac{\cos(k\theta d/2)}{(kd/2)\sqrt{\theta_c^2 - \theta^2}}\right] \tag{9.11}$$

and that of the two lateral plane waves, entering into the vacuum gap from the opposite boundaries $y = \pm d/2$ of the WG:

$$\Psi(x, y) \approx \frac{d}{\sqrt{\lambda x}} \exp(i\pi/4 - ik\theta_c^2 x/2) \frac{\cos(k\theta_c y)}{k\theta_c d/2} \exp(ik\theta_c d/2). \tag{9.12}$$

In Eq. (9.11) $\theta = y/x$. The spatial spectral amplitude $\widetilde{\Phi}(k\theta)$ in Eq. (9.11) includes the function $\sin(k\theta d/2)/(k\theta d/2)$, corresponding to the Fraunhofer diffraction of a plane wave from a thin slit, and a correction term due to the material of the walls. The correction term shifts the positions of the angular spectrum maxima toward smaller angles. $\Phi(x, y)$ can be expressed as the linear combination of resonance modes φ_m by

$$\Phi(x, y) = \sum_{m=0}^{m=m_{\max}} c_m(\theta_m)\varphi_m(y), \tag{9.13}$$

where φ_m is given by

$$\varphi_m(y) = \begin{cases} \cos(k\theta_m y), & |y| < d/2 \\ \cos(k\theta_m d/2)\exp[-k\mu(|y| - d/2)], & \text{else,} \end{cases} \tag{9.14}$$

with $\mu_m \approx (\theta_c^2 - \theta_m^2)^{1/2}$ and c_m given by:

$$c_m(\theta_m) = \|\varphi_m\|^{-1} \int_{-\infty}^{+\infty} \varphi_m(y)dy. \tag{9.15}$$

Taking into account the propagation factor $\exp(-i\chi_m x)$ for each mode, where in the parabolic approximation $\chi_m \approx \theta_m^2[k/2 - i(\beta/\delta^{3/2})/(2^{1/2}d)]$,[19]

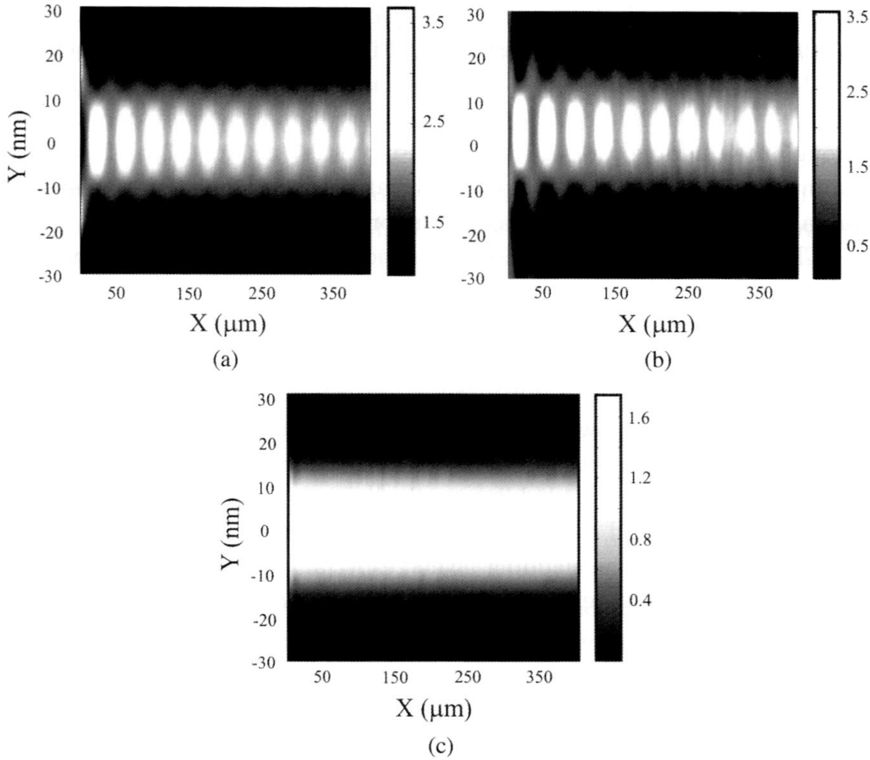

Fig. 9.7. Total field in a WG with Si walls and a 30-nm gap (wavelength $\lambda = 0.1\,\text{nm}$) with a plane wave at the entrance: (a) analytical solution; (b) computer simulation; (c) computer simulation with a step function field ($U = 1$ in the gap, 0 elsewhere) at the entrance (from Ref. 20).

the wave field $\Phi(x, y)$ at any point of the WG is given by

$$\Phi(x, y) = \sum_{m=0}^{m=m_{\max}} c_m(\theta_m)\varphi_m(y)\exp(-i\chi_m x). \tag{9.16}$$

The total field $U(x, y)$ is therefore given by the sum of $\Phi(x, y)$ (Eq. (9.16)) and a function $\Psi(x, y)$, which represents the sum of the two lateral plane waves field (Eq. (9.12)).

In Fig. 9.7, the intensity distribution in the vacuum guiding layer for a 30-nm gap WG with Si walls and photon wavelength $\lambda = 0.1\,\text{nm}$ is shown. The WG supports only one mode.

Figure 9.7(a) depicts the analytical solution given by Eqs. (9.16) and (9.12), and Fig. 9.7(b) represents the result of the computer simulation

Here $\mu_{\text{int}}(y) = 2k_0 \,\text{Im}(\varepsilon)$ is the interference absorption coefficient, $\beta_{gg'} = \sigma_{gg'}/\sigma_{00}$ are the ratios of the partial IS (CS, TDS) cross sections σ_{hh} and σ_{0h} to the cross section σ_{00} of usual incoherent IS far from the diffraction condition. One can immediately recognize in (10.10) the interference term in a form which is typical for any XSW yield curve[17–20] but with the specific coefficients β_{hh} and β_{0h}. These coefficients are determined by the electronic structure of the crystal and represent the key feature of our analysis. In the CS case

$$\sigma_{gg'} = (\mathbf{e}_g\mathbf{e}')(\mathbf{e}_{g'}\mathbf{e}')r_0^2 \sum_n \tilde{\sigma}_{gg'}^{(n)} e^{i(\mathbf{g}-\mathbf{g}')\mathbf{r}_n} e^{-M_{g-g'}^{(n)}} \tag{10.11}$$

$$\tilde{\sigma}_{gg'}^{(n)} = \sum_{i\neq j} f_{ij}^{(n)}(\mathbf{S}_g) f_{ij}^{(n)*}(\mathbf{S}_{g'})\delta(\omega_{ij}^{(n)} - \Omega) \tag{10.12}$$

$$f_{ij}^{(n)}(\mathbf{S}_g) = \int \psi_i^{(n)}(\mathbf{r}) e^{i\mathbf{S}_g\mathbf{r}} \psi_j^{(n)*}(\mathbf{r}) d\mathbf{r}. \tag{10.13}$$

The summation in (10.11) is over all atoms in the unit cell, $\mathbf{S}_g = \mathbf{S}_0 - \mathbf{g}$ are the scattering vectors, $r_0 = e^2/mc^2$, $\mathbf{r}_n$ is the coordinate of nth atom in the unit cell, $M_{g-g'}^{(n)}$ is the thermal (or static) Debye–Waller factor. In (10.12) and (10.13), the summation is over all electrons of the atom of type n, f_{ij} is the form factor (at $i = j$) or the overlap integral (at $i \neq j$) of the electron shells with the binding energies $\hbar\omega_{i,j}$. Note that $\sigma_{0h} = \sigma_{h0}^*$.

There are three factors, differing in their physical nature, that determine the angular dependence of the intensity yield $\kappa(\omega', y)$, Eq. (10.10). The first is the angular dependence of the interference absorption coefficient $\mu_{\text{int}}(y)$, which determines the penetration depth $L(y) = 1/\mu_{\text{int}}(y)$ of the incident beam under conditions of Bragg diffraction. The second factor is the spatial structure of the field in the crystal determined by the amplitude and the phase of reflection coefficient R. The third factor is the specific type of the IS process which depends also on the direction of observation due to the angular dependences of the parameters β_{hh} and β_{0h}.

The penetration depth of the X-ray field is minimal at the center of the total Bragg reflection range: $L(0) = \Lambda/2$, where $\Lambda = \lambda(\gamma_0|\gamma_h|)^{1/2}/\pi C|\chi_{hr}|$ is the extinction length. Using Eq. (10.4) and condition $d\mu_{\text{int}}/d\alpha = 0$ one can see that this depth reaches its maximum L_a (anomalous penetration depth) at the normalized angular deviation from the exact Bragg position of

$$y_m = (1 + b)/(2\sqrt{b}C\varepsilon_h) \tag{10.14}$$

at which $\mathrm{Im}(Y) = 0$. Here $\varepsilon_h = (|\chi_{hi}|/\chi_{0i})\cos(\eta_h - \omega_h)$, where η_h and ω_h are the phases of $\mathrm{Re}(\chi_h)$ and $\mathrm{Im}(\chi_h)$, respectively. At $y = y_m$ we have

$$\mu_{\text{int}}(y_m) = (\mu_0/2)[(1-b) + \sqrt{(1+b)^2 - 4bC^2\varepsilon_h^2}]. \tag{10.15}$$

In the symmetric case ($b = 1$) the depth of the anomalous penetration is $L_a = l_0/(1-\varepsilon_h^2)^{1/2}$, where $l_0 = 1/\mu_0$, while it raises considerably in the strong asymmetric case of $b \gg 1$: $L_a = l_0/(1-\varepsilon_h^2)$. In the latter case, the angular range of the anomalous penetration (10.14) is strongly displaced toward the negative angles. Thus, $y_m = -1.04$ and $L_a = 3.76l_0$ for $b = 1$, and, for example, for $b = 20$, these values are $y_m = -2.43$, and $L_a = 14.1l_0$ (Si, 220 reflection, Cu K_α, σ-polarization, $|\varepsilon_h| = 0.964$). We will see in the next section that using strongly asymmetric reflections with $b \gg 1$ the sensitivity of the XSW technique to the details of the electronic distribution can be greatly enhanced.

The intensity of the CS (10.10) is determined by the sum of three contributions. They are, respectively, CS in the field of the incident beam with the scattering cross section σ_{00}; CS of the diffracted beam quanta with intensity proportional to $b|R|^2\sigma_{hh}$, and the interference term $\sim R\sigma_{0h}$, arising because of the coherence of the reflected and incident radiations.

These features have been clearly observed in the experiment,[8] Fig. 10.1(a): the deep minimum related to the extinction effect (the depth

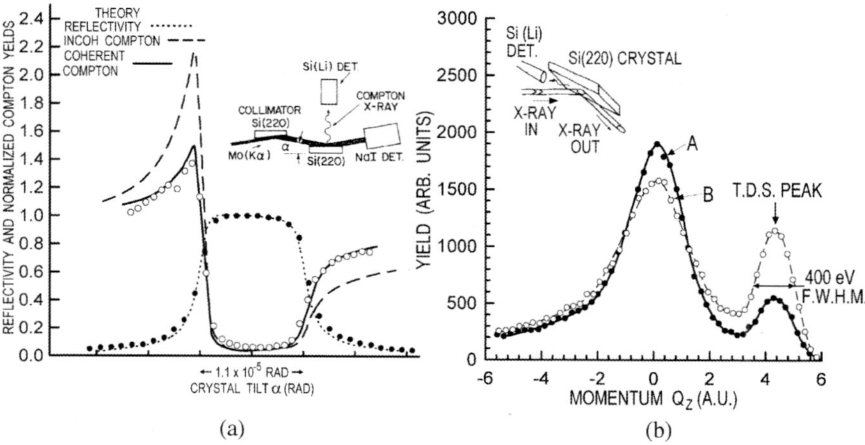

Fig. 10.1. (a) Angular-yield data showing reflectivity (solid circles) and Compton-effect (open circles). Solid line is calculation for coherent CS, dashed line — incoherent (neglecting interference term) CS yield. (b) Spectrum of the scattered radiation: curve A, antinodes between atomic planes; curve B, antinodes on atomic planes, from Ref. 8.

of the CS yield $l'_{\rm CS} = \gamma'/\mu' \gg \Lambda$) and the narrow peak at the lower angle side are clearly visible.

There are two reasons for the enhancement of $\kappa_{\rm CS}$ at $y \approx -1$.

First, in this angular range, the maxima of the field (antinodes) are between the atomic planes, where the density of the valence electrons mostly contributing to the CS is high. Since the valence electrons, due to uncertainty principle, are more delocalized than the core ones, the momentum distribution and therefore the spectrum of the CS by valence electrons should be narrower. Indeed, the curve A on Fig. 10.1(b) in the Compton region is higher and narrower, than the curve B.

Second, in the angular range $y \approx -1$ an abnormally deep penetration of the field into the crystal takes place and therefore the scattering volume increases and so does the CS intensity. Decrease of $\kappa_{\rm CS}$ at $\theta > \theta_B$ is related to the fact that the field's antinodes are now on the atomic planes, i.e., in the region of strongly bound electrons only weakly participating in the CS. Besides, the interference absorption coefficient is still high here: $\mu_{\rm int} > \mu_0/\gamma_0$. Schülke[12] completely confirmed this consideration by the detailed calculation of contributions of various groups of electrons ($1s^2$, $2s^2 2p^6$, $3s^1 3p^3$) to the CS at $y \gg 1$ (no diffraction), $y = -1.1$ and $y = 1.1$. On the angular dependence of the CS intensity, Fig. 10.1(a), the dashed curve corresponds to the case when the third interference term in (10.10) is disregarded, i.e., $\sigma_{0h} = 0$, the case considered by Annaka *et al.*[6]

For the TDS, which is caused mostly by strongly bound electrons, the situation is completely opposite and in many respects similar to the fluorescence from the matrix atoms: the yield at the left edge of the extinction minimum ($y = -1$, nodes on atomic planes) is lower than at the right edge ($y = 1$, antinodes on atomic planes). This is exactly the behavior first observed by Annaka[7] (see also Fig. 10.1(b), where the curve A lies below the curve B in the region of TDS peak).

10.3. Coherent Compton Effect and Electron Density Distribution

As we have seen in the previous section, the coherent CS may serve as an effective tool to study the electronic structure of single crystals: by changing $\Delta\theta = \theta - \theta_B$ the nodes and antinodes of the XSW can be scanned through the inhomogeneous electron density distribution $\rho(\mathbf{r})$, thus revealing the details of this distribution. Unfortunately, due to extinction effect only the "tails" of the $\kappa_{\rm CS}$ angular curve contain information about the electron density structure. In this section we will discuss in more detail how the CS cross-sections entering Eq. (10.10) can be calculated based on the electronic density distribution. We will show also how strong asymmetry of the Bragg reflection may be utilized to significantly enhance the sensitivity of this technique.

The main contribution to CS comes from transitions of atomic electrons into continuous spectrum; therefore, the wave function of the excited electron state looks like the plane wave $\psi_j(\mathbf{r}) = V^{-1/2}\exp(i\mathbf{p}'\mathbf{r})$. Let us replace the summation over j in (10.12) by integration over $\mathbf{p}'$ and substitute $\mathbf{p} = \mathbf{S}_g - \mathbf{p}'$. Assume now that $\Omega \gg \omega_i$, i.e., the energy transferred to electron is much higher than the binding energy, and the transferred momentum $\mathbf{S}_g \gg \mathbf{p}$, i.e., $S_g a \gg 1$, where $a \sim 1/p$ is the radius of an orbital. This is the so-called impulse approximation,[21,22] which is commonly used for calculations of Compton profiles. As a result, for the double differential CS cross-sections we receive[12]:

$$\widetilde{\sigma}_{gg'} = \sum_i \int d^3\mathbf{p}\langle\mathbf{p}|\mathbf{p}+\mathbf{g}'-\mathbf{g}\rangle_i \delta\left(\frac{\eta S_g^2}{2m} - \frac{\eta\mathbf{p}\mathbf{S}_g}{m} - \Omega\right)$$
$$|\mathbf{p}\rangle_i = \psi_i(\mathbf{p}) = \int \psi_i(\mathbf{r})e^{-i\mathbf{pr}}d\mathbf{r}. \tag{10.16}$$

Here $\langle\mathbf{p}|\mathbf{p}+\mathbf{g}\rangle$ is the one-electron density matrix in momentum space. The function σ_{00} represents the well-known kinematical cross-section of CS, corresponding to inelastic and incoherent scattering of the incident quantum $\hbar\mathbf{k}_0$ to the CS-mode $\mathbf{k}'$. From the argument of the δ-function in (10.16), the known dispersion law, Eq. (10.2), in the case of CS on weakly bound electrons follows. Performing the integration in (10.16) for the isotropic electron density distribution one receives

$$\sigma_{00}(p_z) = \frac{m}{\hbar S_0}J(p_z), \quad J(p_z) = 2\pi\sum_i \int_{|p_z|}^{\infty} p|\psi_i(\mathbf{p})|^2 dp. \tag{10.17}$$

The function $J(p_z)$ is the Compton profile. The knowledge of the spectrum σ_{00} enables to determine the probability of finding an electron with momentum $\mathbf{p}$:

$$|\psi(\mathbf{p})|^2 = \left|\frac{1}{2\pi p_z}\frac{dJ}{dp_z}\right|. \tag{10.18}$$

Because of the δ-function in (10.16), the three-dimensional integration over $\mathbf{p}$ is reduced to the integration in a plane perpendicular to the scattering vector $\mathbf{S}_0$, situated at a distance p_z from the origin of coordinates in momentum space. Thus, the spectrum σ_{00} provides information about the one-dimensional (projected onto $\mathbf{S}_0$) momentum distribution function $\langle\mathbf{p}|\mathbf{p}\rangle$, i.e., about the diagonal element of the one-particle momentum density matrix. Similarly, cross-section σ_{hh} is the incoherent CS cross-section of the diffracted beam quanta $\hbar\mathbf{k}_h$ into the direction of observation $\mathbf{k}'$.

Cross-sections with $g \neq g'$ in (10.16) describe the coherent (interference) effects in those states in which the system "crystal-radiation" can be found as a result of both the CS of the incident quantum $\hbar\mathbf{k}_0$, and the CS of the diffracted quantum $\hbar\mathbf{k}_h$. Thus, the non-diagonal cross-section σ_{0h} defines the contribution to the total cross-section from the scattering by the electronic state representing the joint probability of finding electron in states $|\mathbf{p}\rangle$ and $|\mathbf{p}+\mathbf{h}\rangle$. For the first time, this important property has been emphasized in Refs. 9 and 12 (see also Refs. 14 and 23).

The significance of the cross-section σ_{0h} is due to the fact that it is defined by non-diagonal elements $\langle\mathbf{p}|\mathbf{p}+\mathbf{h}\rangle$ of the density matrix. Measured in a diffraction experiment, the atomic factors f_h do not provide such information, because, as it was shown in Ref. 24, $f_h = \int\langle\mathbf{p}|\mathbf{p}+\mathbf{h}\rangle d^3\mathbf{p}$ where the integration is performed over all $\mathbf{p}$-space instead of over a plane, as in (10.16).

The most striking feature of the spectrum σ_{0h} (10.16) is that the energy shift of its maximum differs from the known Compton shift Ω_0 (10.2). The reason is that in the process of the coherent CS the part of the recoil momentum equal to the reciprocal-lattice vector $\hbar\mathbf{h}$ is transferred to the crystal lattice as a whole.[13,23,25] Formally, it can be explained by the fact that, according to the conservation law (δ-function in (10.16)), both the momentum $\hbar\mathbf{h}$ and the electron momentum $\mathbf{p}$ appear on the same basis.

Consider now the CS cross sections integrated over energy. As it follows from (10.12), in the Waller–Hartree approximation the integral cross-sections of the CS by one atom in a unit cell can be written as[14]

$$N_{gg} = Z - \sum_{ij} |f_{ij}(\mathbf{S}_g)|^2, \quad N_{0h} = f_h - \sum_{ij} f_{ij}(\mathbf{S}_0) f^*_{ij}(\mathbf{S}_h). \tag{10.19}$$

Here $f_h = \sum_i f_{ii}(\mathbf{h})$ is the atomic scattering factor. It should be noted that in contrast to integral cross-sections obtained by integrating of CS cross-sections (10.16) over energy in the impulse approximation, the results of Eq. (10.19) are exact.

At large enough scattering vectors $\mathbf{S}_g$, i.e., hard X-ray or gamma radiation and large scattering angles, $f_{ij} \approx 0$ in (10.19). As a result, the diagonal cross-sections N_{gg} tend toward the atomic charge Z. The non-diagonal cross-section N_{0h} at the same time tends toward the atomic factor of the elastic scattering f_h. In the $\mathbf{S}_0 = 0$ limit, the quantities $f_{i \neq j} = 0$, $f_{ii} = 1$; therefore, $N_{00} = N_{0h} = 0$, while $N_{hh}(\mathbf{S}_h) = N_{00}(-\mathbf{h})$. If the CS is observed in the diffraction plane at the angle $\Psi = 2\theta_B$, then $\mathbf{S}_0 = \mathbf{h}$, $\mathbf{S}_h = 0$, and $N_{hh} = N_{0h} = 0$. Since the atomic scattering factor $f_h < Z$, then at small enough scattering vectors $\mathbf{S}_0$ the interference cross-section N_{0h} can become even negative.

These features are clearly seen in Fig. 10.2, where the theoretical angular dependences of the CS integral cross-sections $N_{gg'}$ are shown for Si, reflection (220), Cu K$_\alpha$ radiation. To illustrate the sensitivity to the electron density distribution, the calculations have been performed for two simplest models. In the first model, electron wave functions of free atoms have been used. Note that the majority of the calculations of atomic scattering factors and IS cross-sections were performed in this approximation. In the second model, valence electrons have been treated as a free electron gas. The last model for such crystals as Si and Ge was confirmed by experiments on low-angle IS of fast electrons[1,26] and X-rays.[27–29] Integral cross-sections have been calculated by using Eq. (10.19) and Hartree–Fock integrals f_{ij} for free atoms and Si^{+4} ions published by Freeman.[30] In the model of "ion core + free electrons" (dashed lines) the diagonal cross-sections N_{gg} are equal to the sum of cross-sections of the ion core Si^{+4} and the valence electrons

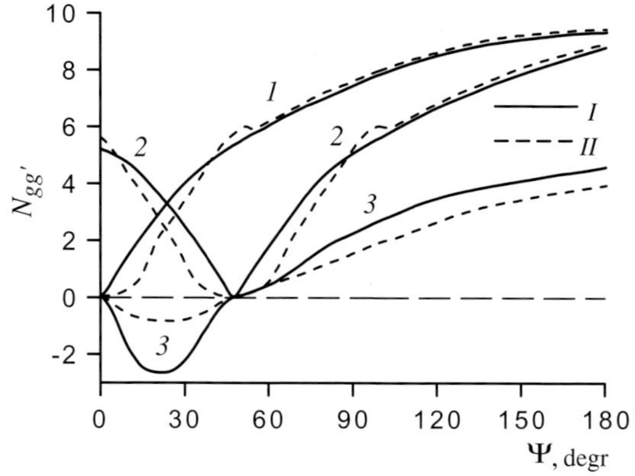

Fig. 10.2. Theoretical angular dependences of the CS integral cross-sections N_{00} (curve *1*), N_{hh} (*2*) and N_{0h} (*3*) in silicon crystal within the models of free atoms (solid curves *I*) and "ion core + free electrons" (dashed curves *II*); Cu K_α radiation, (220) reflection, registration in a plane of diffraction along the vector $\mathbf{h}$(220).[14]

uniformly distributed in a unit cell and therefore not contributing to the interference cross-section N_{0h}. The contribution of the valence $3s^23p^2$-electrons was calculated by using the theory of electron liquid,[31] considering correlation between electrons through Coulomb and Fermi interactions.

Consider now the CS detected along the reciprocal-lattice vector $\mathbf{h}(220)$ as in Ref. 16 (silicon, 220 reflection, Cu K_α radiation, the CS angle Ψ in this case does not depend on the asymmetry factor b and equal to $\Psi = 113.6°$). Then, in the free atoms model $\beta_{hh} = 0.768$ and in the model of "ion core + free electrons" $\beta_{hh} = 0.764$, i.e., diagonal parameters β_{hh} differ by only 0.5%. Contribution of the $3s^23p^2$-electrons into the CS total cross-sections N_{00} and N_{hh} are 48% and 62%, respectively. Atomic scattering factors of $f_h = 8.717$ and 8.673 for two models differ by only $\approx 0.5\%$. On the contrary, the difference in the interference parameter β_{0h}, 0.394 versus 0.287, is more than 37%, which is 10 times higher than the relative change of the diagonal elements.[14]

The angular dependences of the CS yield calculated by using Eq. (10.10) are shown in Fig. 10.3 for symmetric ($b = 1$) and highly asymmetric ($b = 41$) Bragg reflections and the two models described above. As we discussed earlier, the coherent CS peak in the highly asymmetric case is strongly shifted away from the extinction minimum toward the lower angles, leading to the enhancement of the CS peak. Clearly, the κ_{CS} curves in the case of $b \gg 1$ are much more sensitive to the value of the interference parameter β_{0h} and to the electron distribution model. The effect of the angular shift and the enhancement of the CS peak was observed for the first time by Bushuev *et al.*[16] and Afanasev *et al.*[32] (see also Refs. 33 and 34 in which the grazing Bragg–Laue geometry was used to implement highly asymmetric diffraction).

The TDS angular dependence κ_{TDS} in the highly asymmetric case (Fig. 10.3, right bottom panel) demonstrates very interesting behavior with an additional minimum and two maxima. Since for the TDS the parameters $\beta_{hh} = 0.85$ and $\beta_{0h} = 0.83$ are close to each other and to unity (see Refs. 14 and 16 for more details), the shape of the TDS curve is similar to the fluorescence yield from the bulk atoms experimentally observed in Refs. 35 and 36.

Experimental data for the conditions described above is shown in Fig. 10.4 (Si, (220) reflection, Cu K_α radiation, highly asymmetrical monochromator crystal with $b_m = 1/41$) for the sample asymmetry factors of $b = 1$ and 41, Ref. 16. The axis $[1\bar{1}0]$ was parallel to the vertical (rotation) axis of the goniometer and the measurements of the IS were performed in

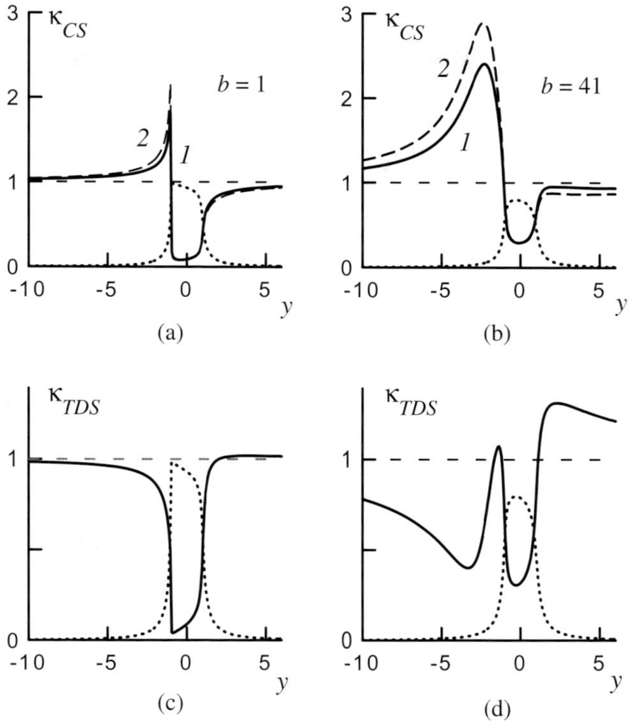

Fig. 10.3. Theoretical angular dependences of the CS (a, b), TDS (c, d), and X-ray reflection $|R|^2$ (dotted curves) in the (220) diffraction of σ-polarized Cu K_α radiation in a silicon crystal for symmetric (a, c) and highly asymmetric (b, d) Bragg diffraction with $b = 1$ and $b = 41$, respectively. For the CS (a, b) solid curves *1* are the free atoms model, and dashed curves *2* are the model of "ion core + free electrons" (from Ref. 14).

the equatorial (horizontal) plane along the h(220) reciprocal-lattice vector by the scintillation counter. By using this geometry and Cu K_α radiation the contribution of the TDS into the measured signal was minimized.

Theoretical curves were calculated by taking into account the angular dependences of both CS and TDS (see Fig. 10.3). The CS cross-sections were calculated for the free atom model. The TDS contribution originated from the vicinity of the (333) reciprocal-lattice point from the phonons with their wave vectors essentially parallel to the [110] axis and nearly touching the boundary of the Brillouin zone ($q_{\min} = 0.97(\pi/a_{110})$). Calculations of the structure amplitudes $F(333)$ and $F(113)$ and single-photon TDS cross-sections for the acoustic and optical phonons yields $\beta_{hh} = 0.85$ and $\beta_{0h} = 0.83$. The TDS cross-section, σ_{00}, is 32% of the CS cross-section, in

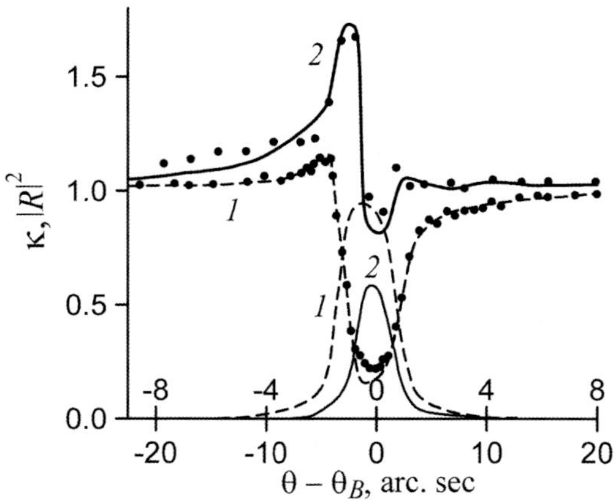

Fig. 10.4. Experimental curves of the angular dependence of the intensity of IS $\kappa(\theta)$, (primarily CS; solid circles), and reflection curves $|R|^2$. (1) Symmetric diffraction ($b = 1$, the lower abscissa scale); (2) asymmetric diffraction ($b = 41$, the upper abscissa scale). The solid and dashed lines are calculations for free atoms model (Si, 220 reflection, Cu K_α), from Ref. 16.

a very good agreement with the experimental value of $30 \pm 3\%$ measured by using crystal analyzer.

Experimental data for the highly asymmetrical case of $b = 41$ confirms the main features presented in Fig. 10.3, such as the enhancement of the IS (with major contribution from the CS) peak. Note that the experimental data at large negative angles are significantly, by 13%–17%, higher than the theoretical curve. We attribute this discrepancy to the fact that the Hartree–Fock approximation slightly overestimates the CS cross-section N_{0h}, as the wave functions of the Bloch electrons in crystal are different from those of free atoms. Indeed, the "ion core + free electrons" model for this angular range, Fig. 10.3(b), gives about 15%–20% higher yield than the free atoms model. For the symmetrical case $b = 1$ both models fit experimental curve equally well confirming our conclusion of much higher sensitivity of the CS-XSW curves to the electron distribution under conditions of highly asymmetrical Bragg diffraction.

10.4. Coherent Compton Effect in the Laue Geometry

Most of the coherent CS studies have been performed in the Bragg case of diffraction, which is a traditional diffraction geometry for the XSW method. Obviously, the deep minimum in the CS yield curves due to extinction effect significantly reduces the sensitivity of the technique even in the highly asymmetric diffraction. From this perspective, the Laue diffraction geometry (cf. Chapter 2) presents a promising alternative approach.[25,37,38] The other important advantage of this geometry is the possibility to measure the CS at the small scattering angles at which the relative contribution of valence electrons is maximum.[38,39]

Let us consider the angular dependence of the CS intensity in the Laue case.[25,37,38,40] Substituting (10.3) and (10.7) into the general expression (10.9), the CS yield in the Laue geometry can be written as:

$$\kappa(\omega', y) = \frac{1}{4(1+y^2)l_\infty}[A_{11}l_{11} + A_{22}l_{22} + 2\,\mathrm{Re}(A_{12}l_{12})] \tag{10.20}$$

where

$$A_{\nu\nu} = T_\nu^2 + b\beta_{hh} + 2\sqrt{b}T_\nu\,\mathrm{Re}\,\beta_{0h}, \quad T_{1,2} = y \mp \sqrt{1+y^2}$$
$$A_{12} = 1 - b\beta_{hh} - 2y\sqrt{b}\beta_{0h}.$$

For the CS measured from the exit crystal surface,

$$l_{\nu\nu'} = \frac{\exp(-\mu_{\nu\nu'}l) - \exp(-\mu' l/\gamma')}{\mu'/\gamma' - \mu_{\nu\nu'}}$$
$$\mu_{\nu\nu} = 2k_0\,\mathrm{Im}(\varepsilon_\nu), \quad \mu_{12} = (\mu_0/2\gamma_0)(1+b) - i(2/\Lambda)\sqrt{1+y^2}. \tag{10.21}$$

The interference absorption coefficients $\mu_{\nu\nu}$ describe the anomalously weak ($\nu = 1$) and strong ($\nu = 2$) absorption, μ_{12} corresponds to the Pendellösung beats under conditions of interbranch scattering with the period defined by the extinction length Λ, $l_\ni = l_{\nu\nu}(|y| \gg 1)$. Analysis of Eq. (10.20) shows that the sensitivity of the coherent CS to the distribution of electron density increases with the thickness of the crystal, with the increase of the asymmetry factor, and with the decrease of the CS angle.

Experimental CS yields from the exit surface of Si crystals under Laue (111) diffraction of Mo K_α radiation are shown in Fig. 10.5 for two crystal thicknesses l corresponding to the values of $\mu_0 l/\gamma_0 = 0.95$ and 4.29. The CS was detected by a semiconductor detector with an energy resolution of 230 eV in the diffraction plane at the angle $\varphi = 60°$ from the exit crystal surface on the reflected beam side (Fig. 10.6). The $\kappa_{\mathrm{CS}}(\Delta\theta)$ curve for thin crystal (curves *a*) shows a slightly asymmetric minimum, which results from the contribution of both the weakly and strongly absorbing fields. As the crystal thickness increases, the κ_{CS} curve acquires a clearly visible maximum at the negative angles (curves *b*; see also Ref. 37). This maximum originates from the anomalous X-ray transmission (Borrmann) effect due

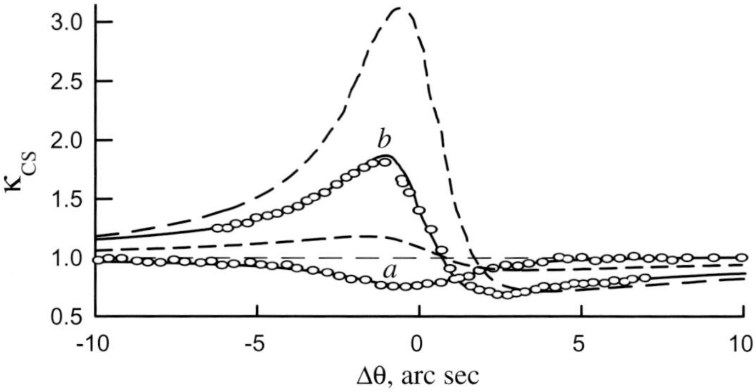

Fig. 10.5. CS yield under conditions of Laue diffraction measured from the exit surface (Si, (111) reflection, Mo K_α radiation). Open circles — experimental data; solid and dashed lines — theoretical predictions for coherent and incoherent CS, respectively. (a) The crystal thickness l is 0.49 mm; (b) $l = 2.2$ mm (from Ref. 38).

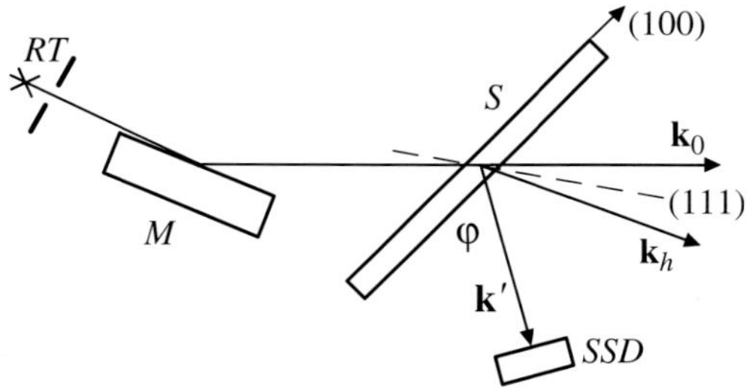

Fig. 10.6. Experimental setup for the coherent CS in the Laue geometry. RT is the X-ray tube, M is the monochromator, S is the sample, SSD is the solid state detector, $\mathbf{k}_0$, $\mathbf{k}_h$ and $\mathbf{k}'$ are the transmitted, reflected and CS wave vectors, respectively (from Ref. 25).

to the standing wave near the exit surface with the antinodes between the atomic planes. Since the main fraction of the valence electrons is located between the atomic planes, we may expect a high sensitivity to valence electron distribution for a thick crystal with $\mu_0 l/\gamma_0 \gg 1$.

Based on the experimental data presented in Fig. 10.5, the contribution of valence electrons to the elastic scattering atomic factor f_h was determined as $f_h^\nu = 1.27 \pm 0.15$.[25] This value agrees well with the value of $f_h^\nu = 1.30$ determined in Ref. 15 from the measurement of the non-diagonal Compton profile of valence electrons integrated over energy. It also agrees well with the difference between the experimental atomic scattering factor $f_h(\mathrm{Si}) = 10.739$,[41] and the calculated value based on the data of Freeman[30] for the ion core $f_h(\mathrm{Si}^{+4}) = 9.439$. For comparison, the dashed lines in Fig. 10.5 are the calculations for the case when one neglects the angular dependence of the CS diagonal cross-sections and, even more importantly, when the coherent nature of the CS is ignored, i.e., if we assume (as in Ref. 32) that $\beta_{hh} = 1$ and $\beta_{0h} = 0$.

As an interesting further application of this approach, we may suggest to study the temperature dependence of the coherent CS intensity as a potential way to differentiate the contribution of valence electrons from that of the ion core to the thermal Debye–Waller factor.

Acknowledgments

The author is very grateful to Alexander Kazimirov for helpful discussion.

References

1. D. Pines, *Elementary Excitations in Solids* (Benjamin, New York, 1963).
2. A. H. Compton and S. K. Allison, *X-rays in Theory and Experiment* (Van Nostrand, New York; reprinted by Van Nostrand, Princeton, 1963).
3. M. J. Cooper, *Rep. Progr. Phys.* **48** (1985) 415.
4. V. A. Bushuev and R. N. Kuz'min, *Usp. Fiz. Nauk.* **122** (1977) 81 (Engl. Transl. *Sov. Phys.-Usp.* **20** (1977) 406).
5. S. Annaka, S. Kikuta and K. Kohra, *J. Phys. Soc. Jpn.* **20** (1965) 2093.
6. S. Annaka, S. Kikuta and K. Kohra, *J. Phys. Soc. Jpn.* **21** (1966) 1559.
7. S. Annaka, *J. Phys. Soc. Jpn.* **24** (1968) 1332.
8. J. A. Golovchenko, D. R. Kaplan, B. Kincaid, R. Levesque, A. Meixner, M. F. Robbins and J. Felsteiner, *Phys. Rev. Lett.* **46** (1981) 1454.
9. W. Schülke, U. Bonse and S. Mourikis, *Phys. Rev. Lett.* **47** (1981) 1209.
10. Z. G. Pinsker, *Dynamical Scattering of X-rays in Crystals* (Springer, Berlin, 1978).
11. A. Authier, *Dynamical Theory of X-Ray Diffraction* (Oxford University Press Inc., New York, 2001).
12. W. Schülke, *Phys. Lett. A* **83** (1981) 451.
13. V. A. Bushuev and R. N. Kuz'min, *Zh. Tekh. Fiz.* **44** (1974) 2568 (Engl. Transl. *Sov. Phys. Tech. Phys.* **19** (1975) 1590).
14. V. A. Bushuev, *Zh. Tekh. Fiz.* **58** (1988) 800 (Engl. Transl. *Sov. Phys. Tech. Phys.* **33** (1988) 487).
15. W. Schülke and S. Mourikis, *Acta Crystallogr.* **A42** (1986) 86.
16. V. A. Bushuev, A. G. Lyubimov and R. N. Kuz'min, *Pis'ma Zh. Tekh. Fiz.* **12** (1986) 141 (Engl. Transl. *Sov. Tech. Phys. Lett.* **12** (1986) 60).
17. A. M. Afanas'ev and V. G. Kohn, *Zh. Eksp. Teor. Fiz.* **74** (1978) 300 (Engl. Transl. *Sov. Phys.– JETP* **47** (1978) 154).

18. A. M. Afanas'ev, P. A. Aleksandrov and R. M. Imamov, *X-ray Diffraction Diagnostic of Submicron Layers* (Nauka, Moscow, 1989) (in Russian).
19. M. V. Kovalchuk and V. G. Kohn, *Usp. Fiz. Nauk* **149** (1986) 69 (Engl. Transl. *Sov. Phys.-Usp.* **29** (1986) 426).
20. I. A. Vartaniants and M. V. Kovalchuk, *Rep. Prog. Phys.* **64** (2001) 1009.
21. P. M. Platzman and N. Tzoar, *Phys. Rev.* **139** (1965) A410.
22. P. Eisenberger and P. M. Platzman, *Phys. Rev. A* **2** (1970) 415.
23. V. A. Bushuev and A. O. Ait, *Vestnik Moscow Univ. Ser. 3*, **27**(5) (1986) 61 (Engl. Transl. *Vestn. Mosk. Univ. Fiz.* **41**(5) (1986) 76–82.
24. R. Benesch, S. R. Singh and V. H. Smith, Jr., *Chem. Phys. Lett.* **10** (1971) 151.
25. V. A. Bushuev, A. Yu. Kazimirov and M. V. Kovalchuk, *Phys. Status Solidi B* **150** (1988) 9.
26. P. M. Platzman and P. A. Wolff, *Waves and Interactions in Solid State Plasmas* (Academic Press, New York and London, 1973).
27. R. J. Weiss, *Philos. Mag.* **26** (1972) 153.
28. W. A. Reed and P. Eisenberger, *Phys. Rev. B* **6** (1972) 4596.
29. Yu. A. Rozenberg, V. F. Karpenko and L. I. Kleshchinski, *Fiz. Tverd. Tela*, **18** (1976) 1841 (Engl. Transl. *Sov. Phys. Solid State* **18** (1976) 1073).
30. A. J. Freeman, *Acta Crystallogr.* **12** (1959) 929.
31. D. N. Tripathy, B. K. Rao and S. S. Mandal, *Solid State Commun.* **22** (1977) 83.
32. A. M. Afanas'ev, R. M. Imamov, E. Kh. Mukhamedzhanov and Li Kong Kui, *Dokl. Akad. Nauk SSSR* **288** (1986) 847 (Engl. Transl. *Sov. Phys. Dokl.* **31** (1986) 492).
33. A. M. Afanas'ev, R. M. Imamov, E. Kh. Mukhamedzhanov and Le Cong Qui, *Dokl. Akad. Nauk SSSR* **295** (1987) 839 (Engl. Transl. *Sov. Phys. Dokl.* **32** (1987) 650).
34. A. M. Afanasev, R. M. Imamov, E. Kh. Mukhamedzhanov and A. A. Nazlukhanyan, *Phys. Status Solidi A* **104** (1987) K73.
35. A. M. Afanas'ev, R. M. Imamov, E. Kh. Mukhamedzhanov and V. N. Peregudov, *Phys. Status Solidi A* **98** (1986) 367.
36. A. M. Afanas'ev, R. M. Imamov, E. Kh. Mukhamedzhanov, Le Kong Kui and V. N. Peregudov, *Dokl. Akad. Nauk SSSR* **289** (1986) 341 (Engl. Transl. *Sov. Phys. Dokl.* **31** (1986) 562).
37. V. A. Bushuev and A. G. Lyubimov, *Pis'ma Zh. Tekh. Fiz.* **13**, 744 (1987) (Engl. Transl. *Sov. Tech. Phys. Lett.* **13** (1987) 309).
38. V. A. Bushuev, A. Yu. Kazimirov and M. V. Koval'chuk, *Pis'ma Zh. Eksp. Teor. Fiz.* **47** (1988) 154 (Engl. Transl. *JETP Lett.* **47** (1988) 187).
39. W. Schülke, *Solid State Commun.* **43** (1982) 863.
40. V. A. Bushuev and A. O. Ait, *Vestnik Moscow Univ. Ser. 3*, **28**(2) (1987) 69 (Engl. Transl. *Vestn. Mosk. Univ. Fiz.* **42**(2) (1987) 83).
41. P. J. E. Aldred and M. Hart, *Proc. Roy. Soc. (London) A* **33** (1973) 223–239.

Chapter 11

THEORY OF PHOTOELECTRON EMISSION FROM AN X-RAY INTERFERENCE FIELD

IVAN A. VARTANYANTS

Deutsches Electronen Synchrotron, DESY,
Notkestraße 85,
D-22607 Hamburg, Germany
National Research Nuclear University, "MEPhI",
115409, Moscow, Russia

JÖRG ZEGENHAGEN

European Synchrotron Radiation Facility,
6 rue Jules Horowitz,
F-38000 Grenoble, France

In this chapter, we will present the theory for the photoelectron emission from an X-ray interference field. The dipole approximation holds astonishingly well, even for hard X-rays, as far as the magnitude of the transition matrix element is concerned. However, the forward–backward asymmetry caused already by higher order multipole terms cannot be neglected when the photoelectron is emitted by the coherent action of two X-ray waves travelling in different directions. We will explicitly elaborate the underlying theory and how corresponding data are analyzed. Furthermore, we will briefly describe the theory behind the X-ray standing wave excited photoemission of valence band electrons.

11.1. Introduction

During the last decade it has become clear by theoretical considerations[1,2] and experimental investigations[3] that an accurate description of the angular resolved hard X-ray photoelectron spectroscopy (HAXPES)[4] has to go beyond the traditionally used dipole approximation and higher-order multipole terms in the photoelectron emission (PE) process have to be included. However, it was not immediately realized how important these non-dipole contributions will be in the case of certain applications of the X-ray standing wave (XSW) technique.[5,6]

If the dipole approximation is valid, the intensity of the PE from an atom exposed to an X-ray interference field (XIF) will be exactly proportional to the intensity of the standing wave at the position of the atom.[6,7] By normalization, the proportionality factor cross-section cancels out. With the position of the standing wave field known, atomic positions can be deduced from the characteristic yield, e.g., employing the dynamical theory of X-ray diffraction for single crystals. However, this simple relationship fails in certain cases. First, if angular resolved PE and π-polarized radiation are considered,[8,9] two E-field vectors are not collinear and consequently the dipole emission lobes from the two X-ray waves do not point in the same direction any longer. Second, and this is the topic of this chapter, when non-dipole terms in the photoelectric process are appreciable, the emission lobes are no longer symmetric around the E-field vectors. Thus, higher-order terms have to be taken into account and the formalism for the XSW analysis has to be revised.

In the last years, several publications addressed this questions from a theoretical[9–12] and experimental[13–17] point of view. If non-dipole terms are taken into account, the expression for the photoelectron yield describing the emission from a particular atomic site has to be generalized and can no longer be described by the well-known expression of the intensity of the XSW.[18] For certain cases it became also clear that these non-dipole contributions are by no means negligible and must be taken into account if the XSW analysis is not to be flawed. However, on the other hand, if multipole contributions are significant (more than several percent) they can directly be determined by the XSW technique and unique information about the scattering process can be extracted.

The XSW method is particularly powerful for the analysis of the structure of adsorbates on crystalline substrates,[6] and photoelectron spectroscopy is here an important tool. Thus, it is important to know how non-dipole terms are to be taken into account in order to obtain the correct positions of adsorbates. These questions will be addressed in the present chapter and we will present an overview of theoretical considerations and experimental studies of non-dipole contributions in the angular resolved PE in the presence of an XSW.

11.2. Photoelectron Scattering Process by a Single Electromagnetic Wave

Within the frame of non-relativistic quantum mechanics for a one-electron, one-photon linear process in first-order perturbation theory the interaction Hamiltonian between the electron and the electromagnetic field, responsible for the photo-absorption, can be written as,

$$\hat{\mathcal{H}}_{\text{int}} = -\frac{1}{c}\int \hat{\mathbf{j}}(\mathbf{r})\hat{\mathbf{A}}(\mathbf{r})dV. \tag{11.1}$$

Here $\hat{\mathbf{A}}(\mathbf{r})$ is the vector potential of the electromagnetic field and $\hat{\mathbf{j}}(\mathbf{r})$ is the current operator

$$\hat{\mathbf{j}}(\mathbf{r}) = \frac{e_0}{2m}[\hat{\mathbf{p}}\delta(\mathbf{r}-\mathbf{r_e}) + \delta(\mathbf{r}-\mathbf{r_e})\hat{\mathbf{p}}],$$

where e_0, m and $\mathbf{r_e}$ are the charge, mass, and coordinate, respectively, of an electron and $\hat{\mathbf{p}} = -i\hbar\nabla$ is the momentum operator.[a]

In the analysis we will use a one-electron central potential model for the description of the photo-excitation process. In case of the photoionization of a many-electron atom, we have to sum the expression for the current operator $\hat{\jmath}(\mathbf{r})$ over the coordinates of all electrons and to chose a many-electron antisymmetric wave function for an atom. Though simplified, a one-electron model gives qualitatively, and even quantitatively, a good description of the process at high energies and, even more essential for the further considerations, the general properties of the photoelectron angular distribution in the central potential model are conserved in the many-electron description.[19] However, for open-shell atoms and photoelectron energies near threshold, when the interaction between the photoelectron and the residual ion is dependent on the ionic term level as well as on the orbital and spin angular momentum coupling of the ion-electron system, more sophisticated theories have to be applied.[20]

In case of the photoelectric process, a photon of energy $E_\gamma = \hbar\omega$ is absorbed by an atom and the energy E_γ is transferred to an electron, which is excited from the ground state $|i\rangle$ to a final state $|f\rangle$. In the following, we restrict ourselves to the case that this final state is unbound. In the non-relativistic limit the differential cross-section for the absorption of a photon

---

[a]For our future treatment it is more convenient to use the electric field vector $\mathbf{E}$ rather then the vector potential $\mathbf{A}$. In case of the Coulomb gauge of the electromagnetic field, they are connected by the relation, $\mathbf{E} = -\frac{1}{c}\frac{\partial \mathbf{A}}{\partial t}$.

plane wave $\mathbf{E} = \mathbf{e}E_0 \exp(i\mathbf{kr} - i\omega t)$ by an atom and the concomitant ejection of a bound electron into a continuum state is determined by the matrix element M_{fi} of the process and can be written in the following form,[21]

$$\frac{d\sigma(\vartheta_p, \phi_p)}{d\Omega} \propto |M_{fi}|^2, \tag{11.2}$$

$$M_{fi} = \langle f| \exp(i\mathbf{k} \cdot \mathbf{r})(\mathbf{e} \cdot \hat{\mathbf{p}})|i\rangle. \tag{11.3}$$

Here ϑ_p and ϕ_p are the spherical angles of the direction of the escaping photoelectron $\mathbf{k}_p$ (see Fig. 11.1). The incident photon is described by the wave vector $\mathbf{k}$. We define here its magnitude different from the definition in other chapters as $|\mathbf{k}| = 2\pi/\lambda$ (it simplifies the wave equations $\exp(i\mathbf{kr})$ saving the factor 2π in the exponential). The polarization direction of the electric field vector is described by the unit vector $\mathbf{e}$. The total cross-section is obtained from (11.2) and (11.3) by integrating over the whole spatial angle Ω of the escaping photoelectron.

11.2.1. *Non-dipole contributions*

In the long wavelength limit $\lambda \gg a$ (where a is the average size of the electron-bound state orbital) it is clear that the retardation factor $\exp(i\mathbf{kr})$ of the matrix element M_{fi} (11.3) can be expanded in a Taylor series

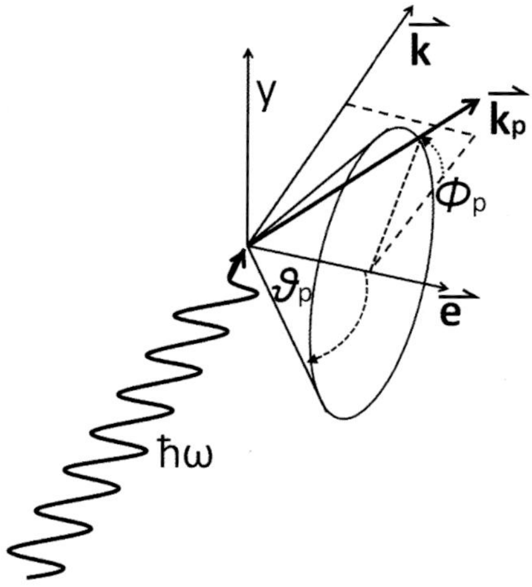

Fig. 11.1. Geometric relationships for the PE process from an atom for an electromagnetic wave with the wave vector $\mathbf{k}$ pointing in the x-direction and the polarization vector $\mathbf{e}$ pointing in the z-direction. The direction of the escaping photoelectron with the propagation vector $\mathbf{k}_p$ is characterized by the spherical angles ϑ_p and ϕ_p.

$\exp(i\mathbf{kr}) = 1 + i\mathbf{kr} - \frac{1}{2}(\mathbf{kr})^2 - \cdots$ which is equivalent to a multipole expansion. The first term gives electric dipole ($E1$) and the second term electric quadrupole ($E2$) and magnetic dipole ($M1$) transitions. We will neglect magnetic transitions for the photo-emission process in the following[b] and take into account only electric ($E1, E2, E3$, etc.) transitions. Neglecting also octupole ($E3$) and higher contributions, we get for the matrix element in (11.3)

$$M_{fi}(\mathbf{s}) = M_{fi}^{D} + M_{fi}^{Q}(\mathbf{s}), \tag{11.4}$$

where $\mathbf{s} = \mathbf{k}/|\mathbf{k}|$ is the unit vector in the direction of the photon propagation. Only the second term corresponding to quadrupole transitions depends on the propagation vector $\mathbf{s}$ but both terms are in general complex numbers and can interfere with each other. Keeping just the main terms we obtain for the photoelectron cross-section,

$$\frac{d\sigma(\vartheta_p, \phi_p)}{d\Omega} \propto \left|M_{fi}^{D}\right|^2 + 2\,\mathrm{Re}[(M_{fi}^{D})^* M_{fi}^{Q}(\mathbf{s})]. \tag{11.5}$$

It was shown in (11.2) that for linearly polarized photons, this general expression for the photoelectron angular distribution may be conveniently parametrized in the following way:

$$\frac{d\sigma(\vartheta_p, \phi_p)}{d\Omega} = (\sigma/4\pi)\{1 + \beta P_2(\mathbf{e}\cdot\mathbf{n}_p) + [\delta + \gamma(\mathbf{e}\cdot\mathbf{n}_p)^2](\mathbf{s}\cdot\mathbf{n}_p)\}, \tag{11.6}$$

where $P_2(z) = 0.5(3x^2 - 1)$ is the second-order Legendre polynomial, and $\mathbf{n}_p = \mathbf{k}_p/|\mathbf{k}_p|$ is a unit vector along the direction of the ejected electron characterized by the momentum $\mathbf{k}_p$ with the magnitude $k_p = |\mathbf{k}_p|$. In Eq. (11.6), the parameter β is the dipole asymmetry parameter, and γ and δ are two additional parameters that take into account non-dipole contributions. The values of parameters β, γ and δ depend on the type of atom, sub-shell and photoelectron energy. The term $\mathbf{s}\cdot\mathbf{n}_p$, which is caused by the quadrupole contribution, introduces a forward–backward asymmetry in the PE process with respect to the photon direction $\mathbf{s}$. Values of parameters β, γ and δ are given for the photoelectron energies E_{ph} from 100 to 5000 eV in Ref. 22.

[b]For a central-field model and a one-electron approximation, the magnetic transition probability for the photo-effect is equal to zero due to the orthogonality of the initial and final state radial wave functions. If core relaxations are taken into account, the probability for magnetic transitions does not vanish but is much smaller than for electric quadrupole contributions.[1,2]

by the condition $\int_0^\infty R_{k'_p l}(r)R_{k_p l}(r)r^2dr = 2\pi\delta(k'_p - k_p)$ and having the asymptotic form[27] for $r \to \infty$

$$R_{k_p l}(r) \sim \frac{2}{r}\sin\left(k_p r - \frac{\pi l}{2} + \delta_l\right). \tag{11.24}$$

Each of the outgoing partial waves of the electron characterized by the angular momentum quantum number l experiences a phase shift δ_l, which is directly related to the form of the scattering potential of the bound electron. It is an important parameter in the theoretical description of the scattering process (see, for e.g., Ref. 27). According to (11.24) the asymptotic (for $r \to \infty$) behavior of the lth partial wave of the emitted electron is determined by δ_l. Corresponding theoretical values of the phase shifts δ_l are available from the NIST data base.[28]

Using Eqs. (11.22)–(11.24) we obtain for the electric multipole moment of the transition (11.21),

$$Q_{LM} = \frac{e_0}{\sqrt{2L+1}}\sum_{l'm'} a^*_{l'm'}\rho^L_{l'}\langle l'm'|Y_{LM}(\mathbf{n})|lm\rangle, \tag{11.25}$$

where

$$\rho^L_{l'} = \int_0^\infty R_{k_p l'}(r)r^L R_{nl}(r)r^2dr \tag{11.26}$$

is the radial integral for the 2^L-pole electric transition and the matrix elements $\langle l'm'|Y_{LM}(\mathbf{n})|lm\rangle$ determine the *transition rules* for the 2^L-pole transitions.

The matrix elements M_{fi} needed for the parameters $S_{\alpha\beta}$ in Eq. (11.16) are now defined by Eqs. (11.20) and (11.21) and (11.25) and (11.26) to *any multipole order*. However, the product of the matrix elements (11.20) contains not only "pure" terms that correspond to dipole–dipole ($E1$–$E1$), quadrupole–quadrupole ($E2$–$E2$), etc. transitions, but also cross terms, especially dipole–quadrupole ($E1$–$E2$) transitions. For the case of the angular-resolved photoelectric process, these are mainly responsible for non-vanishing multipole contributions to the differential cross-section.[1,2]

11.5. Integral Photoelectron Emission from an Interference Field

The total yield of the photoelectrons, Auger electrons, or fluorescence photons emitted from an atom in an XIF is determined by Eqs. (11.15) and (11.17) integrated over all the directions of the escaping photoelectron $\mathbf{n_p}$.

The integration essentially simplifies the solution. Integrating the product of the matrix elements (11.20) $M_{fi}(\mathbf{s}_\alpha)^* M_{fi}(\mathbf{s}_\beta)$ over the whole spatial angle Ω and averaging the result over all the initial magnetic states m yields the following expression[9]

$$\int \langle M_{fi}(\mathbf{s}_\alpha)^* M_{fi}(\mathbf{s}_\beta)\rangle d\Omega \propto \sum_L \sigma^L p_L(\alpha, \beta), \tag{11.27}$$

where indices $\alpha, \beta = 0, h$. In Eq. (11.27), the coefficients σ^L are the total cross-sections of the photoelectron excitation from the subshell n, l corresponding to different multipole transition terms

$$\sigma^L = \frac{A_L}{2l+1} \sum_{l'} |a_{l'}|^2 (\rho_{l'}^L)^2 (l'||Y_L||l)^2. \tag{11.28}$$

The coefficients $p_L(\alpha, \beta)$ in (11.27) determine the *angular dependence* on the polarization vectors $\mathbf{e}_{\alpha\mathbf{s}}, \mathbf{e}_{\beta\mathbf{s}}$

$$p_L(\alpha, \beta) = \frac{8\pi}{2L+1} \sum_M (\mathbf{e}_{\alpha\mathbf{s}} \mathbf{Y}_{L'M'}^{(e)}(\mathbf{s}_\alpha))(\mathbf{e}_{\beta\mathbf{s}} \mathbf{Y}_{LM}^{(e)*}(\mathbf{s}_\beta)), \tag{11.29}$$

where index $s = \sigma, \pi$ denotes σ, or π-polarization (see Fig. 11.2). In (11.28), $(l'||Y_L||l)$ are the reduced matrix elements.[29,30]

Taking for example only dipole transitions into account, only one term with $L = 1$ remains in the sum (11.27). Due to the dipole transition rules $l' = l \pm 1$ one obtains from (11.28) the well-known expression[20] for the dipole cross-section,

$$\sigma^D = \frac{A_1}{2l+1} [(l(\rho_{l-1}^D)^2 + (l+1)(\rho_{l+1}^D)^2], \tag{11.30}$$

where $\rho_{l\pm1}^D = \int_0^\infty R_{kl\pm1}(r) r R_{nl}(r) r^2 dr$ is the dipole integral ($\rho_l^D \equiv \rho_l^1$) (11.26). In the same way we obtain from (11.28) the total cross-sections for quadrupole, octupole, etc. transitions.

As (11.27) shows, after integration over all angles of the escaping photoelectron and averaging over the initial states only "pure" transitions (dipole, quadrupole, etc.) are contributing to the integral PE yield and there are no mixed terms.

Directly from the relations (11.27)–(11.29) one obtain for the parameters $S_{\alpha\beta}$,

$$S_{\alpha\beta} = \sum_L \sigma^L p_L(\alpha, \beta), \tag{11.31}$$

or more explicitly

$$S_{00} = S_{hh} = \sigma^D + \sigma^Q + \sigma^O + \cdots,$$
$$S_{0h} = C^D\sigma^D + C^Q\sigma^Q + C^O\sigma^O + \cdots. \tag{11.32}$$

Here C^D, C^Q and C^O are polarization coefficients corresponding to each multipole term. They can be calculated from (11.29) and are equal to[9]

$$\begin{aligned} C^D &= p_1(0,h) = \mathbf{e_{0s}e_{hs}}, \\ C^Q &= p_2(0,h) = (\mathbf{e_{0s}e_{hs}})(\mathbf{s_0s_h}) + (\mathbf{e_{0s}s_h})(\mathbf{e_{hs}s_0}), \\ C^O &= p_3(0,h) = 1/4\{5[(\mathbf{e_{0s}e_{hs}})(\mathbf{s_0s_h})^2 \\ &\quad + 2(\mathbf{e_{0s}s_h})(\mathbf{e_{hs}s_0})(\mathbf{s_0s_h})] - (\mathbf{e_{0s}e_{hs}})\}. \end{aligned} \tag{11.33}$$

From expression (11.33) for each (σ- and π-) polarization, explicit angular dependence from the scattering angle θ are obtained, i.e.,

$$C^D = \begin{cases} 1, & \sigma\text{-polarization;} \\ \cos 2\theta, & \pi\text{-polarization.} \end{cases}$$

$$C^Q = \begin{cases} \cos 2\theta, & \sigma\text{-polarization;} \\ \cos 4\theta, & \pi\text{-polarization.} \end{cases} \tag{11.34}$$

$$C^O = \begin{cases} 1/8[5\cos 4\theta + 3], & \sigma\text{-polarization;} \\ 1/16[15\cos 6\theta + \cos 2\theta], & \pi\text{-polarization.} \end{cases}$$

For the normalized integral intensity of the photoelectron yield from an atom within the XIF according to Eq. (11.17) one, finally, obtains (in this case $S_R = 1$),

$$Y_A^h = 1 + R + 2\tilde{C}\sqrt{R}\cos(v + \mathbf{h}\cdot\mathbf{r_a}), \quad \psi \equiv 0, \tag{11.35}$$

where

$$\tilde{C} \equiv |S_I| = \frac{C^D\sigma^D + C^Q\sigma^Q + C^O\sigma^O + \cdots}{\sigma^D + \sigma^Q + \sigma^O + \cdots}. \tag{11.36}$$

For most applications the contribution from the octupole term can be neglected (see for e.g., calculations in Refs. 31 and 32). In this case we obtain for the coefficient $\tilde{C}$,

$$\tilde{C} = C^D(1 - Q), \tag{11.37}$$

where the quadrupole contribution is given by

$$Q = (\sigma^Q/\sigma)[1 - C^Q/C^D], \tag{11.38}$$

and $\sigma = \sigma^D + \sigma^Q$ is the total cross-section. For the two polarization states, Eqs. (11.34) yield,

$$Q = \begin{cases} 2(\sigma^Q/\sigma)\sin^2\theta, & \sigma\text{-polarization;} \\ (\sigma^Q/\sigma)\,[\cos 2\theta - \cos 4\theta]\cos 2\theta, & \pi\text{-polarization.} \end{cases} \tag{11.39}$$

So far, there are already two important conclusions. First, in the case of the integral photoelectron yield, according to (11.33), the parameters $S_{\alpha\beta}$ are real numbers. Since $S_{00} = S_{hh}$, additional multipole terms change only the amplitude of the third, the interference term in Eq. (11.35) and multipole contribution can be expressed by one parameter $\tilde{C}$. Second, only in the dipole approximation when ($\sigma^Q = \sigma^O = \cdots = 0$) the photoelectron yield intensity Y_A^h (11.35) features exactly the X-ray wave field intensity (XSW) (11.10) at the particular chosen location in space. Every multipole term in Eq. (11.36) exhibits in fact its own dependence on the scattering angle θ (compare with Ref. 31). This angular dependence allows in principle to determine the multipole terms directly from an XSW experiment.

Interesting results are obtained when analyzing the contribution of the quadrupole term for different scattering angles. For example, an X-ray wave diffracted with a Bragg angle of $\theta = 45°$ is usually regarded as "purely" σ-polarized. However, if quadrupole contributions are taken into account, a "weak" diffracted beam is observed for π-polarization as well. In this case we can obtain from Eq. (11.39): $\tilde{C} = -(\sigma^Q/\sigma)$. Thus, in the case of diffraction for $\theta = 45°$ the interference term in Eq. (11.35) is directly proportional to the value (σ^Q/σ). However, since the reflectivity is (extremely) small in this case, this effect will be difficult to observe.

In the case of the *back reflection*, i.e., for a Bragg angle of $\theta = 90°$ we get from (11.39): $\tilde{C} = 1 - 2(\sigma^Q/\sigma)$ and thus the influence of the quadrupole term is at *maximum*.

In summary, quadrupole terms in the integral cross-section influence the *amplitude* of the cosine function in Eq. (11.35). Thus, the integral photoelectron, Auger electrons, or fluorescence yield depends (due to the different angular dependence of the multipole term) on the value of the

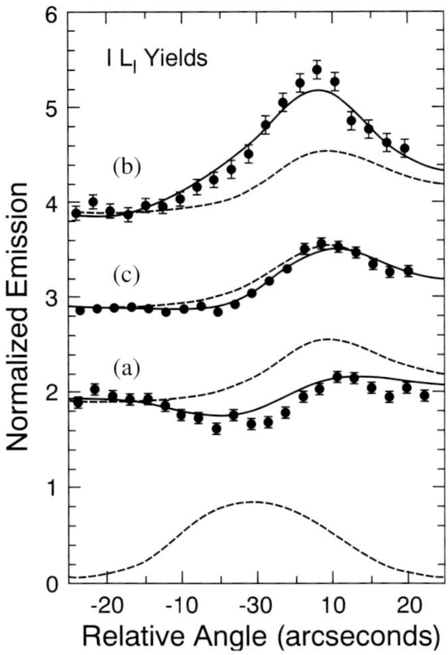

Fig. 11.4. The iodine L_I photoelectron (kinetic energy 1560 eV) yield as a function of glancing angle around the maximum reflectivity of the Ge(111) Bragg reflection for a π-scattering geometry, recorded with different adjustments of the CMA angular aperture: for curve (a), the aperture selects photoelectrons emitted almost in the direction of the polarization vector of the incident beam; for curve (b), the aperture is adjusted to select photoelectrons emitted almost in the direction of the polarization vector of the reflected beam; for curve (c), the aperture was removed, and a wider range of emission angles was selected. The lowest curve is the Ge(111) reflectivity. The theoretical yields shown as solid curves assume an $s \rightarrow p$ dipole transition (cf. Eq. (11.45)); the dashed curves follow the electric field intensity (Eq. (11.10)), i.e., their functional form is given by Eq. (11.17) with $S_R = |S_I| = 1$ and $\psi = 0$. From Ref. 8.

This analysis shows that the shape of the angular-resolved photoelectron yield curves expected from an XSW experiment in the dipole approximation is, in fact, determined by the value of the "asymmetry" parameter β. This value can be different for different subshells and for a given subshell it varies with the photon energy.

11.7. Angular-Resolved Photoelectron Emission in the Dipole–Quadrupole Approximation

As next, the angular distribution of the photoelectron yield from an atom in the XIF is considered, when in addition to the dipole term the quadrupole term has to be taken into account. It is assumed in the following that the initial state of the electron in the atom is an *s-state*, before considering the practically important case of an initial *p-state* of the photoelectron.

11.7.1. *s-initial state*

The photoelectron yield $Y(\Omega)$ (11.15) from an atom in the XSW field is determined by the matrix element defined by Eq. (11.20). Considering dipole ($s \to p$) and quadrupole ($s \to d$) transitions, after straightforward calculations[9] we obtain for the parameters $S_{\alpha\beta}$ in Eq. (11.16)

$$\begin{aligned} S_{00} &= 3(\sigma^D/4\pi)(\mathbf{e_0 n_p})^2[1+\gamma'(\mathbf{s_0 n_p})], \\ S_{hh} &= 3(\sigma^D/4\pi)(\mathbf{e_h n_p})^2[1+\gamma'(\mathbf{s_h n_p})], \\ S_{0h} &= 3(\sigma^D/4\pi)(\mathbf{e_0 n_p})(\mathbf{e_h n_p})\left\{1+\frac{1}{2}[\widetilde{\gamma}^*(\mathbf{s_0 n_p})+\widetilde{\gamma}(\mathbf{s_h n_p})]\right\}. \end{aligned} \tag{11.47}$$

The complex parameter $\widetilde{\gamma}$ is defined as

$$\widetilde{\gamma} = \gamma' + i\gamma'' = \gamma_0 e^{i\Delta}, \quad \gamma_0 = k\frac{\rho_d^Q}{\rho_p^D}, \quad \Delta = \delta_d - \delta_p. \tag{11.48}$$

It is determined by the values of the dipole $\rho_p^D = \int_0^\infty R_{k_p l'=1}(r) r^3 \times R_{nl=0}(r)dr$ and quadrupole $\rho_d^Q = \int_0^\infty R_{k_p l'=2}(r) r^4 R_{nl=0}(r)dr$ integrals (11.26) and the partial phase shifts δ_p, δ_d of the final electron p- and d-state (see Eq. (11.24)). In the case of a single propagating wave, the obtained results are reduced to the ones described by Eq. (11.8) with the parameter γ defined as $\gamma = 3\gamma'$. Furthermore, for a pure dipole transition $\widetilde{\gamma} = 0$.

In case of quadrupole contributions to the photoelectron yield excited by an interference field, generally $S_{00} \neq S_{hh} \neq S_{0h}$. The parameter S_{0h} is now a *complex* number with amplitude and *phase* depending on the photon energy and the scattering geometry. The phase ψ in the expression for the photoelectron yield (11.17) appears only for the angular-resolved PE when the quadrupole term in the multipole expansion is taken into account and it vanishes in the integral PE case (see Sec. 4).

For *σ-polarization* and the particular geometry of the experiment shown in Fig. 11.2, the expressions (11.47) are further simplified. Assuming that the photoelectron detector is located in the (z, y) plane we obtain,[9]

$$S_R = S_{00}/S_{hh} = \frac{1+\gamma'(\mathbf{s_0 n_p})}{1+\gamma'(\mathbf{s_h n_p})} = \frac{1+C_p\gamma'}{1-C_p\gamma'}, \quad S_I = \frac{1+iC_p\gamma''}{(1-C_p\gamma')}. \tag{11.49}$$

Here, for the scattering geometry of Fig. 11.2, the parameter C_p has the value $C_p = \cos\theta_p \sin\theta$, where θ_p is the angle between the z-axis and the propagation vector of the photoelectron (we assume that $\theta_p < \pi/2$). In case of a small multipole contribution ($\widetilde{\gamma} \ll 1$) we obtain for the parameters in Eq. (11.49) $S_R \simeq 1 + 2C_p\gamma'$, $|S_I| \simeq 1 + C_p\gamma', \tan\psi \simeq C_p\gamma''$.

In the *normal incidence geometry*, which is used in many XSW experiments, non-dipole contributions are maximized in the photoelectron yield and cannot be neglected. In this case $\mathbf{s_0 n_p} = -\mathbf{s_h n_p}$ and for the parameter C_p in Eq. (11.49), one obtains $C_p = \cos\theta_p$, where θ_p is an angle between the direction of the escaping photoelectron $\mathbf{n}_p$ and the wave vector $\mathbf{k}_h$. For the normalized photoelectron yield $Y_T^h(\Omega)$ the general expression (11.17) will be valid with parameters

$$S_R = \frac{1+Q'}{1-Q'}, \quad S_I = \frac{1+iQ''}{1-Q'}, \tag{11.50}$$

where $Q' = \gamma' \cos\theta_p$ and $Q'' = \gamma'' \cos\theta_p$. The angular distribution of the photoelectron yield $Y(\Omega)$ calculated according to Eq. (11.17) with and without quadrupole contribution and for different positions of the XSW field is presented in Fig. 11.5.

Directly from (11.50) (this is valid also for the conventional Bragg scattering geometry described by Eq. (11.49)), it can be shown that the three non-dipole parameters are not independent. The following relation is valid

$$|S_I| = 0.5(S_R + 1)\sqrt{1+\tan^2\psi}, \tag{11.51}$$

and two non-dipole parameters only allow to retrieve the structural XSW parameters F^h and P^h (*cf.* Eq. (11.17)).

We will consider now the system composed of an ensemble of atoms absorbed on a crystal surface. The non-dipole parameter S_R can be determined by preparing and measuring an incoherent film, i.e., when $F^h = 0$. In case of an ordered coherent film or a monolayer, when one wants to determine the coherent fraction F^h and the coherent position P^h (*cf.* Eq. (11.17)) the parameter $|S_I|$ and the phase ψ are still unknown. However, taking into account Eq. (11.51), it is sufficient to know either $|S_I|$ or ψ. In the case of an initial s-state we can utilize the relationship between the phase ψ and the partial phase shift $\Delta = \delta_d - \delta_p$ of the electron final p-

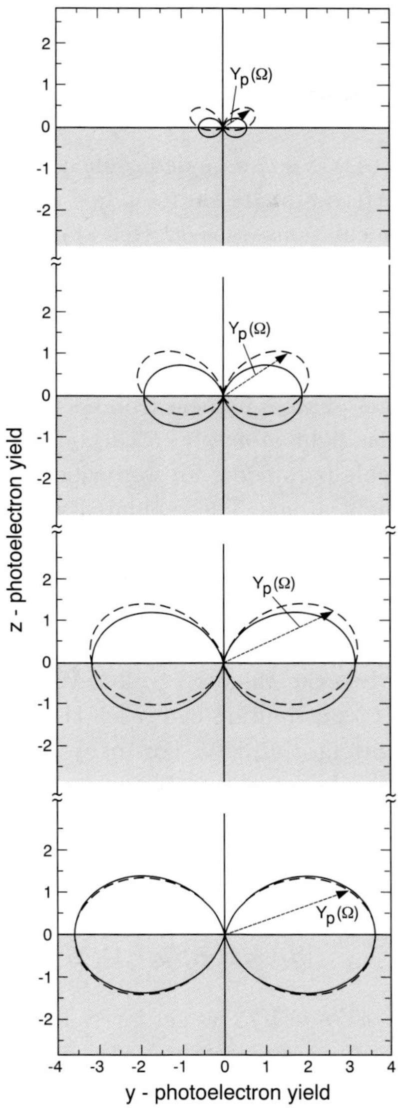

Fig. 11.5. The angular distribution of photoelectrons excited by the XSW field in the backscattering geometry without (solid line) and with (dashed line) quadrupole contribution, according to Eq. (11.17) for different values of the intensity ratio (or reflectivity) $R = |E_H|^2/|E_0|^2$ and phase $\Phi = v - \mathbf{hR_A}$. From top to bottom: $R = 0.95$, $\Phi = -3\pi/4$, $R = 0.9$, $\Phi = -\pi/2$, $R = 0.85$, $\Phi = -\pi/4$, $R = 0.8$, $\Phi = 0$. The beam is polarized in the y-direction and incident on the crystal (shaded region, with the surface at $z = 0$) in $-z$ direction. From Ref. 9.

and d-states. From (11.48) and (11.50) we obtain

$$\tan\psi = \frac{S_R - 1}{S_R + 1}\tan\Delta, \tag{11.52}$$

and applying Eq. (11.51), $|S_I|$ can be determined.

Values of the scattering phase shifts δ_p and δ_d can be obtained from relativistic *ab initio* calculations (see e.g., Ref. 28). These values are needed and used, e.g., in bulk and surface-sensitive structural X-ray methods, which employ electron scattering such as extended X-ray absorption fine structure (EXAFS) and X-ray absorption near edge structure (XANES).[34]

In Figs. 11.6, 11.7 and 11.8, the calculated values of parameters S_R, $|S_I|$ and $\tan\psi$ as a function of the photoelectron energy from 100 to 5000 eV are presented. Different light atoms from carbon to neon are considered. The values of non-dipole parameter S_R were calculated using Eq. (11.50) for $1s$-state of these light atoms. The tabulated values[22] of the parameter γ were used in these calculations. The XSW phase $\tan\psi$ was calculated by expression (11.52) in which the phase shift difference $\Delta = \delta_d - \delta_p$ for each atom and energy was obtained from Ref. 28. The values of parameter $|S_I|$ were determined from Eq. (11.51). In all calculations, the photoelectron escape angle was fixed to the value $\vartheta_p = 45°$. We can see from Fig. 11.8 that the XSW phase ψ contribution can reach the value up to 0.1 rad for light atoms and energies up to 5 keV. For many XSW measurements, the error in determining the phase $2\pi P^h + \psi$ is frequently larger than 0.1 rad. In this case, the contribution of the phase ψ to the XSW photoelectron yield as expressed in Eqs. (11.17) and Eq. (11.51) can frequently be neglected. With this approximation we obtain

$$|S_I| \approx 0.5(S_R + 1). \tag{11.53}$$

For the typical values of $S_R \approx 1.75$ we get for $|S_I| \approx 1.38$, which underlines the importance of these parameters and the necessity of using for the XSW analysis an expression in the form of Eq. (11.17). This approach to determine Δ and consequently the XSW phase ψ was used in Refs. 16 and 35.

It follows from our analysis that non-dipole contributions change the form of the total photoelectron yield curve described by Eq. (11.17). Consequently, if both parameters γ' and γ'' are determined from the experiment, then the values of the dipole and quadrupole integrals and the values of the partial phase shifts δ_p and δ_d for the p- and d-asymptotic

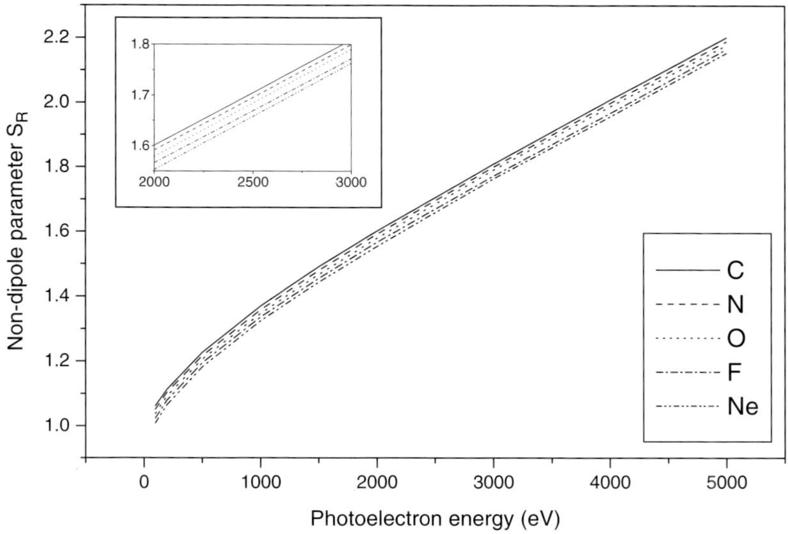

Fig. 11.6. Non-dipole parameter S_R calculated by Eq. (11.50) with parameter γ obtained from the tables.[22] Photoelectron energy range from 2 to 3 keV is shown in the inset.

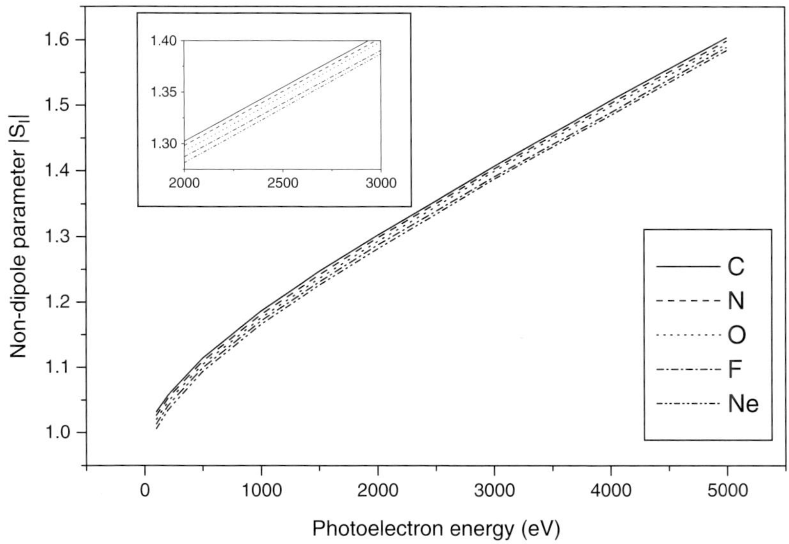

Fig. 11.7. Non-dipole parameter $|S_I|$ calculated according to Eq. (11.51). Photoelectron energy range from 2 to 3 keV is shown in the inset.

of two coherently coupled X-ray beams

$$\frac{d\sigma}{d\Omega} \sim u_a^2|V_a(\mathbf{s_0}) + (E_h/E_0)e^{i\mathbf{h}\cdot\mathbf{r}_a}V_a(\mathbf{s_h})|^2$$
$$+ u_c^2|V_c(\mathbf{s_0}) + (E_h/E_0)e^{i\mathbf{h}\cdot\mathbf{r}_c}V_c(\mathbf{s_h})|^2$$
$$+ u_a u_c e^{i(\mathbf{k}_0-\mathbf{k}_f)\cdot(\mathbf{r}_c-\mathbf{r}_a)}[V_a(\mathbf{s_0}) + (E_h/E_0)e^{i\mathbf{h}\cdot\mathbf{r}_a}V_a(\mathbf{s_h})]^*$$
$$\times [V_c(\mathbf{s_0}) + (E_h/E_0)e^{i\mathbf{h}\cdot\mathbf{r}_c}V_c(\mathbf{s_h})] + c.c. \qquad (11.68)$$

Here, *c.c.* denotes the complex conjugate. Note, that the equation conventionally used in the studies of the X-ray PE from a crystal-valence band from a single propagating electromagnetic beam[39,40] is obtained from Eq. (11.68) by setting $E_h = 0$.

As valence electrons have negligible binding energies ($\varepsilon_b \sim 0$), the product $|(\mathbf{k}_0 - \mathbf{k}_f)\cdot(\mathbf{r}_c - \mathbf{r}_a)|$ will be much greater than 1 at X-ray energies; consequently, the phase factor $e^{i(\mathbf{k}_0-\mathbf{k}_f)\cdot(\mathbf{r}_c-\mathbf{r}_a)}$ will be highly oscillatory, and, therefore, one can neglect the cross-terms compared to the first two terms of Eq. (11.68).[39–41] This approximation leads us to the differential cross-section for two independent mixed-site emitters in the presence of two coherently coupled X-ray beams

$$\frac{d\sigma}{d\Omega} \sim u_a^2|V_a(\mathbf{s_0}) + (E_h/E_0)V_a(\mathbf{s_h})|^2$$
$$+ u_c^2|V_c(\mathbf{s_0}) + (E_h/E_0)e^{i\mathbf{h}\rho}V_c(\mathbf{s_h})|^2. \qquad (11.69)$$

As we are interested in the integral photoeffect, we will be integrating Eq. (11.69) over all solid angle $d\Omega$. It is also possible to take into account non-dipole contributions to the XSW yield. According to the results obtained in Sec. 11.5 (see Eq. (11.27)), integration of matrix elements in Eq. (11.69) in terms of their dipole and quadrupole contributions is given by

$$\int |V_{a,c}(\mathbf{s_{0,h}})|^2 d\Omega \propto \sigma^D_{a,c} + \sigma^Q_{a,c},$$
$$\int V^*_{a,c}(\mathbf{s_0})V_{a,c}(\mathbf{s_h})d\Omega \propto C^D\sigma^D_{a,c} + C^Q\sigma^Q_{a,c}, \qquad (11.70)$$

where $\sigma^D_{a,c}$ and $\sigma^Q_{a,c}$ are the total dipole and quadrupole cross-sections for the anion and cation atoms, and C^D and C^Q are the dipole and quadrupole polarization parameters defined in Eqs. (11.33) and (11.34). From Eqs. (11.69) and (11.71), we now obtain the intensity of the total valence-electron emission from the two coherently coupled beams

$$Y_T \sim u_a^2\sigma_a^T[1 + R + 2C^D(1 - Q_a)\sqrt{R}\cos(v + \mathbf{h}\cdot\mathbf{r}_a)] \\ + u_c^2\sigma_c^T[1 + R + 2C^D(1 - Q_c)\sqrt{R}\cos(v + \mathbf{h}\cdot\mathbf{r}_c)], \quad (11.71)$$

where $\sigma_{a,c}^T = \sigma_{a,c}^D + \sigma_{a,c}^Q$, and $Q_{a,c}$ are the quadrupole contributions for the anion and cation sites, respectively (see Eqs. (11.38) and (11.39)). Note that the inclusion of the quadrupole contributions to the integral photoelectron yield breaks the strict proportionality between the electric-field intensities and the valence-electron emissions by reducing the amplitudes or coherent fractions of the XSW modulations. The coherent position will also be affected, but only in cases where Q_a and Q_c are large and differ by a significant amount.

For the total valence band photoelectron yield in the dipole approximation, i.e., if the quadrupole contributions can be neglected ($Q_a = Q_c = 0$), Eq. (11.71) simplifies to

$$Y_T \sim u_a^2\sigma_a^T[1 + R + 2C^D\sqrt{R}\cos(v + \mathbf{h}\cdot\mathbf{r}_a)] \\ + u_c^2\sigma_c^T[1 + R + 2C^D\sqrt{R}\cos(v + \mathbf{h}\cdot\mathbf{r}_c)]. \quad (11.72)$$

Comparison with Eq. (11.10) shows that in dipole approximation the valence-electron emission associated with each atom is directly proportional to the electric-field intensity at the location of its electronic core.

Expression (11.71) for the total photoelectron yield can be written in the usual parametrized form of the XSW yield from an ensemble of atoms

$$Y_T = 1 + R + 2C^D\sqrt{R}F^h\cos(v + 2\pi P^h), \quad (11.73)$$

where the parameters P^h and F^h are the coherent position and coherent fraction, respectively. Different from core-emission XSW measurements, the photo-emitted electron can originate from different atomic species and elements. Thus, for valence band or conduction band emission P^h and F^h may be described as the phase and amplitude of the hth Fourier coefficient of the $\mathbf{h}$ residence probability function of the valence electrons, weighted, however, by specific cross-sections.

Using a trigonometric identity, Eqs. (11.71) and (11.73) render

$$F^h e^{2\pi i P^h} = \frac{(1 - Q_a)u_a^2\sigma_a^T e^{i\mathbf{h}\cdot\mathbf{r}_a} + (1 - Q_c)u_c^2\sigma_c^T e^{i\mathbf{h}\cdot\mathbf{r}_c}}{u_a^2\sigma_a^T + u_c^2\sigma_c^T}, \quad (11.74)$$

which may be generalized for a unit cell with an arbitrary number of i atoms

$$F^h e^{2\pi i P^h} = \frac{\sum_i (1 - Q_i) u_i^2 \sigma_i^T e^{i\mathbf{h}\cdot\mathbf{r}_i}}{\sum_i u_i^2 \sigma_i^T}. \tag{11.75}$$

Equation (11.75) is the XSW *structure factor for valence-electron emission.* Note that both the bond polarities and photoelectron cross sections appear in Eq. (11.75).

The main difference in the expression for the coherent fraction and coherent position in the case of the photo-excitation from the valence band comparing to the photoexcitation from the core is coming from the contribution of the *product* of the cross-sections $\sigma_{a,c}^T$ and parameters $u_{a,c}^2$ that are determined by the bond polarity value α_p (11.59). Substituting these values into expression (11.74) we have for the coherent fraction and position for an arbitrary reflection **h**

$$\begin{aligned} F^h &= \sqrt{\gamma_a^2 + \gamma_c^2 + 2\gamma_a\gamma_c \cos[\mathbf{h}(\mathbf{r}_c - \mathbf{r}_a)]}, \\ \tan 2\pi\ P^h &= \frac{\gamma_a \sin(\mathbf{h}\mathbf{r}_a) + \gamma_c \sin(\mathbf{h}\mathbf{r}_c)}{\gamma_a \cos(\mathbf{h}\mathbf{r}_a) + \gamma_c \cos(\mathbf{h}\mathbf{r}_c)}, \end{aligned} \tag{11.76}$$

where parameters γ_a and γ_c are determined by

$$\begin{aligned} \gamma_a &= (1 - Q_a)\frac{(1 + \alpha_p)\sigma_a^T}{(1 + \alpha_p)\sigma_a^T + (1 - \alpha_p)\sigma_c^T}, \\ \gamma_c &= (1 - Q_c)\frac{(1 - \alpha_p)\sigma_c^T}{(1 + \alpha_p)\sigma_a^T + (1 - \alpha_p)\sigma_c^T}. \end{aligned} \tag{11.77}$$

For the case of a heteropolar bond, the anions and cations have different atomic cross-sections, and charge is transferred from the less electronegative cation to the more electronegative anion. As the positions of the anion and cation are known ($\mathbf{h} \cdot \mathbf{r}_a = -\pi/4$ and $\mathbf{h} \cdot \mathbf{r}_c = \pi/4$ for the (111) reflections), we may use the XSW structure factor of Eq. (11.74) in the dipole approximation to derive an expression for the coherent positions and fractions for the XSW valence electron yield of the zinc blende semiconductors. Due to the symmetry of the zinc blende structure, the (-1-1-1) reflections will have the same coherent fractions as the (111) reflections, but the coherent positions will be of opposite sign. For the (111) reflection the result is

$$F^{111} = \frac{\sqrt{(1 + \alpha_p)^2 \left(\sigma_a^T\right)^2 + (1 - \alpha_p)^2 (\sigma_c^T)^2}}{(1 + \alpha_p)\sigma_a^T + (1 - \alpha_p)\sigma_c^T} \tag{11.78}$$

and

$$\tan 2\pi\ P^{111} = -\frac{(1+\alpha_p)\sigma_a^T - (1-\alpha_p)\sigma_c^T}{(1+\alpha_p)\sigma_a^T + (1-\alpha_p)\sigma_c^T}. \tag{11.79}$$

This is an important result. We see, that in dipole approximation there are two independent parameters, the bond polarity coefficient α_p and the cross-section ratio σ_c^T/σ_a^T, that are contributing to the expression for the coherent fraction and coherent position. If the bond polarity value α_p is known from theory, then the cross-section ratio σ_c^T/σ_a^T can be determined from the second equation (11.79) and vice versa. The valence band of GaAs, InP and NiO crystals has been analyzed in this way and the results have been published in 2001.[42]

11.9. Summary

The matrix element for the PE is usually treated in a first-order approximation (dipole approximation). It holds astonishingly well even for hard X-rays with a wavelength much shorter than atomic dimensions as far as the magnitude of the transition matrix element is concerned. However, higher-order multipole terms introduce a deviation from the highly symmetric dipole emission profile. The forward–backward asymmetry caused already by the second-order multipole terms (dipole–quadrupole) cannot be neglected when the photoelectron is emitted by the action of an XIF, i.e., by the coherent action of two X-ray waves traveling in different directions. It has to be taken into account for the XSW analysis when analyzing differential, non-angularly integrated photoelectron signals.

To identify which atom of a crystal contributes to the density of states of a valence band, resonant photoelectron spectroscopy is traditionally used. However, the corresponding theory is not simple. Using the site-specific XSW technique is an elegant alternative and has been successfully employed several times for binary (see Chapters 12 and 26) and ternary compounds.[43]

References

1. A. Bechler and R. H. Pratt, *Phys. Rev. A* **39** (1989) 1774; **42** (1990) 6400.
2. J. W. Cooper, *Phys. Rev. A* **42** (1990) 6942; **45** (1992) 3362; **47** (1993) 1841.
3. B. Krässig, M. Jung, D. S. Gemmell, E. P. Kanter, T. LeBrun, S. H. Southworth and L. Young, *Phys. Rev. Lett.* **75** (1995) 4736; M. Jung, B. Krässig, D. S. Gemmell, E. P. Kanter, T. LeBrun, S. Southworth and L. Young, *Phys. Rev. A* **54** (1996) 2127; B. Krässig *et al. Phys. Rev. A* **67** (2003) 022707.

emission is found to be extremely localized: It originates only from the spatial regions of the crystal surrounding the atomic cores. This finding is a direct consequence of the dipole approximation for the photoelectron process, and it allows the delineation of the crystal valence band into its site-specific components. The site-specific PES technique therefore has the unique ability to measure the individual atomic contributions to the total electronic structure of a crystal as demonstrated for the heteropolar crystal GaAs.

Acknowledgments

This research was performed at the Stanford Synchrotron Radiation Laboratory and at the National Synchrotron Light Source, which are supported by the United States Department of Energy. The author is indebted to Dr. Leeor Kronik and Dr. Erik Nelson for their contributions to all aspects of this work.

References

1. W. Braun, A. Goldmann and M. Cardona, *Phys. Rev. B* **10** (1974) 5069; T.-U. Nahm, M. Han, S.-J. Oh, J.-H. Park, J. W. Allen and S.-M. Chung, *Phys. Rev. Lett.* **70** (1993) 3663.
2. N. W. Ashcroft and N. D. Mermin, *Solid State Physics* (Holt, Rinehart and Winston, New York, 1976).
3. J. C. Woicik, E. J. Nelson, D. Heskett, J. Warner, L. E. Berman, B. A. Karlin, I. A. Vartanyants, M. Z. Hasan, T. Kendelewicz, Z. X. Shen and P. Pianetta, *Phys. Rev. B* **64** (2001) 125115.
4. D. P. Woodruff, D. L. Seymour, C. F. McConville, C. E. Riley, M. D. Crapper, N. P. Prince and R. G. Jones, *Surf. Sci.* **195** (1988) 237.
5. A. Einstein, *Ann. Phys.* **17** (1905) 132.
6. W. E. Spicer, *Phys. Rev.* **112** (1958) 114.
7. B. W. Batterman and H. Cole, *Rev. Mod. Phys.* **36** (1964) 681.
8. G. A. Burdick, *Phys. Rev.* **129** (1963) 138.
9. H. A. Bethe and R. W. Jackiw, *Intermediate Quantum Mechanics* (Benjamin Cummings, Reading, New York, 1986).
10. H. A. Bethe and E. E. Salpeter, *Quantum Mechanics of One- and Two-Electron Atoms* (Plenum, New York, 1977).
11. In the core region, the Born approximation may not be valid; however, the standard description of the dipole approximation applies.
12. I. Adawi, *Phys. Rev.* **134** (1964) A788.
13. J. C. Woicik, E. J. Nelson and P. Pianetta, *Phys. Rev. Lett.* **84** (2000) 773.
14. W. A. Harrison, *Electronic Structure and the Properties of Solids* (Freeman, San Francisco, 1980).

15. J. R. Patel and J. A. Golovchenko, *Phys. Rev. Lett.* **50** (1983) 1858.
16. J. R. Chelikowsky and M. L. Cohen, *Phys. Rev. B* **14** (1976) 556.
17. R. G. Cavell, S. P. Kowalcyk, L. Ley, R. A. Pollak, B. Mills, D. A. Shirley and W. Perry, *Phys. Rev. B* **7** (1973) 5313; I. Abbati, L. Braicovich, G. Rossi, I. Lindau, U. del Pennino and S. Nannarone, *Phys. Rev. Lett.* **50** (1983) 1799.
18. J. C. Woicik, E. J. Nelson, T. Kendelewicz, P. Pianetta, M. Jain, L. Kronik, and J. R. Chelikowsky, *Phys. Rev. B* **63** (2001) 041403(R).
19. M. L. Cohen and J. R. Chelikowsky, *Electronic Structure and Optical Properties of Semiconductors*, Springer Series in Solid-State Sciences, Vol. 75, (Springer-Verlag, Berlin, 1988).
20. L. Ley, R. A. Pollak, F. R. McFeely, S. P. Kowalczyk and D. A. Shirley, *Phys. Rev. B* **9** (1974) 600.
21. S. Hufner, *Photoelectron Spectroscopy: Principles and Applications*, 2nd edn. (Springer-Verlag, Berlin, 1996).

Chapter 13

EXPERIMENTAL BASICS

ALEXANDER KAZIMIROV

Cornell High Energy Synchrotron Source (CHESS),
Cornell University, Ithaca, NY 14853, USA

JÖRG ZEGENHAGEN

European Synchrotron Radiation Facility,
6 rue Jules Horowitz, F-38000 Grenoble, France

In this chapter we will explain the experimental essentials for performing X-ray standing wave experiments. The experimental set-ups used for the X-ray standing wave technique have evolved in parallel with the evolution of X-ray light sources, the properties of which will be discussed briefly well.

13.1. Introduction

Every experimental set-up used for X-ray standing wave (XSW) measurements has individual requirements depending on the light source used. The characteristics vary of course also depending on the nature of information a researcher wants to obtain from a particular sample. Furthermore, it also matters in which particular environment the sample under study is placed. There are, nonetheless, some main common features of the experimental set-up which include: (1) an incident X-ray beam produced by an X-ray source and properly conditioned in terms of the monochromaticity $\Delta E/E$ and angular divergence $\Delta\vartheta$; (2) a sample stage allowing positioning of the sample in a chosen diffraction (scattering) geometry at or close to the Bragg angle; (3) a possibility to scan the sample through the Bragg reflection either by rotating the sample or by scanning the energy of the incident beam; (4) an X-ray detector to monitor

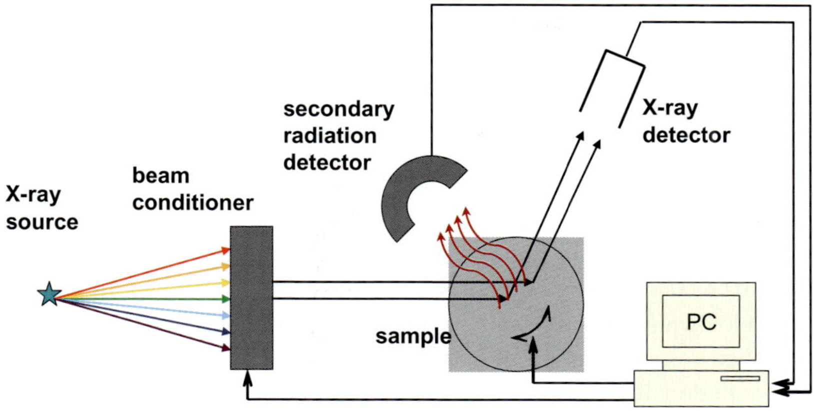

Fig. 13.1. Schematic of a typical XSW set-up. The beam conditioner prepares the incident beam with required degree of monochromaticity $\Delta\lambda/\lambda$ and angular collimation $\Delta\theta$. Computer control is used to perform angular or energy scans and collect information from the detectors for the secondary radiation and the reflected X-rays.

intensity of the beam diffracted or scattered from the sample; (5) secondary detector(s) to register photo-excited radiation coming from the sample while scanning it through the Bragg region. All these parts are shown schematically in Fig. 13.1, and in the rest of the chapter we will discuss them more in detail concentrating on XSW experiment using a perfect sample crystal in the Bragg diffraction geometry.

13.2. X-Ray Sources

13.2.1. *X-ray tubes*

The first demonstration of creating an XSW field in Bragg geometry was performed by using a standard X-ray tube and detecting the X-ray fluorescence signal from the host atoms of a germanium single crystal.[1,2] X-ray tubes were used later in first proof-of-principle experiments on localization of impurities[3] in Si and chemically absorbed atoms on a silicon surface.[4] They are used nowadays for preliminary XSW measurements and also for original research when the secondary radiation yield is not flux limited.

An X-ray tube as we know it today was first introduced in 1913 by William Coolidge, a researcher working for General Electric, and represented a significant improvement over discharge tubes similar to those used by Röntgen in his pioneer experiments. In a Coolidge tube,[5] a hot tungsten filament emits electrons, which are accelerated under applied high voltage towards a water-cooled metal anode. X-rays are emitted from the anode into the solid angle of 2π. All parts are in vacuum inside a sealed glass tube. In modern X-ray tubes electrons are focused on the anode and the X-rays escape through thin Be windows at a certain angle with respect to the anode surface thus defining the source size, typically from few mm down to 0.1 mm. The spectrum of X-rays emerging from the anode consists of two distinct parts. The first one is a continuous spectrum originating from the electrons de-accelerated and eventually stopped in the anode material, which is known as *Bremsstrahlung*. The high energy cutoff is determined by the applied voltage. Superimposed on the *Bremsstrahlung* are characteristic X-ray emission lines of the atoms of the anode material and which are corresponding to electron transitions from the outer shell to the holes in the inner shells produced by the impinging electrons. The characteristic lines are very sharp (few eV), with the peak intensity per line width of about 10^4 higher than the *Bremsstrahlung* and an integral intensity (for K-lines) of about 60% of the total intensity. Figure 13.2 shows a typical spectrum from a Mo tube operating under 40 kV with strong Mo $K_{\alpha_1\text{-}\alpha_2}$ doublet, which can be easily resolved by a perfect crystal analyzer, and K_β characteristic lines.

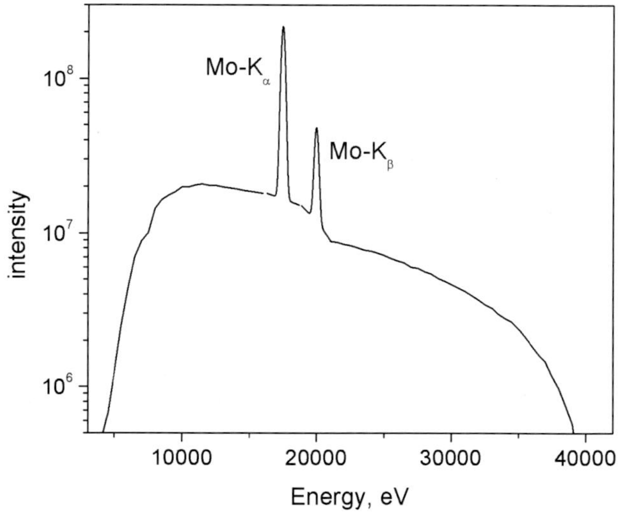

Fig. 13.2. Typical spectrum from the Mo tube operating at 40 keV. Two characteristic K_α and K_β lines are sitting on top of the continuous *Bremsstrahlung* background.

The maximum input power of a sealed tubes is about 2 kW, with only about 0.3% of it converted to X-rays. The maximum power is limited due to the high heat load created in the anode by the bombarding electrons and the limited thermal conductivity. Thus, a design with a rotating anode offers more than 10 times gain in power due to more efficient dissipation of the heat load since the electron beam spot moves on the surface of the water cooled anode. The intensity of the X-ray beam is controlled by adjusting the filament current and high voltage. An X-ray tube represents an almost monochromatic source for an XSW experiment and the size of the X-ray beam, which is emitted in a wide solid angle, will be determined by the size of the slit in front of the beam conditioner (monochromator).

13.3. Synchrotron Radiation

13.3.1. *Introduction*

Charged particles which are accelerated, e.g., by forcing them to move along a curved path, emit electromagnetic waves according to Maxwell equations. Cosmic radio waves and in some instances (Crab nebula) optical emission from high energy electrons moving in magnetic fields naturally occur in our universe.[6] The list of prominent physicists who established the theoretical basis for what is nowadays known as synchrotron radiation (SR) includes the names of Larmor (1897), Lienard (1898), and Schott (1907) (for references on early works, see Refs. 7 and 8). The interest reawakened in 1930–1940s with the work on first circular accelerators, betatrons, and synchrotrons. It was shown in 1944[9] that the losses due to synchrotron radiation were the main factor limiting the maximum energy attainable in a betatron and soon the detailed theory of synchrotron radiation has been developed.[10,11] Synchrotron radiation was first observed at the General Electric 70 MeV electron synchrotron as visible light coming from the transparent electron tube[12] (see Refs. 7, 8, 13–15 for detailed historical accounts). In subsequent years, detailed characterization of SR has been performed both in the US, at the GE synchrotron and later at Cornell,[16] and in the Soviet Union at the 250 MeV synchrotron at the Lebedev Institute in Moscow and remarkable agreement with theoretical predictions has been established. The pioneering work by Tomboulian and Hartman[16] performed at the Cornell 300 MeV synchrotron, in which absorption spectra from Be and Al had been measured in a spectral range inaccessible before, mark the beginning of exploitation of SR. In the following years, research programs based on the unique properties of SR have been established in the US, Europe, and Asia. The era of synchrotron radiation experiments had begun.

In the 1950s–1960s, the storage ring technology was developed as an efficient way to store energetic charged particles mostly for collision experiments. While SR was the unwanted by-product of high-energy physics, research scientists began to utilize it "parasitically" at some synchrotrons and storage rings. These facilities are called the first-generation SR sources. Usually, high-energy machines store two types of beams with opposite charge, typically electrons and positrons, and opposite velocity in order to collide them at the position of a particles detector. The desired high collision probability leads necessarily to a rather rapid decline of the beam current, i.e. to a rather short beam lifetime, requiring frequent, every 1–2 hours, refilling.

In the 1970s–1980s, with a surge of interest in applications of synchrotron radiation, storage rings dedicated to production of SR have been constructed around the world. These are the second-generation SR sources such as SRS in the UK, Photon Factory in Japan, NSLS in the USA, and BESSY in Germany. They produce synchrotron radiation mainly by bending magnets and with some insertion devices such as wigglers and undulators. These facilities typically have a large number of beamlines (about 80 at the NSLS) and employ different parts of the spectrum, from IR via VUV to hard or even high energy X-rays, and use a variety of techniques for numerous applications. Improved beam lifetimes of more than 10 hours and reliable operation allowed obtaining reliable data even in experiments requiring long data acquisition.

The need for higher flux and brightness led to the development of the third-generation SR sources. They are characterized by a very small e-beam emittance that made undulators the most attractive sources of SR. In-between the electron optical elements, the so-called lattice, are many several-meter-long strait sections that can accommodate long undulator or even two or three shorter undulators in a tandem. Three major third-generation hard X-ray SR sources, the European Synchrotron Radiation Facility, ESRF, in France, Spring-8 in Japan, and the Advanced Photon Source, APS, in the USA have been commissioned in the 1990s. They have large circumferences of about 1 km and operate at high electron beam energy of 6 to 8 GeV. They have been joined recently by the PETRA III SR laboratory, located at a sector of the 2.3 km long, rebuilt PETRA III storage ring in Hamburg, Germany. Other third-generation SR sources with slightly lower electron beam energies at around 3 GeV have recently come into operation (e.g., DIAMOND in the UK and SOLEIL in France) or are under construction (e.g., NSLS-II in the USA).

X-ray free electron lasers (XFEL) currently operating in the soft X-ray range (FLASH facility at DESY, Germany) and hard X-ray range (LCLS Stanford, USA) or being under construction in Hamburg (European XFEL) and other countries represent the next, fourth-generation SR sources. They are based on long, up to several kilometers, linear accelerators and long, up to more than hundred meters, undulators which provide the laser like beam of synchrotron radiation. The radiation at the exit of the long undulator is fully coherent up to an X-ray energy of 8–12 keV. The other unique property of the XFEL is a very short bunch length leading to radiation flashes of less than 100 fs. The XFEL beam has an extremely high brilliance (≡ spectral brightness = photons/second/area/solid angle/bandwidth), many orders of magnitude higher than the third-generation sources and an extremely high peak power because of the high flux in a single bunch.[17] Utilizing XFEL radiation will present new challenges and exciting opportunities for the SR community.

Energy Recovery Linacs (ERLs) such as the one proposed at the Cornell University, are the new type of SR sources based on an idea proposed originally in 1965 as an alternative to conventional storage rings for the needs of high energy physics.[18] Different from the storage rings, in an ERL electron bunches circulate only once, they are not at equilibrium and their emittance is determined by the injector. An ERL is characterized by an extremely small emittance because of the very small (few micron) and circular source size, a flexible time structure with a bunch length as short as a few hundred fs and a high repetition rate of up to 2 GHz. It is expected that the brilliance of an ERL photon beam will be more than two orders of magnitude higher than at the third-generation sources with full coherence up to 8 keV.[19]

In developing the XSW technique further, the properties of the laboratory sources (e.g., small brilliance) have been recognized as a strongly limiting factor. In particular, the unique capabilities of the XSW method in localizing adsorbate atoms on clean surfaces required energy tunability and X-ray beams of much higher intensity. First SR-based XSW experiments performed at HASYLAB[20] and at CHESS[21] demonstrated great advantages of using synchrotron radiation and helped to establish the XSW technique as a unique research tool.

13.3.2. *Properties of synchrotron radiation*

The main properties of the synchrotron radiation are: (1) a broad energy spectrum, either continuous (bending magnet, wiggler) or consisting of discrete harmonics (undulator); (2) strong angular collimation; (3) high degree of polarization: in case of a bending magnet (and commonly used wigglers and undulators) linear in the orbit plane and elliptical with opposite helicity above and below the plane; (4) time structure (pulses) determined by the bunch pattern in the ring. In general, the properties of synchrotron radiation can be understood by considering relativistic Doppler (time squeezing) effect, which is due to the fact that the relativistic electron moves with the speed which is only slightly lower than the speed of the photons it emits, and applying it to transform the electron motion from the moving frame of reference to the laboratory (observer) frame. Below we present only a brief description of the properties of SR generated by bending magnet, undulator and wiggler and refer for further reading to original articles,[11] reviews,[22,23] and books[24,25] for a detailed discussion.

Bending magnet. Bending magnets radiate a continuous spectrum with the maximum intensity close to the critical energy

$$E_c = \frac{3\hbar\gamma^3 c}{2\rho} = 0.665 \cdot E_e^2[\text{GeV}] \cdot B[\text{T}], \tag{13.1}$$

where c is the speed of light,

$$\gamma = \frac{E_e}{mc^2} = 1957 \cdot E_e[\text{GeV}], \tag{13.2}$$

E_e is the energy of the electron beam, m is the electron mass, and ρ is the radius of curvature

$$\rho = \frac{E_e}{ecB}, \tag{13.3}$$

where B is the magnetic field. A typical spectrum is shown in Fig. 13.3. The critical energy E_c divides the energy spectrum in two equal parts in terms of the amount of power generated above and below E_c. The angular divergence at the critical energy is

$$\sigma \approx 0.64/\gamma \tag{13.4}$$

(for a Gaussian $2.35 \cdot \sigma_{\text{rms}}$ = full width a half maximum (FWHM)) both in vertical and horizontal planes and it decreases with increasing energy.

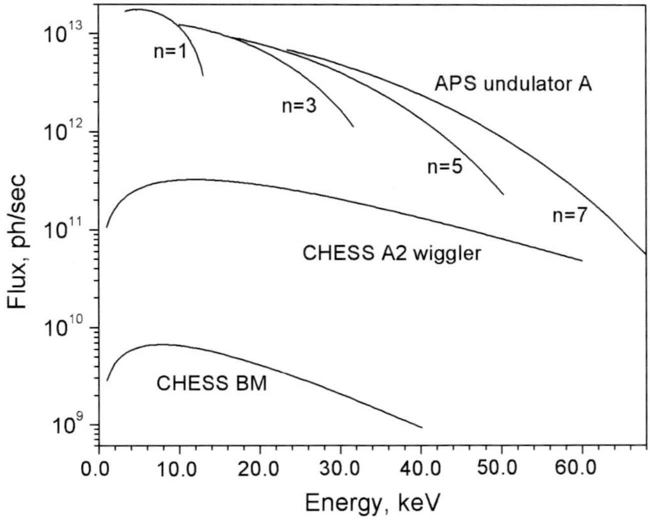

Fig. 13.3. Photon flux through a 0.2 mm (vertical) × 0.5 mm (horizontal) slit at 40 m from the source for a CHESS hard bend magnet, 49 poles CHESS wiggler and the APS type *A* undulator (current 200 mA, beam energy $E_e = 5.3\,\text{GeV}$ for CHESS and $E_e = 7.0\,\text{GeV}$ for APS).

Undulator. An undulator is an insertion device with a periodic sinusoidal magnetic field perpendicular to the direction of the electron beam. It can be characterized by a period λ_u, a number of periods N and a deflection parameter $K = eB\lambda_u/2\pi mc = 0.934\lambda_u[\text{cm}]B[\text{T}] \ll 1$.[26] At $K \ll 1$ the maximum angular slope of the electron trajectory $\delta = K/\gamma$ is much smaller than the angular divergence of the bending magnet radiation γ^{-1}. When $K \gg 1$ the insertion device is called a wiggler. The undulator generates a set of intense odd harmonics with the first (the fundamental) harmonic determined by the undulator equation:

$$\lambda_1 = \frac{\lambda_u}{2\gamma^2}\left(1 + \frac{K^2}{2} + \gamma^2\theta^2\right), \qquad (13.5)$$

where θ is the horizontal observation angle. Since the apparent electron motion in the laboratory reference frame is not pure sinusoidal, higher-order odd harmonics $\lambda_n = \lambda_1/n$ are generated with their intensity growing approximately with the square of K for small K.

As one can see from the undulator equation (13.5), the maximum harmonic energy is on-axis (at $\theta = 0$) and the energy decreases with the observation angle θ as the time squeezing effect is getting weaker. For a single electron the energy bandwidth is getting narrower with increasing the number of periods N and harmonic number n as:

$$\left(\frac{\Delta\lambda}{\lambda}\right)_n \approx \frac{1}{nN}. \tag{13.6}$$

The energy spread in the electron beam $\Delta E_e/E_e$ and a finite emittance of the storage ring ε contribute to the bandwidth and are the main factors determining the energy bandwidth of high order harmonics.

The undulator radiation is collimated within a narrow cone θ_{cen} defined as an angular spread of the photons with the maximum, on-axis, energy (or, alternatively, as an angle containing all energies within the energy bandwidth determined above)

$$\theta_{\text{cen}} \approx \frac{\sqrt{1+K^2/2}}{\gamma\sqrt{N}}, \tag{13.7}$$

with a typical value of few μrad for the fundamental harmonic at the third-generation SR sources. A finite emittance contributes to the broadening of the central cone as:

$$\theta_X = \sqrt{\theta_{\text{cen}}^2 + \sigma_X'^2}, \quad \theta_Y = \sqrt{\theta_{\text{cen}}^2 + \sigma_Y'^2}, \tag{13.8}$$

where σ_X', σ_Y' are the angular divergence of the source in the X (horizontal) and Y (vertical) directions determined by the emittance of the ring ε and and the β-function at the particular location

$$\sigma_X' = (\varepsilon_X/\beta_X)^{1/2}, \qquad \sigma_Y' = (\varepsilon_Y/\beta_Y)^{1/2}. \tag{13.9}$$

For a third-generation source with $\varepsilon_X \approx 7 \times 10^{-9}$ and $\varepsilon_Y \approx 7 \times 10^{-10}$ mrad the typical undulator cone opening (FWHM) is $\sim$20 μrad vertically and $\sim$50 μrad horizontally. Future improvements in the storage ring technology will be directed toward achieving lower horizontal emittance in the 10^{-9} mrad range. It is expected that linac based future sources such as the ERL[19] will have $\varepsilon_X = \varepsilon_Y \approx 0.1 \times 10^{-9}$ mrad and less.

The power radiated into the central cone by the fundamental harmonic can be estimated in practical units as

$$P_{\text{cen}} \cong \frac{(5.69 \cdot 10^{-6}[W]) \cdot \gamma^2 \cdot I[A] \cdot K^2}{\lambda_U[\text{cm}] \cdot (1+K^2/2)^2} \cdot f(K), \tag{13.10}$$

where $f(K)$ is a correction factor close to 1 for small K, giving a typical value of $\sim$10 W. Surprisingly, this value does not depend on the length of the undulator (number of periods N) as the central cone is getting narrower with N. For comparison, the total power radiated by an undulator into all harmonics over all angles is

$$P_{\text{tot}}[\text{kW}] = 0.633 \cdot E^2[\text{GeV}] \cdot B_0^2[\text{T}] \cdot L[\text{m}] \cdot I[A], \tag{13.11}$$

giving several kW and it depends linearly on the undulator length L. Only very small part of this power contained within the central cone is used in the experiment. The entrance slit in front of the monochromator is usually adjusted around the central cone cutting off the rest of the radiation and reducing the thermal load on the first crystal.

The energy spectrum of the flux through an aperture of 0.2 mm (vertical) $\times$ 0.5 mm (horizontal) located at the distance of 40 m from the source at a fixed $K = 2.24$ is shown in Fig. 13.3. One can see that the intensities of the first and the other odd harmonics are much higher than for the even harmonics. For a single electron, the intensity of the even harmonics on-axis is zero. They appear because of the finite size of the aperture and the random distribution of motion of individual electrons in a bunch (finite emittance).

In an XSW experiment typically the first or the third harmonics are utilized. Tuning the energy of a harmonic to a desired value is achieved by tuning the magnetic field B by changing the undulator gap and thus adjusting the deflection parameter K. As one can see from the undulator equation (13.5), opening the undulator gap and reducing the magnetic field shifts harmonic energy toward higher value. On the other hand, increasing the intensity of the high-energy part of the spectra, i.e., of higher-order harmonics, requires increasing K. Undulator tuning curves for the APS type A undulator are shown in Fig. 13.4 for the first four odd harmonics. When properly designed, undulator odd harmonics smoothly cover a wide energy range, thus allowing experiments at any energy.

Improvements in machine technology over the last decade lead to a significant reduction in emittance and combined with progress in undulator technology resulted in many orders of magnitude gain in spectral brightness (brilliance) of the third-generation SR sources. X-ray sources based on linear accelerator technology such as the XFEL and ERL utilize undulators exclusively.

Wiggler. Undulators become Wigglers at high K values. With increasing K to the level where the electron deflection angle $\delta = K/\gamma$ is larger than γ^{-1} the time-squeezing (Doppler) effect becomes highly non-uniform and only small parts of the trajectory tangent in the direction of observation

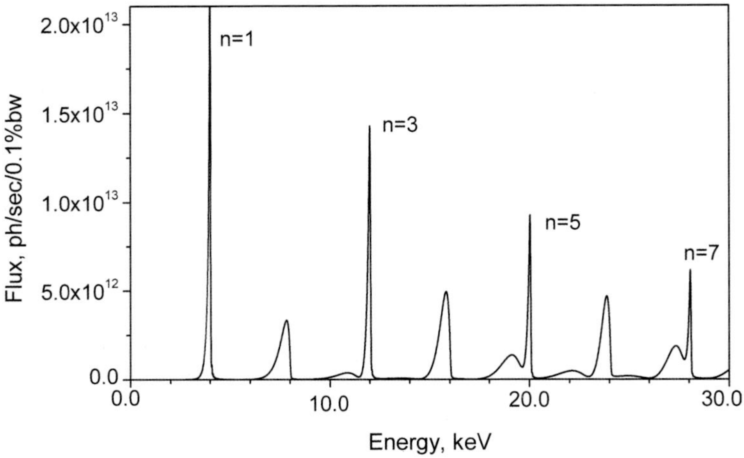

Fig. 13.4. Energy spectrum of the X-ray flux through an aperture of 0.2 mm (vertical) $\times$ 0.5 mm (horizontal) located at a distance of 40 m from the source for an APS type A undulator, $K = 2.24$.

contribute to the radiated intensity. These parts can be approximated by segments of a circle with the radius $\rho = \gamma\lambda_u/2\pi K$. Coherent interference of radiation from all parts of the electron trajectory at $K \ll 1$ transforms into an incoherent sum of intensities from $2N$ tangent points at $K \gg 1$. Thus, at high K undulators act as a sequence of $2N$ bending magnets.

As one can see from the undulator equation (13.5), the energy position of the undulator harmonic and the separation between odd harmonics on-axis can be written as:

$$E_n \approx \frac{4\pi\hbar nc\gamma^2}{\lambda_u(1 + K^2/2)}, \tag{13.12}$$

$$\Delta E_n \approx \frac{8\pi\hbar c\gamma^2}{\lambda_u(1 + K^2/2)}. \tag{13.13}$$

With increasing K, all harmonics shift to the lower energy, become more closely spaced, more stronger harmonics emerge at higher energy and the spectrum as a whole exhibits a comb-like structure with the maximum intensity shifted toward higher energy. Eventually, at high K and a finite emittance and acceptance angle, all harmonics merge into a continuous spectrum. Similar to a bending magnet, the wiggler spectrum is characterized by a critical energy which can be expressed through

K as

$$E_c = \frac{3\pi c\hbar\gamma^2 K}{\lambda_u}. \tag{13.14}$$

This energy corresponds to a critical harmonic number

$$n_c = \frac{3K}{4}\left(1 + \frac{K^2}{2}\right) \approx \frac{3K^3}{8}. \tag{13.15}$$

An energy spectrum from the permanent magnet 49 pole CHESS wiggler ($\lambda_u = 12\,\text{cm}$, $B_0 = 0.8\,\text{T}$, $K = 9$) is shown in Fig. 13.3. The critical energy of this wiggler is 14.9 keV which corresponds to $n_c \cong 270$. At high energy $E > 50\,\text{keV}$, the wiggler flux exceeds the flux from the CHESS hard bend magnet by (more than) two orders of magnitude.

13.4. Beam Conditioning

13.4.1. *DuMond diagram*

As for any interference field, the quality of the XSW pattern can be characterized by the visibility of the interference fringes, which strongly depends on the quality of the incident beam. The most important parameters are the monochromaticity $\Delta\lambda/\lambda$ and the angular divergence $\Delta\theta$. They have to be optimized for a particular experiment, i.e., the values of $\Delta\lambda/\lambda$ and $\Delta\theta$ should be small enough not to wash out the interference fringes and, at the same time, conditioning optics should transmit enough intensity to allow to measure the secondary radiation yield with sufficient statistics in reasonable time. These are two opposing requirements and thus a matter of compromise. Developed over the last few decades, perfect crystal optics (for review see e.g., Refs. 27, 28) are used to condition the X-ray beam for a particular experiment. A graphical technique known as DuMond diagram[29] proposed in the early years of X-ray spectroscopy is an excellent tool to analyze and optimize perfect crystal optics.

According to dynamical diffraction theory, the width of the intrinsic, or Darwin, rocking curve for a perfect plane wave incident on a crystal and a symmetric reflection (see Chapter 2 of this book) is

$$\omega_0 = \frac{2\,|P|\,r_e\lambda^2\sqrt{F_H F_{\bar{H}}}}{\pi V \sin(2\theta_B)}, \tag{13.16}$$

where $|P|$ is the polarization factor, r_e is the classical electron radius, F_H is the structure factor of reflection $\mathbf{H}$, V is the unit cell volume, and θ_B is the

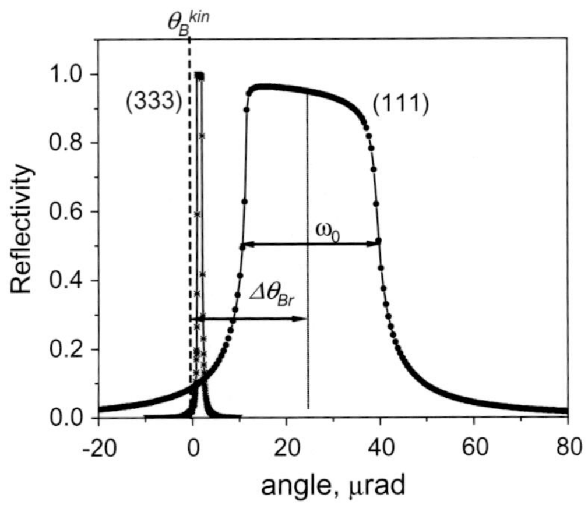

Fig. 13.5. The Si(111) and (333) rocking curves at 10 keV and 30 keV, correspondingly. The zero angle corresponds to the Bragg angle calculated by using Bragg's law (kinematical Bragg angle). Due to refraction, the center of the Darwin curve is shifted toward higher angle.

Bragg angle. An intrinsic rocking curve for the Si(111) reflection at 10 keV is shown in Fig. 13.5. For comparison, the (333) rocking curve at 30 keV is also shown on the same angular scale. The zero of the angular scale is set at the value corresponding to the angle calculated by using Bragg's law (kinematical Bragg angle). Due to refraction, the center of the Darwin curve is shifted from the kinematical Bragg angle toward higher angle and this shift (for symmetrical reflection) can be expressed as:

$$\Delta\theta_B = \frac{r_e \lambda^2 F_0}{\pi V \sin(2\theta_B)}. \tag{13.17}$$

The DuMond diagram is based on a graphical representation of Bragg's law in angular ϑ (horizontal axis) and the wavelength λ (vertical axis) space as shown in Fig. 13.6. The horizontal width ω of the inclined band in Fig. 13.6 is the width of the intrinsic rocking curve. The horizontal $\Delta\lambda$ and the vertical $\Delta\theta$ bands represent the energy and the angular spread of the radiation incident on the crystal. Then, the intersection of all three bands (black diamond area in Fig. 13.6) shows the ϑ- and λ-distribution of the radiation transmitted by the crystal.

In Fig. 13.7, two double crystal arrangements most commonly utilized in X-ray optics are presented. On the left, Fig. 13.7(a), a parallel, or $(+n, -n)$ scheme is shown. The index n in this notation, which was first introduced in Ref. 30, denotes a crystal reflection. The sign combination

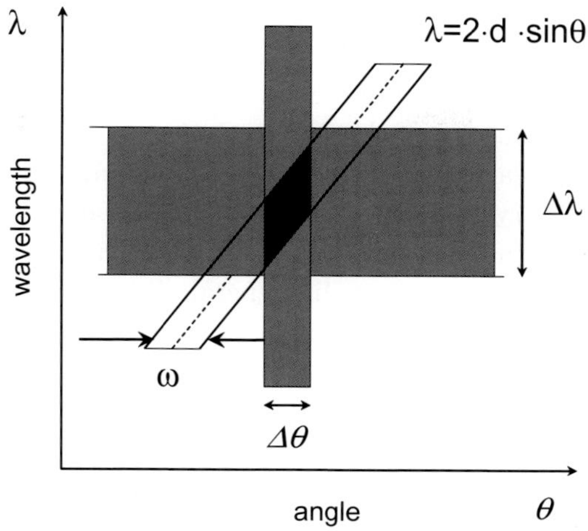

Fig. 13.6. DuMond diagram. The inclined band shows the diffraction region of the crystal. The horizontal width ω is the intrinsic width of the rocking curve. The horizontal $\Delta\lambda$ and the vertical $\Delta\theta$ bands are the energy and the angular spread of the incident radiation. The intersection of all three bands (black diamond) shows the distribution of the radiation transmitted by the crystal.

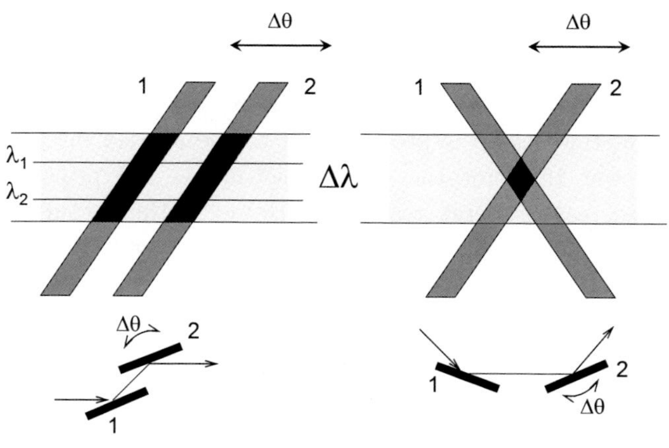

(a) (b)

Fig. 13.7. Parallel $(+n, -n)$ (a) and non-parallel $(+n, +n)$ (b) double crystal arrangements and corresponding DuMond diagrams. See text for the details.

$(+, -)$ means that the beam incident on the first crystal and the beam diffracted from the second crystal are on the opposite sides with respect to the beam between the first and the second crystals. When both crystals are identical and the reflections are the same, the reflection bands in the

DuMond diagram for both crystals are parallel. When one of the crystals is rotated relative to the other one in $\Delta\theta$, all wavelengths incident on the first crystal are reflected by the second one at a given angle. This is the reason why this scheme is called a *nondispersive* one, i.e., the whole energy bandwidth of the beam incident on the first crystal can pass the second crystal. If different crystals and/or different reflections are used, a $(n, -m)$ scheme, the corresponding bands in the DuMond diagram are not parallel any more and this arrangement is, in general, dispersive. Each wavelength will pass the second crystal at a particular angle $\Delta\theta$.

In Fig. 13.7(b), a non-parallel, or dispersive, $(+n, +n)$ scheme is shown. As can be easily seen from the DuMond diagram only a small part of the energy bandwidth, which corresponds to the diamond-shaped intersection area and is determined by the intrinsic width ω_0 and the value of the Bragg angle as

$$\frac{\Delta E}{E} = \frac{\omega_0}{\tan\theta_B}, \tag{13.18}$$

can pass both crystals. As the crystals rotate relative to each other, the energy of the transmitted beam is changing. This set-up serves as a basis for high energy resolution monochromators (see e.g., Ref. 31). Choosing high order reflections at near backscattering condition $\theta_B \lessapprox 90^\circ$ an energy resolution of 1 meV level and better can be achieved.[32] The intensity throughput of this set-up is proportional to the shaded intersection area and much lower than for the parallel set-up — the price one has to pay for the increased energy resolution. However, the throughput can be increased at near backscattering condition practically without loss in energy resolution since the angular acceptance of crystals is strongly enhanced at $\theta_B \approx 90^\circ$.

So far we considered symmetrical reflection, i.e., the incident and exit angle of the X-rays with the surface of the crystal is the same. For an asymmetrical reflection, the width of the rocking curve depends on the asymmetry factor

$$b = \frac{\sin(\theta_B - \alpha)}{\sin(\vartheta_B + \alpha)}, \tag{13.19}$$

where α is the angle between the surface of the crystal and the atomic diffraction planes, as illustrated by Fig. 13.8. Suppose, the X-ray beam is coming from the left side of Fig. 13.8 at a shallow incident angle $\theta_B - \alpha$ to the surface of the crystal, as shown by the black solid arrows. Then the angular

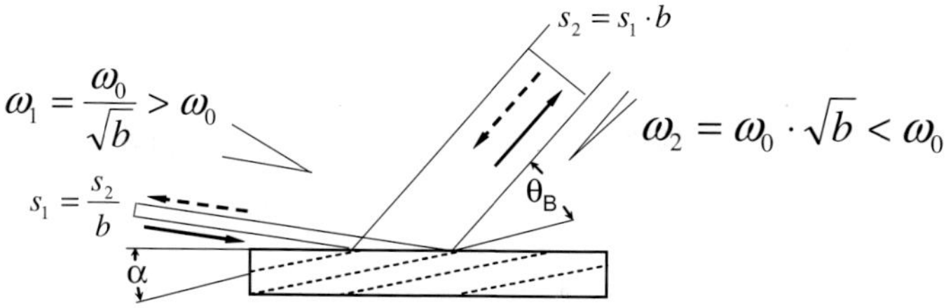

Fig. 13.8. Diffraction from an asymmetrically cut crystal. The solid arrows, from left to right, show the beam path at $b < 1$, the dashed arrows, from right to left, illustrate the case of $b > 1$.

width of the crystal "acceptance" window is $\omega_1 = \frac{\omega_0}{\sqrt{b}} > \omega_0$ and the angular divergence of the "exit" beam is $\omega_2 = \omega_0\sqrt{b} < \omega_0$. This remarkable property of the asymmetric reflection follows directly from the dynamical theory (see, e.g., Ref. 32 or 33 for details) and leads to a very important application in X-ray optics: When a monochromatic and divergent incident beam, e.g., from an X-ray tube, is incident at shallow angle on an asymmetric crystal then the crystal will accept X-rays within a range wider than ω_0 angular range and diffract into a narrower angular range. The size of the diffracted beam is increased by a factor of b. In other words, the beam will be compressed in angle and expanded in size.

This unique property of asymmetric reflection is used frequently to control the angular divergence of X-ray beams. In particular, highly asymmetric crystals with the asymmetry factor $b \leq 0.1$ are used to produce almost perfect plane waves for XSW experiments. If necessary, several subsequent reflections from crystals with $b < 1$ are used to reach the needed degree of collimation. Note that the flux density will be significantly reduced.

The opposite situation happens when the beam is incident on the crystal from the other side, at high incident angle with the surface (from the right side in Fig. 13.8 indicated by the dashed arrows). The acceptance window is now $\omega_2 = \omega_0 \cdot \sqrt{b} < \omega_0$ and the exit divergence is $\omega_1 = \frac{\omega_0}{\sqrt{b}} > \omega_0$, i.e., the crystal accepts X-rays within narrower angular range and diffracts them into a wider angular range. The diffracted beam is compressed in size. Sophisticated combinations of crystals and reflections are used in order to collect all X-ray photons from the X-ray source available in a desired angular and energy range in order to maximize the intensity available for a certain experiment.

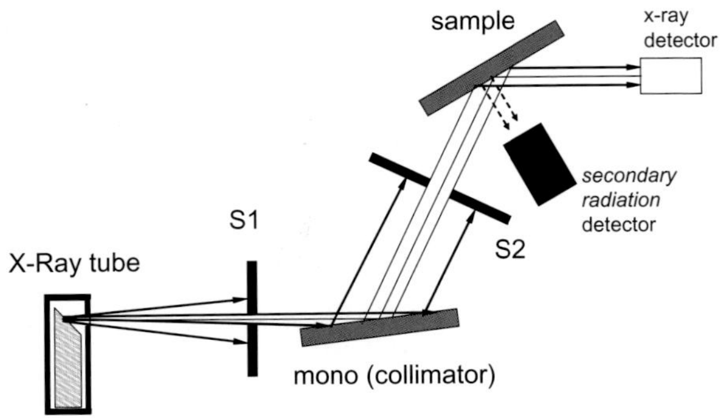

Fig. 13.9. Nondispersive set-up with an asymmetrical monochromator/collimator crystal for laboratory XSW experiments. For details see text and the DuMond diagram in Fig. 13.10.

13.4.2. *Laboratory XSW optical set-up*

A set-up for the XSW measurements by using a laboratory X-ray source, either a sealed tube or a rotating anode, is shown schematically in Fig. 13.9. As we discussed before, an X-ray tube radiates monochromatic characteristic radiation into a wide solid angle of 2π. Angular collimation of the beam is performed by using a perfect collimator crystal which is usually called monochromator crystal (though, in many cases, working with highly monochromatic characteristic radiation it acts like a collimator, and not like a monochromator). The divergence of the beam incident on the monochromator crystal is limited by a slit (S1 in Fig. 13.9) placed in front of the crystal. The distance between the tube and a monochromator is typically 0.1 to 1 m. To avoid or minimize the washing out of the XSW interference fringes by the angular or energy spread of the incident beam, a nondispersive optical setup with an asymmetrical crystal is utilized. This requires using the same material and the same reflection for the monochromator crystal as for the sample. The requirements on perfection of the monochromator crystal are extremely high and fully satisfied by only few crystals such as perfect dislocation free semiconductor crystals (Si, Ge, GaAs) and a few others (diamond, sapphire, etc.) with sufficient crystalline quality. Defects in crystals render the rocking curve broader than that predicted by the dynamic theory either due to mosaic structure or the diffuse scattering of X-rays by individual dislocations and point defects. The search for a monochromator crystal usually starts with Si and Ge: Knowing the d-spacing of the reflection under study one should try to find a Si- or

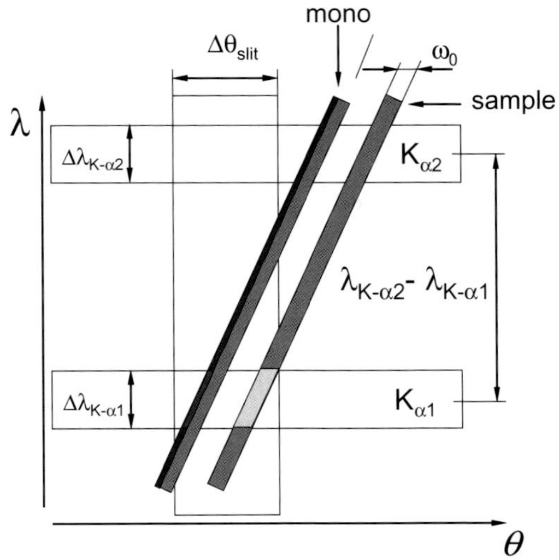

Fig. 13.10. DuMond diagram for the laboratory set-up with an asymmetric collimator crystal (see text for the details).

Ge-reflection with a *d*-spacing which is as closely matched as possible. If the best match results in an unacceptably high dispersion, then other crystals than Si and Ge may need to be considered.

The DuMond diagram illustrating a laboratory set-up is shown schematically in Fig. 13.10. Usually, the K_α radiation is utilized as the strongest component of the spectrum. It consists of two strong lines, $K_{\alpha 1}$ and $K_{\alpha 2}$, with a typical width of a few eV and an energy separation of about 100 eV (the width of the Mo $K_{\alpha 1}$ line is 6.31 eV at an energy of 17.478 keV and for the Mo $K_{\alpha 2}$ line these values are 6.49 eV and 17.373 keV, respectively[34]). The two K-lines are shown in Fig. 13.10 as two horizontal bands. The vertical region shows the angular distribution of the radiation incident on the first crystal, which is defined by the size of the slit S1 and the source size. Usually, the size of the slit S1 of about 0.1–0.2 mm is sufficient to separate $K_{\alpha 1}$ and $K_{\alpha 2}$ lines. The left inclined grey band represents the diffraction region of the first crystal in the case of a symmetrical reflection. The solid black band, narrower in angle, represents the output characteristics of an asymmetrically reflecting crystal. The bright red area shows the resultant distribution (θ, λ) of the radiation from the X-ray tube incident on the sample. This area will be convoluted with the region shown in blue of the diffraction band of the sample as the sample is scanned

the incident angle close to the critical angle at 10 keV leads to a significant reduction of the intensity at 30 keV (about 10^{-3} for the chosen example). The mirror can be inserted in the beam before the monochromator, also helping to reduce a heat load by about a factor of two, or downstream in the monochromatic beam, which is more common for modern undulator beamlines.

The second approach is based on crystal optics. One of the commonly used techniques is a slight misalignment (detuning) of the two monochromator crystals. Because the range of reflectivity for the harmonics is very narrow, the slight misalignment of two crystals, which causes about $20 \div 30\%$ reduction in intensity (overlap of two (111) curves) for the (111) reflection, leads to a strong reduction in overlap for the (333) harmonics and suppression of higher harmonics to less than 1%[36] can be achieved. Technically, this method can be implemented by using a special feedback electronic unit which keeps overlap between two crystals constant by monitoring beam intensity before and after monochromator and adjusting the voltage on a piezo crystal which is installed to change the angle of one, typically the second, monochromator crystal relative to the other one.[37] Another possibility to reduce harmonics would be using a slightly dispersive set-up with two crystals with different lattice constant such as Si and Ge,[38] asymmetrical reflections,[39] Laue–Bragg monochromators,[40] etc. (for an overview see e.g., Ref. 33 Sec. 15 and references therein).

13.5. Detection of Secondary Radiation

A variety of absorption and inelastic scattering channels can be exploited (*cf.* Chapters 1, 10, 11, and 22) in an XSW experiment, each of them yielding unique information about the sample under study. Typically detected are: (i) X-ray quanta, and (ii) electrons. These are the subjects of the well-established areas of X-ray fluorescence and photoelectron spectroscopy with various techniques developed over decades and the reader is referred to appropriate handbooks for details. Below we discuss the basics of fluorescence and photoelectron detection and the specifics related to the XSW experiment.

13.5.1. *Detection of fluorescence radiation*

13.5.1.1. *Introduction*

There are two major techniques to detect X-ray fluorescence radiation. The first one is the wavelength-dispersive technique. It is based on using a crystal (or a multilayer) as an analyzer. The X-ray radiation from the sample is collimated and incident on the analyzer crystal and the intensity of the X-rays diffracted by the analyzer crystal is measured by the scintillation detector. The spectrum is obtained by rotating the analyzer and using Bragg's law to convert angle into wavelength (energy). Thus, tuning spectrometer to a particular energy is performed by adjusting the angle of the analyzer. First wavelength-dispersive spectrometers became available in the 1940s and at present various crystals and multilayer are used to cover an energy range from about 100 eV to 100 keV. The analyzer crystals are usually not perfect and a typical energy resolution of a crystal spectrometer is around 1%. For the most frequently used energy range of 5 to 15 keV, typically an energy resolution of about 20–30 eV is achieved by a proper choice of the analyzer crystal and the spectrometer geometry. Much better resolution can be achieved with specialized instruments. Very often curved crystals and multilayer are used to increase the solid acceptance angle (see, e.g. Ref. 41 for details).

The use of crystal spectrometers in the XSW experiments so far has been very limited because of the needed X-ray flux. As we could see from the previous section, for the XSW measurements, a highly monochromatic and collimated beam is required. The price one has to pay for the beam conditioning is intensity. The situation may change in a near future. New high-brilliance SR sources may have enough intensity in a well-conditioned incident beam to use crystal spectrometers for XSW measurements, especially in the experiments in which high energy resolution is required, e.g., to resolve close in energy characteristic fluorescence lines. The maximum count rate of a crystal spectrometer is determined by the detector, which is of the order of few times 10^5 and higher for a scintillation detector.

The second major approach is based on energy dispersive detectors. They became available in 1960s with the progress in semiconductor technology. We will discuss them more in detail in the next section. We want to mention here that one of the advantages of the energy dispersive detectors is the possibility to measure the whole energy spectrum simultaneously while by using the wavelength-dispersive spectrometers only a small energy range, to which the analyzer crystal is tuned, can be recorded.

13.5.1.2. *Semiconductor detector*

The semiconductor diode structure is the basis of common energy dispersive detectors. High quality silicon, germanium and some other semiconductor materials (CdTe, HgI_2, GaAs, and others) are used. Two *p*- or *n*-type contacts on opposite surfaces are reverse-biased thus creating a depletion region in the bulk of the crystal. In order to apply high bias voltage, the crystal has to have a very low concentration of free carries. This requires either a high purity material, as in the case of germanium (HPGe detectors), or charge compensation of *p*-type silicon with lithium donor atoms in the case of Si(Li) detectors. To reduce thermal generation of carriers, the crystal is cooled by liquid nitrogen. Recently, thermoelectrically cooled (Peltier effect) high count rate silicon drift detectors became available.[42] These detectors require only electric power to provide cooling and they are more compact than traditional detectors with liquid nitrogen cryostat.

The incident X-ray photons interacts with the crystal material by photoelectric absorption creating an energetic photoelectron and an inner-shell vacancy. The photoelectron interacts with the atoms through a series of low-energy ionization events until it is slowed down to drift velocity. The inner-shell vacancy relaxes via a cascade of Auger electrons or low-energy X-rays. The result of all these process is the production of a large number of free electrons and holes that form a cloud inside the depletion region. They are separated by the high electric field of the bias voltage and collected at the opposite contacts, one of which is connected to the gate of the field effect transistor (FET). The result is a voltage pulse with an amplitude proportional to the energy of the incident photon.

The average energy required for an X-ray photon to create one electron-hole pair ε is 3.86 eV for Si and 2.96 eV for Ge. It is easy to calculate that an X-ray photon with an energy of 3.691 keV (Ca-K_α fluorescence) will generate 956 pairs in silicon which, assuming 0.1 pF feedback capacitance in the first stage amplification, will produce a voltage pulse of 1.53 mV.[43] To detect this pulse, an elaborate signal processing system with a very low noise level is required. The pulse height analysis (PHA) processing electronics consists of a preamplifier (the FET is a part of it), a main spectroscopic amplifier with several gain stages and a pulse shaper, an analog-to-digital converter, and a multichannel analyzer. Since the incident photons arrive randomly, there is always a chance of a pileup: if two events occur within a time shorter than the shaping time of the amplifier two pulses overlap producing an erroneous energy signal. The pileup pulse can be rejected by a special rejection electronics by gating the output pulses based on the inspection of the fast discriminator output. This rejection leads to a "dead" time, which is a percentage of time when electronics is unable to process any new incoming signal. Obviously, the shorter the shaping time the lower the dead time and the higher the count rate. Unfortunately, decreasing the shaping time also leads to a decrease in energy resolution. Thus, the choice of the shaping time is always a compromise between the energy resolution and the count rate. Accurate dead time correction is extremely important. Usually it is performed by passing pulses from a random pulse generator at a desired (average) frequency through the signal processing electronics. The pulser peak appears on the energy spectra as an additional "energy" peak and allows to determine the dead time and thus to correct all other peaks in the spectra. Modern spectroscopic units may not have an input for an external pulse. They internally determine the dead time and generate a corresponding signal which can be read out by the controller.

Energy resolution, the ability to resolve individual characteristic lines, is one of the most important characteristics of the detector. The FWHM of an X-ray line is the sum, squared, of the contributions from the detection process ΔE_{det} and the processing electronics ΔE_{elect} as $\Delta E_{\text{total}} = \sqrt{\Delta E_{\text{det}}^2 + \Delta E_{\text{elect}}^2}$. The ΔE_{det} component is determined by the statistics of the free-charge generation and can be expressed as $\Delta E_{\text{det}} = 2.35\sqrt{F\varepsilon E}$, where E is the energy of the X-ray quanta, ε is the energy required to produce one electron-hole pair, and F is a Fano factor which describes the departure from the Poisson statistics due to the fact that the individual events are not fully independent, $F = 0.12$ for Si and 0.08 for Ge. The ΔE_{elect} depends on the shaping time and usually dominates at high count rates.[43]

Detector efficiency is another important parameter. It is determined by many factors such as: the solid angle, which is defined by the size of the crystal (typically, 10–80 mm^2) and the sample to detector distance; the absorption coefficient of the detector material; the thickness of the beryllium window (typically, 25 μm, thinner for the detectors designed for low-energy applications). Usually, the low-energy cut-off is determined by the thickness of the entrance window and the high energy limit is due to the decrease in photoelectric absorption. Thus, Ge detectors are much more efficient than Si detectors at the energies of $\geq$20 keV.

There are several artifacts that can be present in the energy spectrum[43]:

(1) *Escape peaks.* If the energy of the incident photon is higher than an absorption edge of the detector material, the absorption process involves generation of fluorescence inside the crystal. Most of the fluorescence photons will be absorbed in the volume of the crystal. There is, however, a small probability that some of them will escape the crystal, mostly through the front surface, and will not contribute to the registered charge. This energy loss will result in a peak in the spectrum with the energy reduced

by the energy of the characteristic fluorescence line. For Si detectors this reduction is the energy of the Si K_α line, 1.740 keV (the probability of the generation and escaping of the Si K_β line is very low). For Ge detectors, these are Ge K_α, Ge $K_{\beta 1}$, Ge $K_{\beta 2}$ lines with the energies of 9.876, 10.981, and 11.10 keV, respectively. The intensity of the escape peak depends on the proximity to the absorption edge. Thus, for a Si detector it is about 3% for a parent peak of 2–3 keV and about 0.1% for a parent peak of 10 keV. The presence of the escape peaks is a much more serious problem for Ge detectors due to the much higher escape probability because of the much higher absorption length of the Ge fluorescence.

(2) *Sum peaks* are a result of the pileup when two absorption events happen so close in time that the rejection electronics is unable to recognize them as separate events. It will show up in the spectrum as a peak with the energy of the sum of the energies of the separate photons.

(3) *Diffraction peaks*, which are the peaks originating from the photons diffracted from the sample, are common when a broad energy bandpass incident radiation is used. They are usually not observed in the XSW experiment utilizing highly monochromatic and collimated incident beam.

(4) *Contamination peaks* are all spurious peaks originating from the experimental environment but not from the sample. This include notorious lines from Fe, Cr, Cu, and other metals from the slits, parts of the spectrometer, chamber walls, etc. The detector has to be very carefully shielded to suppress or reduce these peaks. The Ar K_α peak is usually present in the spectrum when the XSW experiments is performed in the open air. To suppress it, the sample should be placed in an evacuated or helium-filled container. Parts of the detector may also be the sources of the contamination such as the trace elements in Be windows.

The example of an experimental spectrum measured in an XSW experiment is shown in Fig. 13.14. The spectrum was recorded with the RÖNTEC silicon drift XFlash detector from a Zn-doped InP film epitaxially grown on a InP(001) substrate by using 10.0 keV incident photons.[44] The signal from the pulse generator (peak #1) is placed in the region of the first channels where there are no peaks from the sample. The In L peaks (#5) and P K_α peak (#3) from the substrate and the elastic peak (#8) dominate the spectrum. Two escape peaks can be seen: #2, of about 2% from the parent In L peaks, and #6, of 0.2% from the parent elastic peak. The contamination Ar K_α peak (#4) is seen on the low energy tail from the In L peaks. The peak of interest, Zn K_α (#7) is sitting on a strong background created by Compton scattering from the substrate.

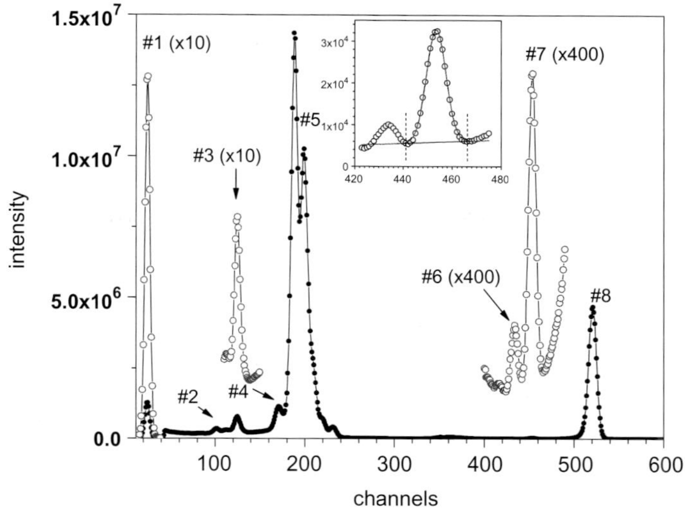

Fig. 13.14. The spectrum from a Zn-doped InP film epitaxially grown on a InP(001) substrate excited by a 10.0 keV synchrotron-X-ray beam. The peak #1 is coming from the random pulse generator used for dead time correction. The peaks from the sample are #3, P K_α, #5, In L lines, #7, fluorescence from the Zn dopant. Artifacts are #2, escape peaks from the In L lines, and #6, escape signal from the elastic peak #8. Peak #4 is the Ar K_α peak from the air. The fit of the Zn K_α peak (#7) by a Gaussian function (line) and the linear background (straight line) is shown in the inset.

Optimization of the fluorescence spectrum is usually performed by optimizing the detection geometry with the goal to maximize the signal of interest and minimize the strongest unwanted peaks. The later ones are usually the elastic peak, the Compton scattering, and the fluorescence from the substrate, they typically dominate the spectrum and contribute the most to the dead time. The effective way to minimize their contribution is to register radiation at the low exit angle to the surface of the sample, thus by minimizing the scattering volume. In experiments with synchrotron radiation, the contribution from the elastic and the Compton scattering can be significantly reduced by registering radiation in the horizontal plane in the direction along the polarization vector, usually the horizontal plane.

13.5.2. *Detection of electrons*

13.5.2.1. *Introduction*

Electrons scatter much stronger than X-rays. At lower electron energies (<1 keV) elastic scattering dominates and at higher kinetic energies the

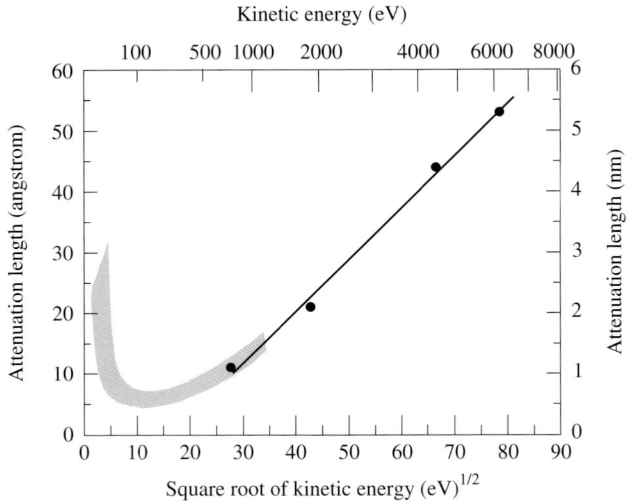

Fig. 13.15. The effective attenuation length (EAL) of electrons. Shown is the "universal" curve (broad band) and the experimentally determined EAL for electrons in GaAs.[46] Since electrons in the eV to keV range are predominantly scattered by valence and conduction electrons, the EAL does not show a very strong dependence on the atomic number.

inelastic scattering is more pronounced. The electrons are losing energy on their path by a multitude of electronic and phononic excitations and this has two consequences: (a) The escape depth of electrons is small. It is expressed by an attenuation lengths l marking the distance at which the electron signal is attenuated by $1/e$. The attenuation length is only weakly energy dependent as shown in Fig. 13.15. (b) Specific electron emission lines are frequently riding on a very high background, which is drastically enhancing the needed data acquisition time and rendering the extraction of net peak integrals more difficult. A typical electron energy spectrum recorded with a concentric hemisphere analyzer (see below) is shown in Fig. 13.16.

Energy discriminating detectors for charged particles are much easier to construct than such devices for X-rays and can nowadays reach an energy resolution down to $\Delta E/E = 10^{-6}$. The simplest electron detectors are Geiger–Müller type gas proportional counters, channeltrons, and channel plates, which offer only very modest energy resolution. High and very high energy resolution is achieved by electrostatic analyzers such as the cylindrical mirror analyzer (CMA) and the concentric hemispherical

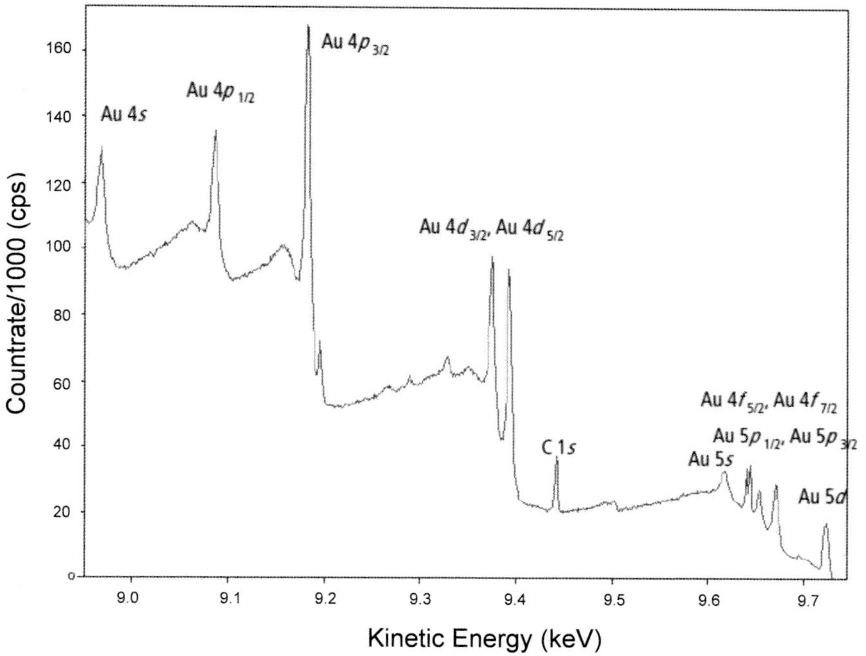

Fig. 13.16. Electron energy spectrum from a gold sample excited by 9.8 keV X-rays. Besides various Au emission lines, the C 1s line from carbon impurities can be distinguished. Other lines, not labeled in the spectrum, result mostly from various Au Auger emission processes. Visible in the spectrum with the highest kinetic energy is the emission from the Au 5d valence electrons.

analyzer (CHA). Less frequently used (at synchrotron sources) are energy-resolving electron analyzers using retarding fields or employing the timing information from the short synchrotron light pulse, so-called time of flight (TOF) spectrometers.

13.5.2.2. *Electron multipliers*

Channeltrons/spiraltrons and channelplates are amplifying electron detectors, i.e., electron multipliers and must be operated in vacuum. Typical electron multipliers consists of a discrete number of dynodes, connected by a cascade of resistors, which is dividing the applied high voltage (order of 1 kV). In case of channeltrons/spiraltrons and channelplates, the multiplier is simply a glass tube, or a large number of them in case of channelplates, internally covered with a resistive coating which also functions as electron emitter. Over the length of the resistive coating, a

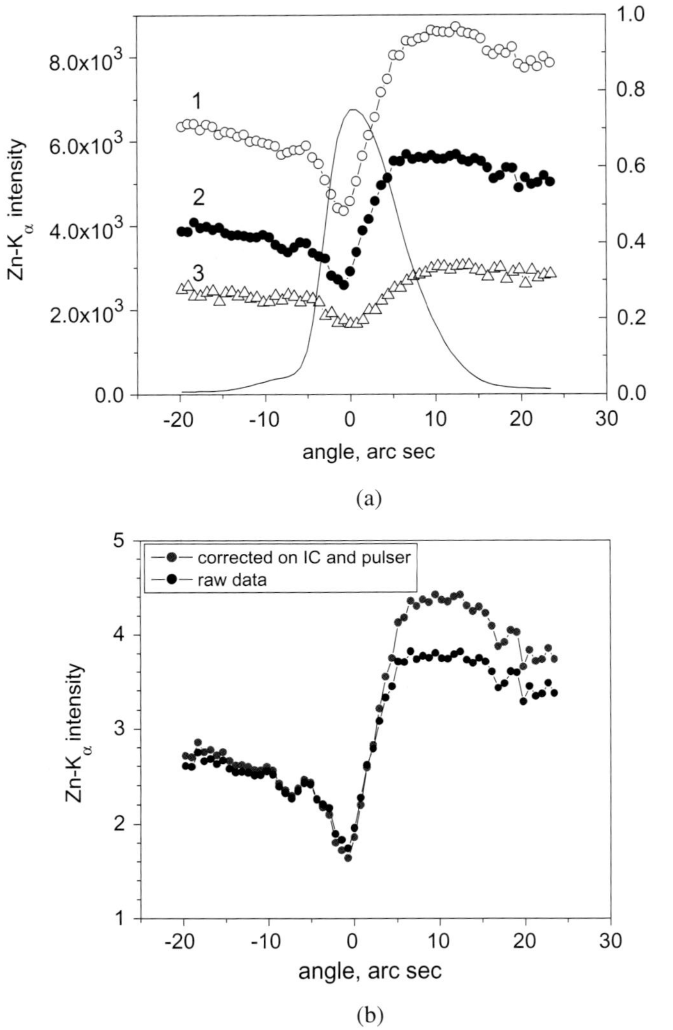

Fig. 13.21. Treatment of XSW yield raw-data. See text for details.

is the angular dependence of the background. Next, the experimental curve (number 2 and lower curve in Fig. 13.21(b)) has to be corrected for dead time by dividing by the pulser counts and by dividing by the intensity monitor signal to account for variations in the incident beam intensity during the measurement. The corrected curve is the upper curve in Fig. 13.21(b). This curve is now ready for fitting by using the XSW formalism as given in Chapter 1 (Secs. 1.8 and 1.9). The as determined coherent fraction f^H and coherent position P^H of the Zn contains the information about the position of Zn atoms in the InP unit cell of the epitaxial layer.

In the example presented in this section, we considered a very common case of recording the fluorescence yield from a small amount of atoms in the Bragg case of diffraction. Other types of corrections may be needed depending on experimental specifics. For example, the change of the beam footprint on the sample surface has to be taken into account in the experiments performed at low incident angles, e.g. TR-XSW (Chapter 5), or in scanning a wide angular range, e.g. when measuring small yield variations from a very thin film (Chapter 16).

13.7. The Beamline ID32 at the ESRF: A Dedicated XSW Station

Only very few beamlines worldwide at synchrotron radiation laboratories were dedicated or even only partially dedicated to XSW measurements. The by-far most productive and fully dedicated of these had been the ID32 insertion device beamline at the ESRF in Grenoble, France, which we will describe in the following in some detail. It serves as an example of a station optimized for XSW measurements, which requires tailored X-ray optics and instrumentation. The ID32 beamline[48] had been in operation since 1995.[49] Since December 2011, the beamline is closed for users and at the time this book goes to print, it will be decommissioned.

The ID32 XSW beamline used a high-β section of the ESRF storage ring. Here, the horizontal angular emittance of the electron beam in the synchrotron is small, at the expense of a larger source (e-beam) size, 30 μrad and 0.9 mm, respectively, in case of ID32. As radiation source, the XSW beamline ID32 used three undulators installed in a 5-m-long straight section, which provided a horizontally, linearly polarized beam with a maximum photon flux of $\approx 1 \times 10^{13}\,\mathrm{ph \cdot s^{-1} \cdot mm^{-1}}$ (at 0.001% bandwidth, 30 m source distance, and 200 mA current in the storage ring). Two of the undulators (U35), with 35 mm period, were used for energies roughly between 2.7 keV and 30 keV. The third undulator (U42) with a period of 42 mm extended the energy range down to 1.4 keV at the minimum gap opening of 11 mm. The beamline was designed for operation without any windows between source and sample. Thus, the X-ray beam was transported in high vacuum (HV) from the ultra HV (UHV) in the storage ring to the

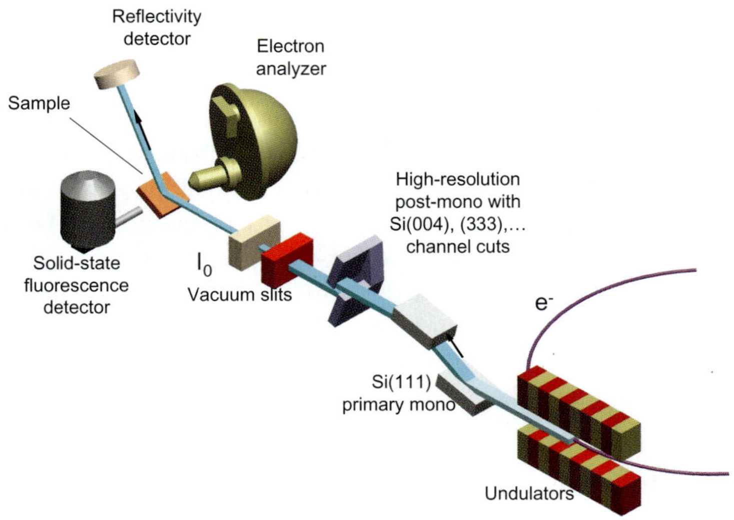

Fig. 13.22. Schematic experimental layout of the ID32 beamline.

sample in the UHV chamber in the second experimental hutch (see further below). The principle layout of the beamline is shown in Fig. 13.22.

Optics hutch 1. In the primary optics hutch (OH1), the first optical element behind a water cooled front end slit system is the double-crystal, "pseudo-channel-cut" monochromator, installed at a distance of about 30 m from the center of the straight section. Two sets of crystals, a set of Si(111) crystals and a set of multilayer, are mounted side by side (see Fig. 13.23) on a common, high precision, stepper motor-driven rotation axis coupled to an angular encoder with 50 nrad resolution. The primary crystals are placed between liquid nitrogen cooled heat exchangers. This allows them to withstand a maximum power of about 100 W without noticeable distortion by the heat load. At 200 mA, e-beam current in the storage ring and the gap of two undulators closed, this power level is reached at a beam size of about 0.5 mm^2. The secondary crystals, also cooled, are mounted on a slide allowing to adjust the distance D to the first crystals. Using the pair of Si(111) crystals, the monochromator is typically operated with the exit beam kept at a fixed height $H = 16$ mm with respect to the incident beam. To achieve this, the secondary crystal is tracking as a function of Bragg angle θ via $H = D\cos(2\theta)$. The angular orientation of the secondary crystals with respect to the primary ones can be adjusted in two Eulerian angles ("pitch" and "roll") over a range of 1° with

Fig. 13.23. Photograph of the interior of the primary monochromator of ID32 showing the two sets of crystals, each 25 mm wide, sandwiched between gold-plated heat exchangers. The crystals hit by the undulator beam are on the bottom and the secondary crystals are above. The Si(001) crystals covered with multilayer are on the left and the primary crystal is 150 mm long because of the grazing angle of incidence.

sub-microrad resolution using stepper motors and over a range of about 0.3 mrad with 10 nrad resolution by piezo crystals. Using the set of Si(111) monochromator crystals, the energies attainable at fixed exit beam position $H = 16\,\mathrm{mm}$ range from 2.05 keV to 28 keV determined by the chosen design parameters of the monochromator. The energy resolution $\Delta E/E$ of the main monochromator utilizing the Si(111) crystals is 1.3×10^{-4} up to about 13 keV at which point the resolution degrades since the source divergence becomes larger than the Si(111) rocking curve width.

The second set of monochromator crystals is a pair of Si(001) substrates covered with two stripes of multilayer with 3 nm and 6 nm d-spacing. This allows to reach energies down to 1.4 keV, the lowest energy of the U42 undulator at the closed gap. Using the multilayer, a very high flux density is transmitted, however, at a very moderate resolution of about 1%. Because of the large d-spacing, the angle of incidence on the multilayer is small. This requires reducing the beam "offset" H. Using $H = 3$ mm and using the multilayer with the smaller d-spacing (3 nm), a maximum energy of 20 keV can be reached.

Two meters downstream of the main monochromator, a set of two-dimensional parabolic beryllium compound refractive lenses (CRL)[50] is installed in a UHV chamber. The lenses are contained in packets (presently, one, two, four, and eight) which can be moved in and out of the beam. As in-line optical element (the beam goes straight) CRLs are very convenient, but they are dispersive, i.e., their focal length is energy dependent. The lenses placed at a distance of 32 m from the source and ≈ 16 m to the sample in the first experimental hutch. Thus, a de-magnifying ratio of the electron beam source of 2:1 is achieved. A total of 34 lenses with two different radii can be inserted using remote control into the beam. By combining 3 lenses of larger radius with 31 lenses of smaller radius, an optimal focus is obtained for 127 discrete energies in the range ≈ 2 keV to 20 keV. With the given electron beam source dimensions in the high-β section, the beam is focused to a size of a few ten micrometers vertically by a few hundred micrometers horizontally.

Optics hutch 2. The second optics hutch (OH2) is housing equipment for further conditioning the monochromatic beam. For high-resolution XSW experiments (and for high-resolution photoelectron spectroscopy experiments) two high-precision post-monochromator (PM) axes are mounted in two UHV chambers (*cf.* Fig. 13.24). Both rotary axes are driven by a stepper motor with sub-microrad resolution. Each axis is equipped with several channel-cut (CC) crystals (two are shown in Fig. 13.25), which can be selected and aligned with high precision in the beam. The CC crystals are mounted on a torsion bearing driven by a piezo crystal and the rotation is controlled by a Heidenhain RPN886 angular encoder

Fig. 13.24. Left: The two UHV chambers containing the two axes of the ID32 post-monochromator. Right: View of the interior of one of the chambers with the high resolution rotary axis, mounted on a kinematic support for y and z-adjustment. By horizontal translation along y, i.e., perpendicular to the beam direction, different channel-cut crystals can be selected.

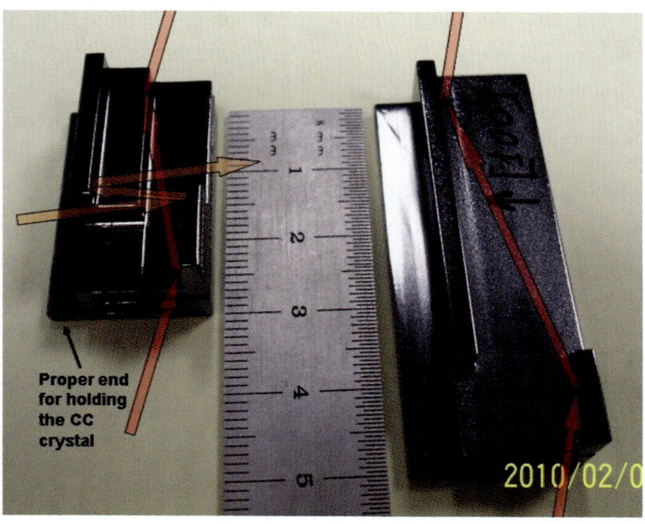

Fig. 13.25. Two channel-cut crystals. The channel-cut crystal on the left is designed to be used with two distances of the "Bragg-mirrors". At small Bragg angles, the CC is used with 5 mm distance between the mirrors, thus keeping the resultant offset of the exit beam smaller than 10 mm. A large Bragg angles close to 90° is used with 10 mm distance between the mirrors providing a large enough beam offset ≈1.5 mm so that the beam reflected from the second mirror can still pass the first one. The corresponding beam path is indicated. The mounting end of the crystal is separated from the "mirrors" by strain-relief-cuts to avoid any strain (introduced by glueing or clamping) propagating to the "mirrors".

with 50 nrad resolution. It is another advantage of the set-up with two axes that the second axis can correct for the beam-offset introduced by the first axis if the CC on the second axis is mounted in a dispersive arrangement (+/−, −/+) (provided that the used CCs and diffraction planes are identical for both axes). Using both axes in this dispersive arrangement, high-energy resolution can be achieved at variable energy, but this may be at the expense of a lower photon flux. To allow maximum intensity throughput, the source divergence needs to be smaller than the angular acceptance of the chosen reflection. When the output beam position is not an issue and the excitation energy is somewhat flexible, a single axis of the post-monochromator and just one CC crystal can be used. However, to achieve the desired energy resolution, the angular acceptance of the used PM CC should not exceed the photon beam divergence from the undulators. With this proper choice of crystals (frequently used are Si(111), Si(220) and Si(311) crystals), the incident photon energy bandpass can be varied largely. For instance at 9.9 keV, it can be varied from 1.3 eV down to 15 meV by using either the Si(111) main monochromator or using e.g. a Si(555) CC in backscattering geometry. A beam intensity monitor mounted on a linear translation is used for the alignment of the CC of the first PM. By the same linear drive, a beam-stop can be moved into place that absorbs the higher harmonics that pass through the first CC. For a high-resolution XSW measurement, the energy band-pass and the angular divergence of the beam must be smaller than the corresponding values for the used single crystal reflection, which may be smaller than 50 meV and 10 μrad, respectively. Achieving such low values for ΔE and $\Delta\theta$ is practically only possible by using two CCs in dispersive arrangement (see discussion above using DuMond diagrams).

A 450-mm-long flat mirror with three different stripes (SiO_2, Ni, Pd) located behind the PM can be inserted into the beam path, acting as a low-pass filter, suppressing higher harmonics. It is mounted on a hexapod and is usually operated with the monochromatic beam incident at a fixed angle of 0.1°. If such a mirror is used when performing XSW experiments it has to be assured that slope errors are much smaller than the beam divergence desired for the measurement.

Experimental hutch 1. The first experimental hutch (EH1) houses a diffractometer which can be used in two different modes. As a six-circle Kappa instrument, it is typically employed for XSW experiments (*cf.* Fig. 13.26) using single crystals and multiple diffraction planes, e.g.,

Fig. 13.26. Photograph of the six-circle kappa diffractometer in EH1 of the ID-32 beamline with a small UHV chamber which containins the sample. The chamber is portable and the sample typically has been prepared in a laboratory not located directly at the beamline. The chamber is inclined, since the sample has been aligned for an XSW scan using a diffraction plane, which is not parallel to the sample surface. Visible is also the detector for the beam which is reflected from the sample in the vertical plane as well as the solid state detector for recording the X-ray fluorescence from the sample.

for XSW imaging. As a so-called 2 + 2 + 2 six-circle instrument, two additional circles with an xyz-stage are mounted (not shown) and it is used typically for surface X-ray diffraction (SXRD) experiments or for XSW measurements employing standing waves created by total reflection from a sample surface or by a multilayer. The diffractometer is mounted on a kinematical support allowing horizontal and vertical alignment and tilting of the whole diffractometer by $\pm 1°$ (e.g., in order to adjust the angle of incidence of the beam on the sample for grazing incidence SXRD experiments). When experiments are carried out in EH1, a UHV transfer tube connecting the vacuum line at the entrance with the vacuum line at the exit of the hutch is removed. The beam position is kept constant on the sample on the diffractometer with an accuracy of about 1 μm with the help of gas filled beam position monitors (I_0 in Fig. 13.22), the signal of which is fed back to the piezo crystals adjusting the pitch (causing vertical beam movement) and roll (causing horizontal beam movement) of the second

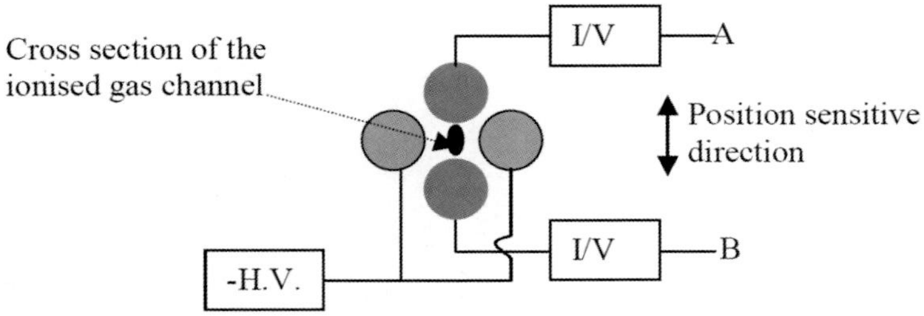

Fig. 13.27. Schematic of the position sensitive quadrupole ionization chamber. Circles representing the cross-section of the cylindrical electrodes. I/V denotes the current-to-voltage converter (picoampermeter). A and B signal values are proportional to the charge fraction measured on the collecting electrodes. The biasing electrodes are placed on a common negative potential. The X-ray beam is perpendicular to the paper surface.

crystal of the monochromator in OH1. These monitors are based on the ionization chamber principle (see Fig. 13.27[51]). For the given distance of the beam position monitor to the monochromator of 16 m, the position stability corresponds to an angular stability of about 50 nrad, which is much smaller than the Si(111) reflection curve width at any energy used at the ID32 beamline ($\approx$10 μrad at 28 keV). Thus, the position feedback mechanism replaces previously used intensity feedback solutions.[37]

For XSW experiments in EH1, fluorescence detection is used as shown in Fig. 13.26. The six-circle Kappa diffractometer allows to align the sample such that any diffraction plane can be used for XSW experiments. Employing diffraction planes which are inclined with the sample surface, it is usually desired to align the sample such that the incident and the exit angles with the surface are equal (symmetrical reflection) which simplifies data analysis. The orientation of the sample in this case is schematically shown in Fig. 13.28. The diffractometer can carry heavy loads up to 12 kg on the kappa stage (and ten times more in the $2 + 2 + 2$ configuration). This allows XSW measurements to be performed with the sample in different environments, i.e., in a UHV chamber, in a cryostat, in a high-temperature furnace, in an electrochemical cell, etc.

Experimental hutch 2: UHV experiments. XSW measurements in UHV and using photoelectron detection are performed (at 55 m distance from the source) in experimental hutch 2 (EH2) with the UHV flight tube in EH1 in place, which thus connects the UHV system(s) installed in EH2 directly to the storage ring vacuum. One UHV chamber had been equipped with a Perkin-Elmer PHI model 10-360 hemispherical analyzer fitted with a multichannel plate detector with 16 channels/electrodes. The instrument

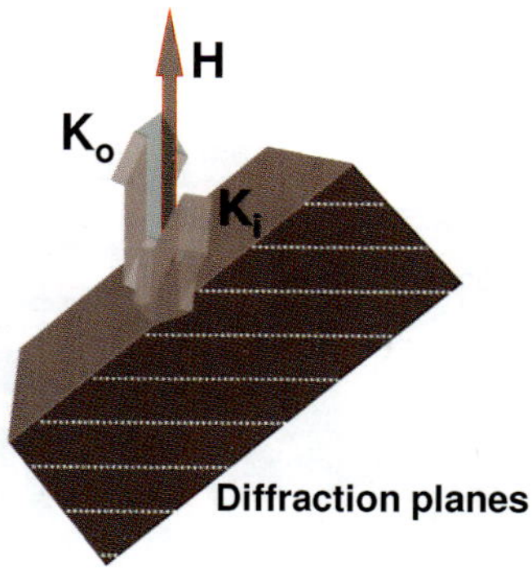

Fig. 13.28. Sketch of the geometry in case a diffraction plane of the sample is used which is inclined with the sample surface. The orientation of the sample is chosen such that (1) the polarization vector of the X-ray beam lies within the chosen diffraction plane and (2) the angle of incidence of the X-ray beam with the surface plane equals its exit angle with the surface plane.

had been in use for more than 10 years. It allowed XPS and XPS/XSW experiments detecting electrons with kinetic energies of up to 4.8 keV. The UHV system was equipped with a flow-through cryostat coupled with a Cu braid to the sample holder allowing to reach temperatures down to 30 K. Standard sample holders allow heating up to about 1000°C. The manipulator (VG OMNIAX) allowed x, y, and z translations, rotation around the vertical manipulator axis and rotation around one orthogonal axis with moderate accuracy and stability. These mechanical deficiencies can be tolerated in case of XSW measurements in back-scattering geometry (NIXSW measurements, see Chapter 4). A sputter gun, an electron gun, reverse reading low energy electron diffraction (LEED) system, evaporators, and residual gas analyzer are available for sample preparation and characterization. This system was was decommissioned in 2010 after being in use for NIXSW experiments for more than 10 years.

The second UHV chamber is made of mu-metal, a magnetically soft material (typically a Ni, Fe, Cu, Mo alloy) with very high permeability to effectively shield the interior of the chamber from external magnetic fields. This is necessary for (very) high-resolution XPS experiments in order to not distort the trajectories of the photoelectrons. The chamber, shown schematically in Fig. 13.29, carries a PHOIBOS 225 HV photoelectron analyzer from the company SPECS, which is able to detect photoelectrons with kinetic energies of up to 15 keV with resolution ΔE better than 50 meV. The analyzer at ID32 is using a 2D delay line detector to record

42. P. Lechner, *et al.*, *Nucl. Instrum. Meth. A* **377** (1966) 346; G. Bertuccio, *et al.*, *Nucl. Instrum. Meth. A* **377** (1966) 352;
43. A. T. Ellis. In *Handbook of X-Ray Spectroscopy*, R. E. Van Grieken and A. A. Markowicz (eds.) (Marcel Dekker, New York, Basel, 2002), pp. 199–238.
44. A. A. Sirenko, A. Augazzaden, A. Kazimirov, *Comp. Mater. Sci.* **33** (2005) 132.
45. M. J. Bedzyk, G. Materlik, M. V. Kovalchuk, *Phys. Rev. B* **30** (1984) 4881.
46. C. Dallera, G. Panaccione, G. Paolicelli, B. Cowie, J. Zegenhagen and L. Braicovich, *Appl. Phys. Lett.* **85** (2004) 4532.
47. Available at: http://www.certif.com/.
48. J. Zegenhagen, B. Detlefs, T.-L. Lee, S. Thiess, H. Isern, L. Petit, L. Andrė, J. Roy, Y. Mi and I. Joumard, *J. Electron Spectrosc.* **178–179** (2010) 258.
49. F. Comin, R. De Martino, K. Fezzaa, V. Formoso, L. Ortega and R. Signorato, *J. Physi. IV* **7** (1997) C2-343.
50. A. Snigirev, V. Kohn, I. Snigireva and B. Lengeler, *Nature* **384** (1996) 49; A. Snigirev, V. Kohn, I. Snigireva, A. Souvorov and B. Lengeler, *Appl. Opt.* **37** (1998) 653.
51. M. Kocsis, private communication.
52. J. A. van Bokhoven, T.-L. Lee, M. Drakopoulos, C. Lamberti, S. Thiess and J. Zegenhagen, *Nat. Mater.* **7** (2008) 551.
53. M. Drakopoulos, J. Zegenhagen, A. Snigirev, I. Snigireva, M. Hauser, K. Eberl, V. V. Aristov, L. Shabelnikov and V. Yunkin, *Appl. Phys. Lett.* **81** (2002) 2279.
54. A. Kazimirov, D. H. Bilderback, R. Huang, A. Sirenko and A. Ougazzaden, *J. Phys. D: Appl. Phys.* **37** (2004) L9.

Part II

INTRODUCTION TO PART 2

ALEXANDER KAZIMIROV and JÖRG ZEGENHAGEN

Chapters 14 to 27 present the reader with a broad range of application highlights. We invited contributions which we believe represent particularly attractive scientific cases and are best suited illustrating the strength of the XSW technique. We hope that you will find these examples instructive, stimulating, and entertaining.

Chapter 14, written by Michael J. Bedzyk and Paul Fenter dwells on an issue which was already touched upon in Chapter 1. Rather straight forward, the XSW technique allows the element specific imaging of atomic distributions and the authors present the principles and some illustrative examples.

In Chapter 15 Terrence Jach elaborates the application of the XSW technique for the analysis of quasicrystals whose crystallographic aspects are also reviewed. Since translational symmetry is missing in quasicrystals, it is *a priori* not obvious that the XSW technique can provide any useful information.

The next two chapters deal with the important class of epitaxial systems. The XSW technique can be used to analyze the structure of the epitaxial overlayer either by creating the XSW by the thin film (Chapter 16) or by the substrate crystal (Chapter 17) as described by Alexander Kazimirov *et al.*

Chapter 18, written by S.I. Zheludeva *et al.* reports about investigating ion species in organic macromolecules and lipid-protein systems on surfaces. Element specific structural information on a length scale of several nm is obtained by utilizing the total reflection XSW method.

Since the XSW technique is absolute element specific and applicable to very dilute systems, it has also found applications in interfacial geochemistry as described by Paul Fenter in Chapter 19.

A smart researcher will not attempt using one technique alone for characterizing and understanding a complex surface structure. In Chapter 20 M. Siebert *et al.* describe how the XSW technique can efficiently be combined with DFT calculations.

Strange, complex structures as result of two-dimensional alloying are created on the (111) surfaces of silicon and germanium by some metal adsorbates. How the XSW technique helps to understand these systems is presented in Chapter 21 by one of the editors.

Jan-Ingo Flege *et al.* describe the virtues and pitfalls of using the XSW technique in combination with X-ray induced ion and/or atom desorption in Chapter 22. This is particularly interesting for obtaining some information about the surface adsorption of very light elements such as hydrogen.

The next two chapters deal with X-ray waveguides, which were treated in Chapter 9. In Chapter 23, Ajay Gupta describes how the enhancement of the XSW inside an X-ray waveguide can be utilized to analyze the elemental distribution of its interior and to study diffusion processes on the nano-scale. In Chapter 24, Liberato de Caro *et al.*, employ the small intense beam exiting the waveguide for coherent diffraction imaging on the nano-scale.

Combining photoelectron spectroscopy (PES) with the XSW technique, structural information about different chemical species of the same element can be obtained, as described in Chapter 25 by D. Phil Woodruff.

Even valence electronic states can be associated with atomic positions, as it was described in Chapter 12 by Joseph Woicik. In Chapter 26 he describes the application of this "site specific valence band analysis" to transition metal oxides such as the highly correlated V_2O_3.

Finally, in Chapter 27, S.-H. Yang *et al.* combine chemical state sensitive PES, valence band PES and X-ray emission spectroscopy to obtain chemical and electronic information about interfaces utilizing XSW produced by Bragg reflection from a multilayer.

The appendix contains reminiscences of some of the prominent practioners of the XSW technique about the early developments of the XSW technique in three continents. It all started at AT&T Bell Laboratories in Murray Hill, New York. Boris W. Batterman, Jene A. Golovchenko, and Walter M. Gibson report about the early ongoings in the United States, but also about what happened in Aarhus, Denmak (appendix A1, A2, and A3). S. Kikuta describes how the Japanese community became involved shortly after the first ground breaking experiments of Batterman became public (A4). Finally Gerhard Materlik reports about his early interest in the XSW technique and developments at the Hamburg Synchrotron Radiation Laboratory (HASYLAB) in Germany (A5).

Chapter 14

XSW IMAGING

MICHAEL J. BEDZYK

*Departments of Materials Science &
Engineering and Physics & Astronomy,
Northwestern University, 2220 Campus Dr.,
Evanston, IL 60208, USA
Argonne National Laboratory,
Argonne, IL 60439, USA*

PAUL FENTER

*Chemical Sciences and Engineering Division,
Argonne National Laboratory, 9700 S.
Cass Avenue, Argonne, IL 60439, USA*

The summation of XSW measured *hkl* Fourier components is used to generate a model-independent real-space map of an XRF-selected atomic distribution. As a demonstration, this Fourier inversion method is used to generate a set of *1D* maps for a set of naturally occurring impurity ions within a mica crystal and a *3D* map for adsorbed Sn atoms on a Ge(*111*) single crystal surface.

14.1. Introduction

In the most general sense, XSW probes the partial element-specific geometrical structure factor, $\mathcal{F}_{\boldsymbol{H}}$ (in both amplitude and phase), in or above a crystal. As already explained in Chapter 1, $f_{\boldsymbol{H}}$ is the amplitude (also known as the coherent fraction) and $P_{\boldsymbol{H}}$ is the phase (also known as the coherent position) of the $\mathbf{H}^{th}$-order Fourier coefficient of the normalized distribution function[1,2]:

$$\mathcal{F}_{\boldsymbol{H}} = \int_{uc} \rho(\mathbf{r}) \exp(2\pi i \mathbf{H} \cdot \mathbf{r}) d\mathbf{r} = f_{\boldsymbol{H}} \exp(2\pi i P_{\boldsymbol{H}}). \tag{14.1}$$

The phase sensitivity of XSW derives from the fact that the detector of the E-field is the fluorescent atom itself, lying within the "near-field" where the X-ray fields coherently interfere. (This is in contrast to conventional diffraction measurements; where the structure factor phase is lost because the scattered intensity is detected in the far-field.)

XSW can therefore be used directly to reconstruct the direct-space structure from a set of Fourier coefficients collected in reciprocal space. Based on the Fourier inversion of Eq. (14.1), the full time-averaged direct-space distribution of the atomic-centers, $\rho(\mathbf{r})$, of each fluorescent atomic species can be synthesized by the discrete Fourier summation[3,4]:

$$\begin{aligned}\rho(\mathbf{r}) &= \sum_{\boldsymbol{H}} \rho_{\boldsymbol{H}} = \sum_{\boldsymbol{H}} f_{\boldsymbol{H}} \exp[2\pi i(P_{\boldsymbol{H}} - \boldsymbol{H}\cdot\mathbf{r})] \\ &= \sum_{\boldsymbol{H}\neq-\boldsymbol{H},\boldsymbol{H}\neq 0} f_{\boldsymbol{H}} \cos[2\pi(P_{\boldsymbol{H}} - \boldsymbol{H}\cdot\mathbf{r})]. \end{aligned} \tag{14.2}$$

The above simplification to a summation of cosine terms makes use of $f_O = 1$ and the symmetry relationship analogous to Friedel's law that makes $f_{\boldsymbol{H}} = f_{-\boldsymbol{H}}$ and $P_{\boldsymbol{H}} = -P_{-\boldsymbol{H}}$. This reconstruction of direct-space profiles is referred to as XSW imaging. The images are obtained without any reference to model structures, thereby representing in direct-space the information obtained by measuring the fluorescent yield modulations within the range of Bragg reflection.

Since the XSW phase is directly linked to the phase of the substrate electronic structure factor,[5] the Bragg XSW positional information is acquired in the same absolute substrate unit cell coordinate system that was chosen when calculating the X-ray structure factors $F_{\mathbf{H}}$.

There are a number of ramifications that derive directly from the fact that the atomic density is determined by discrete Fourier synthesis as described in Eq. (14.2). In particular, the fidelity of the inverted image depends directly on the sampling frequency and the range over which the Fourier components are sampled. In the case of Bragg XSW, the sampling frequency in reciprocal space is limited to the discrete locations corresponding to allowed bulk Bragg reflections. Consequently, the image $\rho(\mathbf{r})$ generated from the XSW measured $f_{\boldsymbol{H}}$ and $P_{\boldsymbol{H}}$ values, is unique only in a volume corresponding to the primitive crystallographic unit cell of the bulk crystal; that is, the derived image is a projection of the full density profile into the unit cell. Consequently, the Bragg-XSW measurement is best-suited to probing structures that are localized over distances that are small with respect to the unit cell size, so that the full distribution can be determined uniquely. The range of reciprocal space over which the set of $\mathcal{F}_{\boldsymbol{H}}$ values is measured determines the spatial resolution of the XSW image. This is evident when considering a one-dimensional reconstruction. The highest spatial frequency of the derived density profile corresponds to a period of $d_{H\text{-}max} = 1/H_{max}$. In particular, the phase sensitivity of XSW allows the two half-periods of the sinusoidal function to be distinguished directly, leading to an expected resolution of $\sim \frac{1}{2} d_{H\text{-}max}$.

This approach is illustrated by simulating the Fourier synthesis of the diamond crystal structure along the *111* direction. As seen by the two vertical lines in Figs. 14.1(a) and 14.1(b), the *1D* projection for this bilayer distribution can be represented by

$$\rho(z) = \frac{1}{2}\left[\delta\left(z + \frac{1}{8}\right) + \delta\left(z - \frac{1}{8}\right)\right], \tag{14.3}$$

where z is the *1D* unit cell fractional coordinate relative to d_{111}. Using Eq. (14.1), the Fourier coefficient for this distribution is:

$$\mathcal{F}_{hhh} = \cos(\pi h/4), \tag{14.4}$$

which is plotted in Fig. 14.1(c). For this case the Fourier coefficient amplitudes are $f_{000} = f_{444} = f_{888} = 1$, $f_{111} = f_{333} = f_{555} = f_{777} = 1/\sqrt{2}$ and $f_{222} = f_{666} = 0$, and the phases are $P_{111} = P_{777} = P_{888} = 0$ and $P_{333} = P_{444} = P_{555} = \frac{1}{2}$. In Fig. 14.1(a) the individual Fourier components ρ_{hhh} are plotted for $h = 0$ to 8 and then summed according to Eq. (14.2) to produce the *1D* image in Fig. 14.1(b). This simple example provides the essential characteristics of the imaging process by discrete Fourier synthesis.

The location of the two halves of the diamond lattice bilayer is clearly resolved in this reconstruction of the density profile, in spite of the fact that maximum of the $\boldsymbol{H}^{th}$ contribution to the density does not in general coincide with the locations of the atomic planes. As expected, the resolution in the reconstructed profile (defined as the FWHM of a discrete feature) in a given $[hkl]$ direction is equivalent to one-half of the smallest d-spacing (corresponding to H_{max}) that is included in the summation. In the example, this results in a spatial resolution of $(d_{111}/8)/2 \sim 0.2$ Å, or $\Delta z = \frac{1}{16}$, as observed in Fig. 14.1(b). Extra oscillations in the reconstructed profile are due to truncating the Fourier sum (effectively assigning zero amplitude to all terms higher than $hhh = 888$) and will be present so long as the amplitude of the missing Fourier components is nonzero (that is, the actual profile is not "resolved" by the data, but instead has a width limited by

impact of the actual sampling of the Fourier components be kept in mind in interpreting the reconstructed images.

14.2. *1D* Profiling of Lattice Impurity Sites

The usefulness of this approach for directly revealing impurity site distributions was first demonstrated through measurements of impurities in muscovite (i.e. mica).[3] Muscovite is a layered silicate with a tetrahedral–octahedral–tetrahedral (TOT) sheet structure (with each sheet containing tetrahedrally coordinated Si and Al in the T layers and octahedrally coordinated Al in the O layer), with monovalent cations in the interlayer between TOT sheets to provide charge-balance (Fig. 14.2). The distribution of impurities (i.e. at *<1%*) cannot be determined directly by X-ray crystallography since they have only a weak affect upon the total electron density. While the local incorporation site might be obtained by EXAFS, the ability of EXAFS to determine the local coordination geometry can be limited when the element is found in multiple sites in a lattice. The use of XSW to directly image the impurity distributions is illustrated in Fig. 14.2.

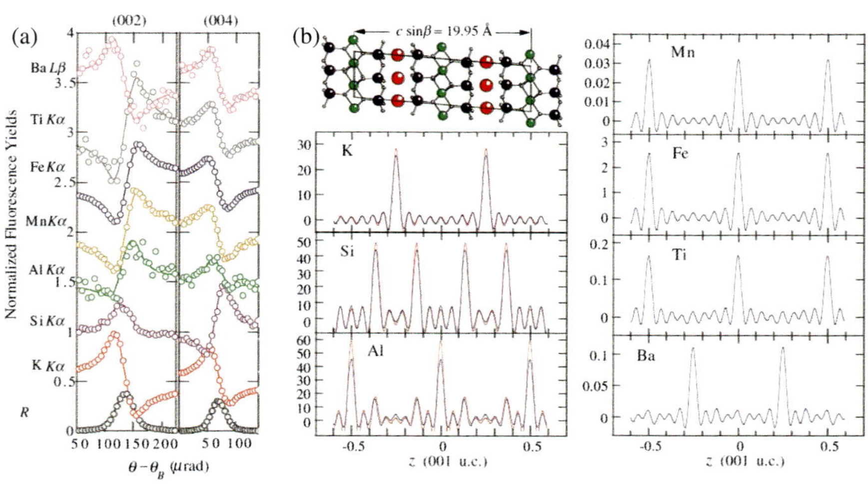

Fig. 14.2. Mica *1D* ionic profiles: (a) Reflectivity and fluorescence modulations for the (*002*) and (*004*) reflections for both majority (Al, Si, K) and minority (Ba, Ti, Fe, Mn) elements in muscovite. (See Ref. 3 for the (*006*) to (*0016*) data and the measured values of f_H and P_H.) (b) The derived *1D* density profiles obtained by discrete Fourier synthesis of the partial element structure factors as described in Eq. (14.2). The mica crystal structure is shown in (b) scaled so that the atom positions correspond directly with calculated profiles.

XSW data for numerous impurities (Mn, Fe, Ti, Ba) are obtained simultaneously through the use of an energy-dispersive XRF detector while exciting the muscovite Bragg reflections. The characteristic XSW modulation of the fluorescent yield is obtained in spite of the fact that muscovite crystals are non-ideal crystals (due to their tendency to bend). By illuminating just a tiny area of the surface with a $\sim 20\text{-}\mu m$-sized X-ray beam with a relatively low photon energy of $7\ keV$, it was however possible to find regions of the crystal where the observed rocking curve width was close to the theoretical value.

The fluorescent modulation for each element was measured for $\mathbf{H} = 00L$ with $L = 2$ through 16 for $L = even$ (due to the absence of odd-order reflections along $00L$). Results for $L = 2$ and 4 are shown in Fig. 14.2(a). These raw XSW data show that the fluorescent yield modulation for Ba follows that of K, while the modulation of Ti, Fe and Mn all follow the behavior of Al. Thus a traditional XSW analysis would conclude that Ba distribution is centered on the interlayer site, while the Ti, Fe and Mn would be centered on the octahedral site (the majority site for Al).

The derived model-independent elemental profiles obtained by Fourier synthesis are shown in Fig. 14.2(b). Only reflections along $\mathbf{H} = 00L$ were measured since the height of a cation is sufficient to distinguish the various lattice sites due to the layered nature of the muscovite structure. From the synthesis of the impurity data in Fig. 14.2, it is immediately obvious that Ba is found exclusively at the interlayer site, while Mn, Fe, and Ti are all found in the octahedral site. This information is model-independent, since only the total electron density of the muscovite crystal (i.e., that determined by crystallography) was used in the dynamical X-ray scattering theory calculations of the rocking curves. Also apparent from these model-independent density distributions is a significant degree of "ringing" due to an abrupt termination of the experimental data. This suggests that the intrinsic distribution width of each element is narrow with respect to the resolution width $d_{001}/16 = 1.2\ Å$. These artifacts could be reduced through the use of "Hanning" windows imposed on the measured coherent fractions, but at some cost of resolution in the images.

Although the distribution of the major elements (Si, Al, K) is well established by crystallography, these distributions were well reproduced by the imaging approach showing that no significant artifacts are introduced into the derived density profiles due to the discrete Fourier synthesis, as seen by comparing to the derived profiles to that calculated using the known crystal structure. These data confirm that Si is found only in the tetrahedral site, K is only in the interlayer site, and Al is distributed between tetrahedral and octahedral sites. Other elemental impurities in muscovite (e.g., Na, Rb...) were not imaged either because the X-ray beam energy was too low to excite measurable X-ray fluorescence, or the fluorescent X-rays were too low in energy to be measured by the fluorescence detector. Through measurement of the elemental composition (e.g., by X-ray fluorescence), these data provide the site-specific composition of the crystal.

We note that it should be possible to obtain the individual profiles of the major elements solely from these XSW data even if the crystal structure were not known *a priori*. For instance, the K and Al distributions are out phase for $L = 2$, but in phase for $L = 4$, with the Al distribution centered on the origin and K displaced by $d_{001}/4$. The modulation amplitude of Al at (004) is reduced with respect to that observed for K suggesting that the Al is distributed unevenly between a majority site near the origin and minority site separated by approximately $d_{001}/8$. In contrast, the (002) modulation of Si has a fluorescence modulation corresponding to $f_{002=0}$, suggesting that there are two equally occupied Si sites displaced by $d_{001/4}$. The Si site is displaced by $d/4$ with respect to the center of the Al distribution since its (004) fluorescence yield modulation is out of phase with respect to that for Al. By continuing along these lines and iterating with dynamical diffraction calculations, it therefore should be possible to determine the major element distributions within a crystal even when the crystal structure is unknown.

14.3. *3D* Map of Surface Adsorbate Atoms

Having demonstrated that XSW imaging works for *1D* profiling of impurity bulk lattice distributions,[3] the method was extended to *3D* mapping of adsorbate atoms and ions on single crystal surfaces. Cases included surfaces prepared by molecular-beam epitaxy,[6,7] cation and anion adsorption from solution,[8,9] atomic-layer deposition,[10] and molecular self-assembly.[11]

One of the UHV surfaces studied was $\frac{1}{3}$ ML of Sn on Ge(111).[6] This surface exhibits a $(\sqrt{3} \times \sqrt{3})R30^\circ$ to (3×3) reversible phase transition at $T_C \sim 210\ K$.[12] For each of these two surface phases a *3D* Sn atomic map was generated by Eq. (14.2) summation of the (111), (333), $(11\,\bar{1})$, and $(33\,\bar{3})$ XSW measured Fourier components and their 3-fold symmetry equivalents. Figure 14.3 shows the image generated by the $\sqrt{3} \times \sqrt{3}$ room temperature XSW data. The top view is a cut at $2.0\,Å$ above the top of the Ge bilayer and the side view is through the long diagonal of the (1×1) unit cell. In these images, the dark red represents the highest Sn density. The small black circles are added to indicate the bulk-terminated Ge atomic positions. The density oscillations that appear in the image are artifacts due to the truncation of the Fourier summation. This method projects the atomic distribution into the substrate (Ge) crystal primitive unit cell. Thus, if there are two distinct heights in the extended *2D* superlattice, their projections will superimpose to form a combined distribution. (STM and/or SXRD are needed to extend beyond the 1×1 unit cell.) The resolution along the *111* direction corresponds to $\frac{1}{2}d_{333} = 0.5$ Å. Within this resolution, the Sn distribution shows an elongation along *111*.

After refinement of the data (based on a T_4 site model), it was determined that two-thirds of the Sn are at *1.85(5) Å* and one-third *0.45(5) Å* higher. Comparing the XSW results above and below T_C confirmed a previously proposed dynamical fluctuation order–disorder phase transition for this corrugated *2D* system.[13]

In general, this Fourier inversion method is of particular importance when a simple atomic position model cannot be easily found that is consistent with the set of XSW measured Fourier components. This proved to be the case for a differently prepared $\frac{1}{3}$ ML Sn/Ge(*111*) surface. This surface, which was annealed at *573 K* as opposed to *473 K* for the result shown in Fig. 14.3, produced a different set of Fourier coefficients that generated an image with an unexpected (and unwanted) third Sn position corresponding to Sn substituting for Ge in the bottom of the Ge bilayer. Because this third Sn position lacked any long-range order, it was undetectable by diffraction (LEED and SXRD).

As a general principle, XSW imaging (or Fourier inversion) is most powerful when used as the first stage in the XSW analysis process. After producing the model-independent $\frac{1}{2}$-Å-resolution *3D* atomic image, a model can then be proposed that is based on the image. This leads to the second XSW analysis stage corresponding to a χ^2 fit of a set of simultaneous equations in which each XSW measured *hkl* Fourier coefficient is equated to the corresponding parameterized coefficient from the model.

$$f_{\boldsymbol{H}} \exp(2\pi i P_{\boldsymbol{H}}) = \sum_j c_j \exp(i2\pi \boldsymbol{H} \cdot \mathbf{r}_j) \exp(-2\langle u_{H,j}^2 \rangle d_H^{-2}). \qquad (14.7)$$

where the fit parameters are the coordinates $(\mathbf{r}_j)$, occupation fractions (c_j) $(\Sigma c_j = 1)$, and r.m.s. vibrational amplitudes $(\langle u_{H,j}^2 \rangle^{1/2})$ for each of the model proposed sites.

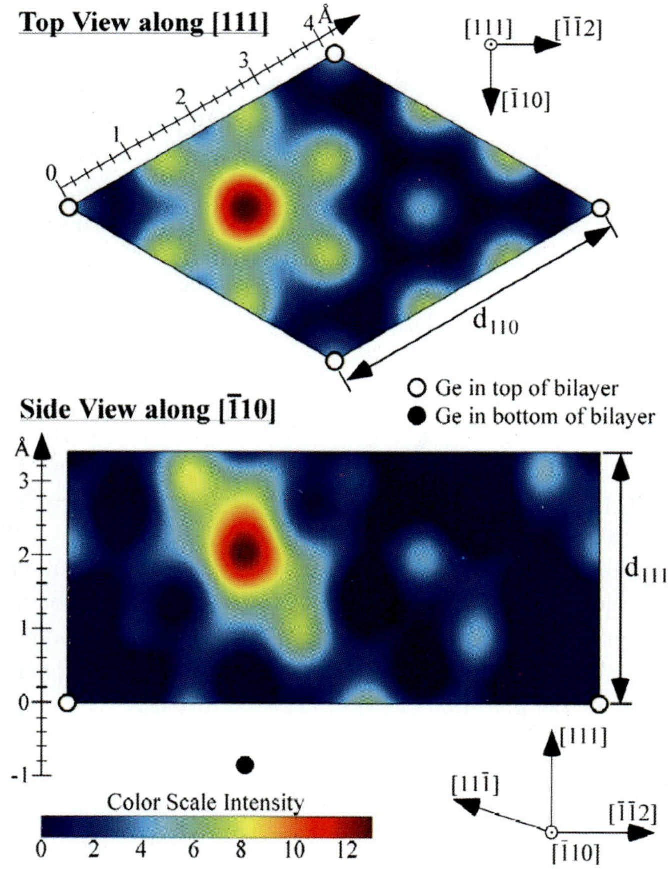

Fig. 14.3. *2D* cuts through the maximum density of the *3D* Sn atomic map generated by inserting XSW measured f_H and P_H into Eq. (14.2) for the $\sqrt{3} \times \sqrt{3}$ phase of $\frac{1}{3}$ ML Sn/Ge(*111*). The top cut corresponds to the Ge(*111*) *1* × *1* surface unit cell. The Sn maximum is located laterally at the T_4 site, which is directly above the Ge atom in the bottom of the bilayer.

14.4. Experimental Description

While a general description of the XSW instrumentation and procedures is discussed in Chapter 13, this Fourier inversion approach has some unique requirements with respect to its implementation. An experimental end-station used for these studies is shown as a photograph and schematic in Fig. 14.4 and further described in Refs. 14 and 15. The end-station uses two non-dispersive double-bounce silicon post-monochromators with a *d*-spacing typically chosen to closely match that of the selected Bragg planes

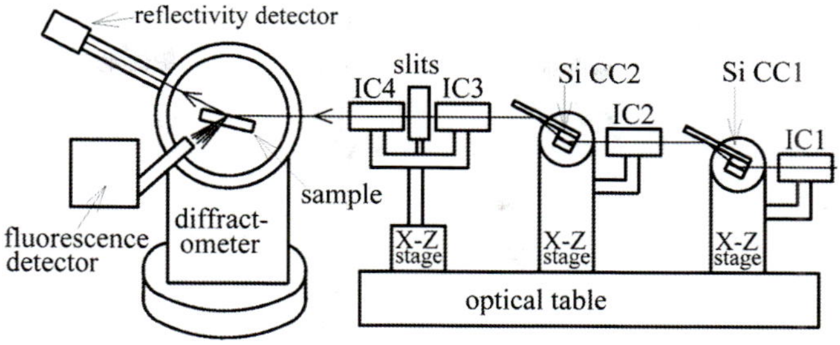

(a)

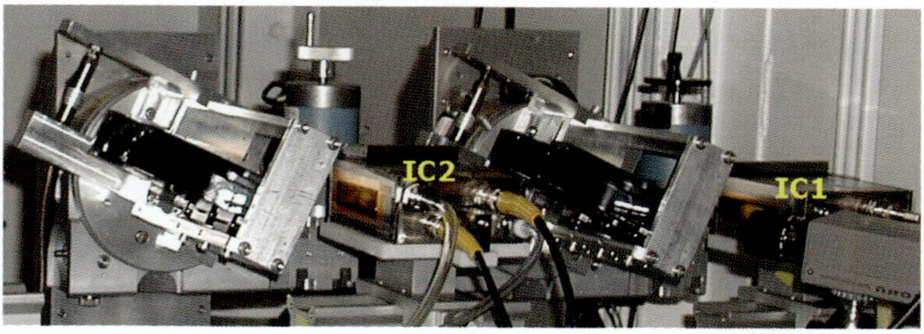

(b)

(c)

Fig. 14.4. Experimental system. (a) Side view schematic of the experimental components: four ion chambers (IC) monitor the X-ray flux after the beam is conditioned by the post-monochromator components. Two separate X-Z-θ stages manipulate Si channel-cut monochromators (Si CC1 and Si CC2) in the beam for the purpose of reducing the energy bandpass and angular divergence of the X-rays incident on the sample. (b) Side view image of the post-monochromator components. (c) Image showing up-stream view of the three channel-cut crystals mounted on a torsion-bearing stage (TS) attached to a Huber 410 rotary stage (HS). A piezo actuator (P) uses feedback from the ion chambers to maintain a fixed detuning value for the Si CC monochromators. The horizontal X translation stage at the base of each Si CC stage is used to select which of the three channel cuts intercepts the beam. Taken from Ref. 15.

of the sample. Each of the two rotational stages is equipped with a set of remotely selectable Si (hhh), $(hh0)$, and $(00h)$ channel-cut crystals. The incident beam channel-cut monochromators are used, either individually or in tandem, to tune the X-ray beam dispersion characteristics and optimize the measurement to best match the various Bragg reflections for the sample. More specifically, it is important to optimize the observed fluorescence modulation to improve the experimental sensitivity for measuring a particular Fourier coefficient. The capability of this type of setup to quickly change the λ versus θ emittance function of the post-monochromator for each reflection of the sample is important because the strength or weakness of a substrate Bragg reflection is, in general, unrelated to the importance of that Fourier component to the elemental density profile to be measured. Consequently, all symmetry inequivalent Fourier coefficients extending over the relevant reciprocal space volume should be measured to obtain an image.

14.5. Conclusion

As shown in this chapter, the Bragg XSW technique can be used to image the *3D* real-space distribution of atoms in an element-specific and model-independent manner. Due to the limited set of allowed *hkl* Bragg reflections, the real-space information is limited to the spatial extent of the substrate unit cell. The XSW imaging technique has become a powerful tool especially with the availability of highly brilliant third-generation synchrotron sources. It is particularly well adapted to adsorbate systems, as shown here, or dopants within the unit cell of host lattices as demonstrated by the *1D* imaging of naturally occurring trace impurities in mica[3] or for *3D* imaging of Mn doped GaAs.[16] For nanocrystals grown on the surface of a single crystal substrate, XSW imaging can be used to measure the correlation between the two lattices, nanoparticle and substrate, revealing important parameters for describing the interfacial structure.[17] As described in Chapters 5 and 19, the Fourier inversion of XSW data collected at the 0th-order Bragg peak (or total external reflection) can be used to produce a model-independent *1D* atomic density profile of the extended structure above a mirror surface[18,19] with a lower-resolution, but longer length-scale; well beyond the reach of this single crystal Bragg XSW method.

Acknowledgments

Colleagues who contributed to the development and demonstration of these ideas include: Likwan Cheng, John Okasinski, Zhan Zhang, Chang-Yong Kim, Anthony Escuadro, Jeffrey Catalano, Donald Walko, Denis Keane, and Neil Sturchio. The results in Figs. 14.2 and 14.3 were obtained at beamlines 12-ID-D and 5-ID-C, respectively, at the Advanced Photon Source, which is supported by the US Department of Energy. The work was also partially supported by the US National Science Foundation MRSEC Program and the Basic Energy Sciences Geoscience Research Program of the US Department of Energy.

References

1. N. Hertel, G. Materlik and J. Zegenhagen, *Z. Phys. B* **58** (1985) 199.
2. M. J. Bedzyk and G. Materlik, *Phys. Rev. B* **31** (1985) 4110.
3. L. Cheng, P. Fenter, M. J. Bedzyk and N. C. Sturchio, *Phys. Rev. Lett.* **90** (2003) 255503.
4. M. J. Bedzyk and L. W. Cheng, *Rev. Mineral Geochem.* **49** (2002) 221.
5. M. J. Bedzyk and G. Materlik, *Phys. Rev. B* **32** (1985) 6456.
6. J. S. Okasinski, C. Kim, D. A. Walko and M. J. Bedzyk, *Phys. Rev. B* **69** (2004) 041401.
7. A. A. Escuadro, D. M. Goodner, J. S. Okasinski and M. J. Bedzyk, *Phys. Rev. B* **70** (2004) 235416.
8. Z. Zhang, P. Fenter, L. Cheng, N. C. Sturchio, M. J. Bedzyk, M. L. Machesky and D. J. Wesolowski, *Surf. Sci.* **554** (2004) L95.
9. J. G. Catalano, Z. Zhang, P. Fenter and M. J. Bedzyk, *J. Colloid Interf. Sci.* **297** (2006) 665.
10. C. Y. Kim, J. W. Elam, M. J. Pellin, D. K. Goswami, S. T. Christensen, M. C. Hersam, P. C. Stair and M. J. Bedzyk, *J. Phys. Chem. B* **110** (2006) 12616.
11. R. Basu, J.-C. Lin, C.-Y. Kim, M. J. Schmitz, N. L. Yoder, J. A. Kellar, M. J. Bedzyk and M. C. Hersam, *Langmuir* **23** (2007) 1905.
12. J. M. Carpinelli, H. H. Weitering, M. Bartkowiak, R. Stumpf and E. W. Plummer, *Phys. Rev. Lett.* **79** (1997) 2859.
13. J. Avila, A. Mascaraque, E. G. Michel, M. C. Asensio, G. LeLay, J. Ortega, R. Perez and F. Flores, *Phys. Rev. Lett.* **82** (1999) 442.
14. D. A. Walko, O. Sakata, P. F. Lyman, T.-L. Lee, B. P. Tinkham, J. S. Okasinski, Z. Zhang and M. J. Bedzyk, *AIP Conf. Proc.* **705** (2004) 1166.
15. M. J. Bedzyk, P. Fenter, Z. Zhang, L. Cheng, J. S. Okasinski and N. C. Sturchio, *Synchrotron Radiation News* **17** (2004) 5.

16. T. L. Lee, C. Bihler, W. Schoch, W. Limmer, J. Daeubler, S. Thiess and J. Zegenhagen, *Phys. Rev. B* **81** (2010) 235202.
17. Z. Feng, A. Kazimirov and M. J. Bedzyk, *ACS Nano* **5** (2011) 9755.
18. M. J. Bedzyk, in *arXiv:0908.2115v1* (2009).
19. V. Kohli, M. J. Bedzyk and P. Fenter, *Phys. Rev. B* **81** (2010) 054112.

Chapter 15

X-RAY STANDING WAVES IN QUASICRYSTALS: ATOMIC POSITIONS IN AN APERIODIC LATTICE

TERRENCE JACH

Material Measurement Laboratory,
National Institute of Standards and Technology,
Gaithersburg, MD 20899-8371, USA
terrence.jach@nist.gov

The observation of the Borrmann effect in quasicrystals demonstrates the presence of dynamical diffraction and, by inference, the existence of X-ray standing waves. In this chapter, we investigate the nature of the atomic order in icosohedral quasicrystals and the manner in which dynamical diffraction comes about. While atoms in quasicrystals do not occupy periodic atomic planes, they occur in distributions that give rise to well-defined X-ray standing wave effects. It is possible to demonstrate that the coherent fraction and coherent position obtained from a specific X-ray reflection correspond to a correlation function and centroid of the atomic distribution relative to that "reciprocal quasi-lattice vector." We show experimental X-ray standing wave effects in the quasicrystal Al-Pd-Mn and their relation to the symmetries of the different elemental distributions.

15.1. Introduction

We normally think of Bragg reflection X-ray standing waves as a consequence of the periodicity of perfect crystals. The interference pattern that is produced by the incident and diffracted beam at the Bragg condition is commensurate with the unit cell of the crystal. The phase of the standing wave is constrained by conditions of dynamical diffraction that allow us to control the position of the X-ray field. The remarkable precision with which this is possible permits us to determine the locations of individual atoms in the unit cell with great accuracy.

The discovery of quasicrystals — materials that diffract but do not possess translational invariance — presented crystallographers with a challenge. The first quasicrystals were identified by the presence of sharp diffraction peaks with icosahedral symmetry.[1] These could not be ascribed to space groups associated with periodic arrays of atoms. It did not take long to identify a new type of crystal lattice which does not require translational invariance.[2] It was found that the Fourier Transform (FT) of the lattice points still resulted in delta functions of intensity, just as in normal crystals. Thus, it is natural to extend the principles of dynamical diffraction to this novel lattice and look at the behavior of those features with which we are familiar. As we shall see, many of the characteristics of dynamical diffraction in perfect crystals, including the consequences of X-ray standing waves, show up in quasicrystals in new and interesting ways.

Let us start by reviewing some of the properties of quasicrystals. Shechtmann, Blech, Gratias, and Cahn observed electron diffraction patterns from a binary metal alloy with two-, three- and five-fold symmetries.[1] The five-fold symmetry, combined with unsuccessful attempts at indexing the diffraction spots,[3] led to the conclusion that the patterns could not correspond to the Fourier transform of a periodic lattice generated by the customary three primitive vectors. It was observed, however, that the observed reflections could be satisfactorily indexed with a finite ($N > 3$) set of vectors.[4] Mathematically, a function is labeled "quasiperiodic" if its Fourier transform results in a set of delta functions (i.e., diffraction peaks) that are not uniformly spaced, but which are spanned in phase space by a finite set of vectors.[5] In the case of the materials that were observed initially, labeled from their symmetries as icosahedral, it was observed that six basis vectors suffice.

The imposition of five-fold symmetry requires basis vectors that are characterized by irrational values. The icosahedral point groups 235 and $m\bar{3}\bar{5}$ in the International Tables of Crystallography require indexing vectors that are linear combinations of rational coefficients and τ, the Golden Mean. The Golden Mean $[\tau = (1 + \sqrt{5})/2]$ shows up characteristically in geometry involving pentagonal symmetry.[4]

A relevant mathematical theorem states that a D-dimensional *quasi-periodic* function requiring a basis of N vectors can be derived from a *periodic* N-dimensional function cut by a D-dimensional plane.[5] This provides a recipe for modeling quasicrystals: we define a *6D* periodic lattice cut by *3D* planes.[6] The intersections of the planes with the higher-order lattice points (to within a small volume that defines acceptance) specify "quasi-lattice" sites in three-dimensional space. As in ordinary Bravais lattices, the primitive quasi-lattice may be dressed with an atom or a combination of atoms. The role of the theorist is to select a *6D* lattice and *3D* planes and acceptance criteria, which result in a realistic assemblage of atoms. Atomic species must be joined by reasonable bond distances and not overlap. The role of the experimentalist is to determine whether these models are consistent with observation. While we shall concern ourselves with icosahedral quasicrystals here, it should be noted that quasicrystals with other symmetries such as decagonal, octahedral, and dodecagonal have been reported.[7–9]

15.2. One-Dimensional Quasi-Lattices

While the reduction of six dimensions to three is difficult to visualize, it is possible to illustrate the projection of a two-dimensional periodic lattice into a one-dimensional quasi-lattice.[10] Figure 15.1 shows a diagram of a *2D* square lattice of unit spacing designated by a coordinate system (X, Y).

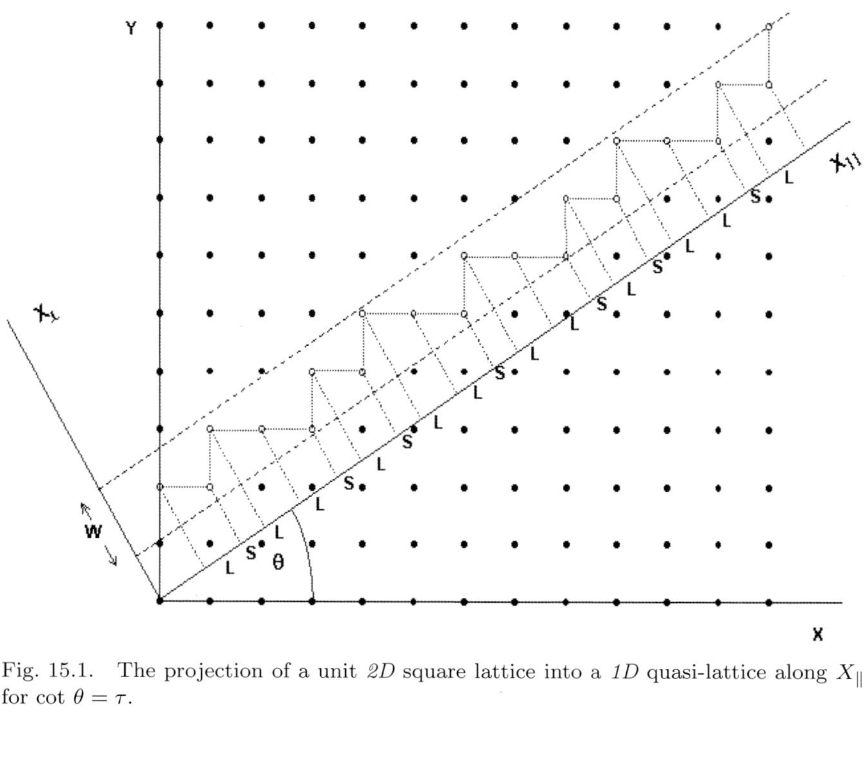

Fig. 15.1. The projection of a unit *2D* square lattice into a *1D* quasi-lattice along $X_\parallel$ for cot $\theta = \tau$.

A second coordinate system ($X_{\|}$, $X_{\perp}$) is shown rotated with respect to the first by the angle θ. We define a narrow region of width W parallel to $X_{\|}$ that determines an "area of acceptance." The coordinates of lattice points within this region (open circles) are noted on $X_{\|}$. For any θ for which $\cot\theta$ is rational, the *1D* lattice of those coordinates will ultimately repeat, defining the periodic lattice of a regular crystal. A trivial example would be $\cot\theta = 1$, which repeats along $X_{\|}$ over a distance $\sqrt{2}$.

The model permits us to see what happens at angles θ with irrational slopes where the lattice does not repeat. For five-fold symmetry, an appropriate angle between the axes is given by $\cot\theta = \tau$. The *1D* quasi-lattice is constructed by projection of lattice points in the acceptance region W on $X_{\|}$. Because of the acute angle between $X_{\|}$ and X, the distance between successive points depends on the projection of vertical (short) or horizontal (long) legs along $X_{\|}$. With S and L being the corresponding two distances, the sequence LSLLSLSLLSLLSLSLLSL... is not accidental, but rather is the geometrical realization of a Fibonacci sequence.

Fibonacci sequences are defined by a recursion relation $F_{n+1} = F_n + F_{n-1}$ $(1, 1, 2, 3, 5, 8, \ldots)$.[11] The series of distances can be created mathematically by the concatenation of two preceding sequences starting with the prescription S; S→L; L→LS ... (S, L, LS, LSL, LSLLS, LSLLSLSL, ...). Each sequence is F_n in length, and the sequence in the limit of $n \to \infty$ is quasiperiodic. For the geometrical construction, the Fibonacci sequence is a mathematical consequence of choosing the slope of the acceptance region parallel to $X_{\|}$ to be a reciprocal of the Golden Mean, since the limit of the ratio of successive terms in the numerical sequence ($F_0 = 0$; $F_1 = 1$) is τ.

Ideal, aperiodic, one-dimensional quasi-lattices are characterized by atomic positions that are Fibonacci sequences in which the interatomic distance is proportional to 1 or τ. This may seem unphysical in one dimension but is enforceable in three dimensions. It is easily shown by a simple geometric argument that unphysical additional spacings $\tau - 1$ and $\tau + 1$ are avoided by selection of the acceptance width, $W = \mathrm{S} + \mathrm{L}$.[12] The ratio of L to S elements in a sequence in the limit $n \to \infty$ is also τ.

The reciprocal lattice, being the Fourier Transform (FT) of the *2D* square lattice in position space, is also a square lattice, as shown in Fig. 15.2. The angular relation of the rotated axes $Q_{\|}$, $Q_{\perp}$ to Q_x and Q_y is retained. Sharp diffraction spots occur at the coordinates of lattice points on $Q_{\|}$.[10] The indexing of the diffraction spots on the quasi-lattice along $Q_{\|}$ again

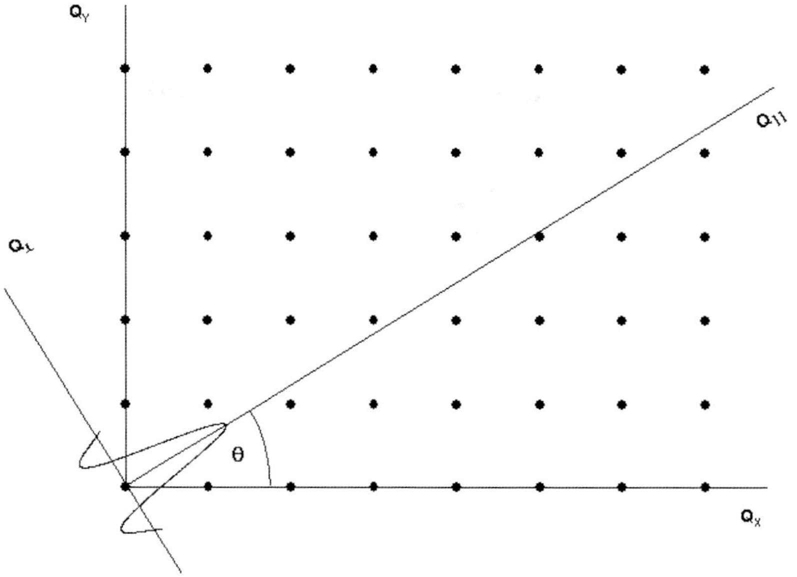

Fig. 15.2. Representation of quasi-lattice axis in momentum space, showing the FT of the acceptance function, which determines the diffraction spot intensity on $Q_{||}$.

requires two spanning vectors proportional to integral multiples of the unit vectors $(1, 0)a$ and $(0, \tau)a$, where $a = (\sin\theta)^{-1}$. Because the entire *2D* lattice in real space (Fig. 15.1) is convoluted with a step function that is 1 for points with $X_\perp$ within W, and 0 otherwise, the amplitude of the diffraction spots in reciprocal space is convoluted with the FT of the real space function, $\sin(W\ Q_\perp)/(W\ Q_\perp)$, which is depicted on the $Q_\perp$ axis. The intensity of diffraction spots projected on $Q_{||}$ is proportional to the amplitude squared, i.e. $I \propto [\sin^2(W\ Q_\perp)/(W\ Q_\perp)^2]$. This explains why there are many visible quasicrystal reflections, but few that show significant intensity.

A related geometric approach to the generation of quasi-lattices is that of tiling. Penrose originally discovered that combinations of two rhombic tiles with a ratio of areas 1: τ could produce infinite *2D* areas that had five-fold symmetry.[13] The vertices of tiles correspond to quasi-lattice points. In the *3D* equivalent, two rhombohedra have been developed by Mackay, which fill space quasiperiodically.[14] The intersection points have been discerned as atomic cluster sites in icosohedral quasicrystals.

The imposition of similar acceptance criteria in the *6D* hyperspace results in a hypercube which projects out a *3D* acceptance domain, a triacontahedron, in the $X_{\perp}^{3}$ space.[6] The triacontahedron can be decomposed into 20 rhombohedra (10 of each type) demonstrating the close connection between the projection method from *6D* to *3D* and the geometric aspects of tiling. In advanced modeling, different acceptance (or occupation) domains may be assigned to different points on the *6D* lattice so as to designate specific atom species or binding sites.[15]

15.3. Dynamical Diffraction from *1D* Quasi-Lattices

The first dynamical theory of *1D* quasicrystals was developed by Berenson and Birman.[16] They determined the scattering factor for a Fibonacci lattice and derived the standard dispersion relation treatment for dynamical diffraction. It was evident that standing wave solutions should exist in this model, and, as a result, anomalous transmission (the "Borrmann effect") was predicted. The derivation assumed a "reciprocal quasi-lattice" vector $\mathbf{H} = (2\pi/\sqrt{5}d)[n + m\tau]$, where n and m are integers, and d is a fundamental atomic distance. They also assumed that $F_{\mathbf{H}} = F_{-\mathbf{H}}$, which implies a centrosymmetric quasicrystal, the meaning of which will be discussed below. Berenson and Birman considered both the two-beam theory and the effects of multibeam corrections.

Chung and Durbin extended the theory with calculations of a hypothetical Fibonacci lattice of $Ga_{\tau}As$, with interplane distances which were a multiple of either 1 or τ.[17] They calculated, using Darwin's method, that the rocking curve would look similar to that of the reflection from a perfectly periodic set of Si(111) planes. They also calculated the absorption of the Ga and As by the X-ray standing wave as the diffraction angle was scanned from one side of the rocking curve to the other. By mapping the displacements of all of the atomic planes in the *1D* lattice, modulo a single cycle of the standing wave (i.e., by mapping the coherent phase position of each atomic plane into a single zone $-\pi < \varphi < \pi$), they demonstrated that scanning the rocking curve would result in lower and higher absorption. The conclusion is that, in recording the X-ray fluorescence from one species of the constituent atoms, a well-defined coherent position with a characteristic coherent fraction should be observed.

An explanation given by Prince is shown in Fig. 15.3.[18] It depicts a modification of the *2D* lattice projection diagram to illustrate the

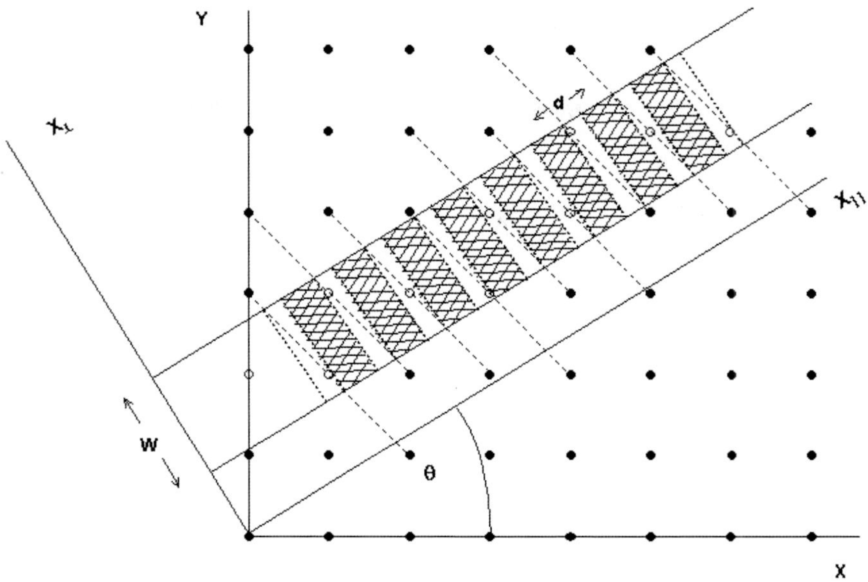

Fig. 15.3. The projected *2D* lattice sites in W reside in periodic zones with respect to the $[1,1]$ direction, which give rise to the charge distribution responsible for the scattering vector $q = 1/d$ (the dotted lines are the projections on the $[1, 1\tau]$ direction).

contribution of reflections by a specific crystal "plane," here indexed as $(1, 1\tau)$, of the *1D* Fibonacci crystal. We see that along $X_{\parallel}$ the contributions of the atoms from those planes within the acceptance region occur within a periodic zone, where the atoms will *always* be found. In contrast, there is a hatched region, also periodic, where the atoms are *never* allowed. Because of the irrational slope by which $X_{\parallel}$ has been rotated, the actual positions of the atoms will vary continuously, filling the allowed region with a uniform distribution.[6,18] The effect seen by the standing waves is equivalent to a periodic rectangular charge density distribution that gives rise to the scattering vector responsible for the Bragg reflection. The complex X-ray susceptibility χ_q of this reflection originates from the lowest Fourier component of the charge distribution.

Thus, while there is not strictly a periodic array of atoms in the quasicrystal for each reflection, there is a real periodic density of charge ρ_q that serves as an absolute reference for phase. The periodic zones that either always or never contain atoms give rise to the two X-ray standing wave solutions corresponding, respectively, to the β- and α-branches of the two-beam dispersion surface in dynamical diffraction.[19]

15.4. Centrosymmetry versus Non-Centrosymmetry

The question of centrosymmetry is as important in quasicrystals as it is for periodic crystals; however, it has a slightly different meaning. Tiling models have shown that there can be a central origin of symmetry (i.e., inversion symmetry) which theoretically is preserved throughout the quasicrystal.[20] If we collapse all the periodic zones depicted in Fig. 15.3 into a single cell modulo the distance d, a schematic distribution of the atomic positions should look something like Fig. 15.4(a). In that case, a uniform distribution of a single atomic species over the allowed zone gives rise to a

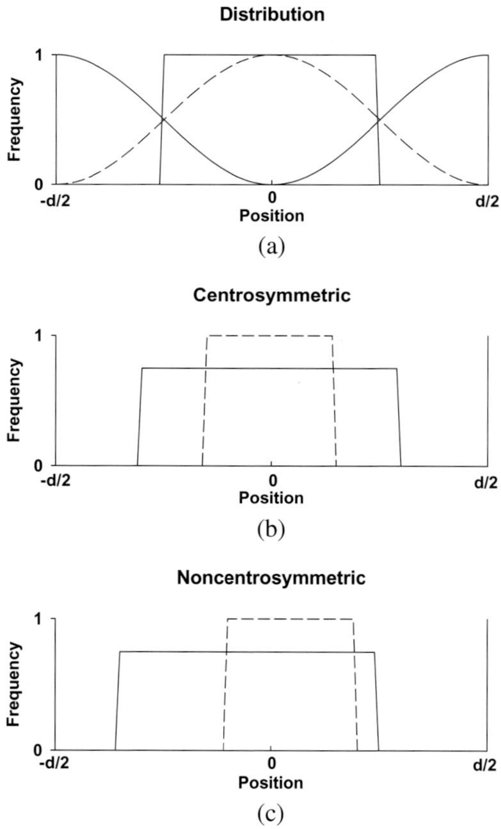

Fig. 15.4. (a) The schematic distribution of atoms in the *1D* Fibonacci lattice collapsed into a single periodic zone of a strong Bragg reflection $q = 1/d$, showing the X-ray standing wave for (solid) α- and (dashed) β-branches. (b) Schematic of centrosymmetric distribution of two atomic species. (c) Schematic of two species in non-centrosymmetric distribution. See Ref. 22.

centrosymmetric cell, since the single species is responsible for the charge distribution. Such a system will show large but finite regions of inversion symmetry no matter where the center is chosen.[21]

The X-ray standing wave, which occurs at the Bragg condition for this reflection will have a periodicity d. The lowest order α-branch X-ray standing wave is shown as a solid line. The β-branch standing wave is shown as a dashed line. Unlike the simplest case in a periodic lattice, the standing wave is never displaced to a position where the atoms are all located at a node in the wave field. The translation of the wave that occurs as we traverse the X-ray rocking curve will excite a fluorescence standing wave curve characterized by a well-defined coherent fraction and coherent position, which are the correlation function and the centroid of the atomic distribution, respectively.[22]

In the case of two atomic species, the charge density ρ_q, which gives rise to the reflection originates from the distribution of both species. Since the atomic form factor $f(q)$ for each species is independent of the other as a function of energy, a centrosymmetric quasicrystal has to have a centrosymmetric distribution of each species. The charge density ρ_q giving rise to χ_q is the result of both species, so if one distribution is non-centrosymmetric, the other must be as well. In the case of three or more atomic species in the quasicrystal, it is possible to have combinations of centrosymmetric and non-centrosymmetric distributions which give rise to χ_q.

15.5. Quasicrystals in Three Dimensions

While it is one thing to justify the structure of quasicrystals mathematically, this does not explain how or why quasicrystals occur in nature. Quasicrystals have been grown by various techniques (rapid solidification, sputtering, laser ablation), but the highest-quality materials are grown from a melt.[23] The tendency of atoms to pack icosahedrally is the energetically favorable solution for small clusters but is normally frustrated beyond the second atomic layer.[24] The tendency to grow into icosahedral quasicrystals is thus a question of icosahedral clusters that are stabilized into a structure that maintains an overall icosahedral symmetry in the bulk. The role of entropy in stabilizing the quasicrystalline structure, in addition to simple energetic considerations, has been treated extensively.[25] Metal alloys with stoichiometries that depart slightly from the quasicrystal ratios tend to form complex unit cells with large repeat distances. They correspond to slight

is unconstrained result in values of f that appear markedly lower than predicted in the model.

The conclusions that we reach from the above results for the two-fold symmetry reflections are that the Pd atoms appear to be located in centrosymmetric sites, but the Al atoms appear to be in sites which deviate from centrosymmetry by about 0.1 d. This would require the Mn atoms to be offset in the opposite direction along the two-fold quasicrystal axis with respect to the Pd atoms since the overall charge contributed by all the atoms determines the phase of the q.

We have also measured XSW effects along a five-fold axis of Al-Pd-Mn.[40] The (42222$\bar{2}$) reflection occurs at a calculated $q_{42222\bar{2}} = 0.4643$ Å^{-1}. Figure 15.9 shows the reflection, at $\theta_B = 90°$, which was observed at a nominal beam energy of 2859 eV.

The Al fluorescence standing wave shows evidence of a slight displacement of Al atoms from the centrosymmetric positions along the

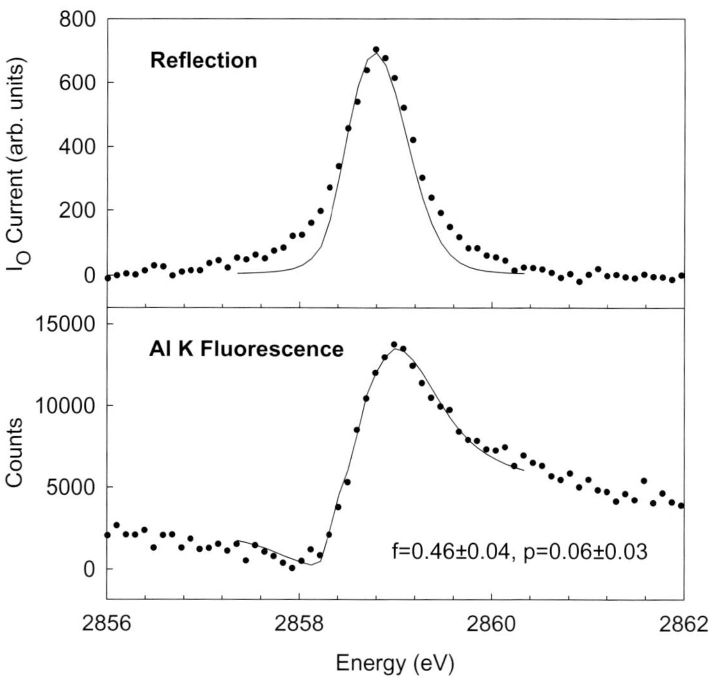

Fig. 15.9. Energy scan of Al-Pd-Mn along the two-fold symmetry axis showing (a) data and fit to the (42222$\bar{2}$) rocking curve for $\theta_B = 90°$, and (b) Al K fluorescence with f, p unconstrained.

five-fold axis. Data recently obtained from powder diffraction support the evidence that the Al-Pd-Mn quasicrystal system is non-centrosymmetric.[41] It is hoped that the availability of atomic models are large enough to calculate χ_q and $\chi_{\bar{q}}$ accurately for the non-centrosymmetric case (where $\chi_q \neq \chi_{\bar{q}}$) and the resulting XSW values of f and p will allow for a careful comparison between the model and the experiment along different symmetry axes.

Since phasons are responsible for the mosaic spread in the diffracted beam, they logically cause a diminution of the coherent fraction in the XSW. This has been studied using a model for the phason distortion in a quasicrystal.[42] The calculation shows that the effect of phasons is to diminish the "empty" part of the collapsed zone. The magnitude of the effect depends on $Q_\perp$, with f decreasing the most for large values of $Q_\perp$, and very little for small $Q_\perp$.

Finally, XSW effects have been observed in X-ray photoemission spectroscopy from Al-Pd-Mn.[43] This method allows one, in principle, to look at photoemission lines from all of the elements, although the photoemission cross-section may still be low at the energy of a Bragg reflection, $\theta_B = 90°$. In addition, the short escape depth of the photoemission electrons makes the resulting XSW curve highly sensitive to sample quality near the surface. Figure 15.10 shows the integrated Al 1s photoemission line intensity that was obtained on beamline ID32 at the European Synchrotron Radiation Facility as the Si(111) monochromator was scanned through the (42222$\bar{2}$) reflection of Al-Pd-Mn. The X-ray photoemission spectrum was obtained in ultra-high vacuum with a hemispherical analyzer electron spectrometer.

Figure 15.10(a) displays the rocking curves as a function of energy from the quasicrystal sample after the surface was sputtered (filled circles), and then after it was annealed (open circles). Figure 15.10(b) displays the integrated Al 1s X-ray photoemission line intensity for the sample as a function of energy. After sputtering (filled circles), although the rocking curve shows a narrow diffraction line, the surface disorder down to the escape depth of the electrons results in a line intensity that merely reflects the enhanced intensity of the Bragg peak without any phase coherence ($f \cong 0$). After annealing the crystal at 850 K to produce an ordered surface (open circles), the intensity in Fig. 15.10(b) shows the onset of an XSW characteristic with $f \neq 0$. X-ray standing waves clearly sample the order of the surface layer, and we are able to see this down to the escape depth of the electrons (here about 24 Å).

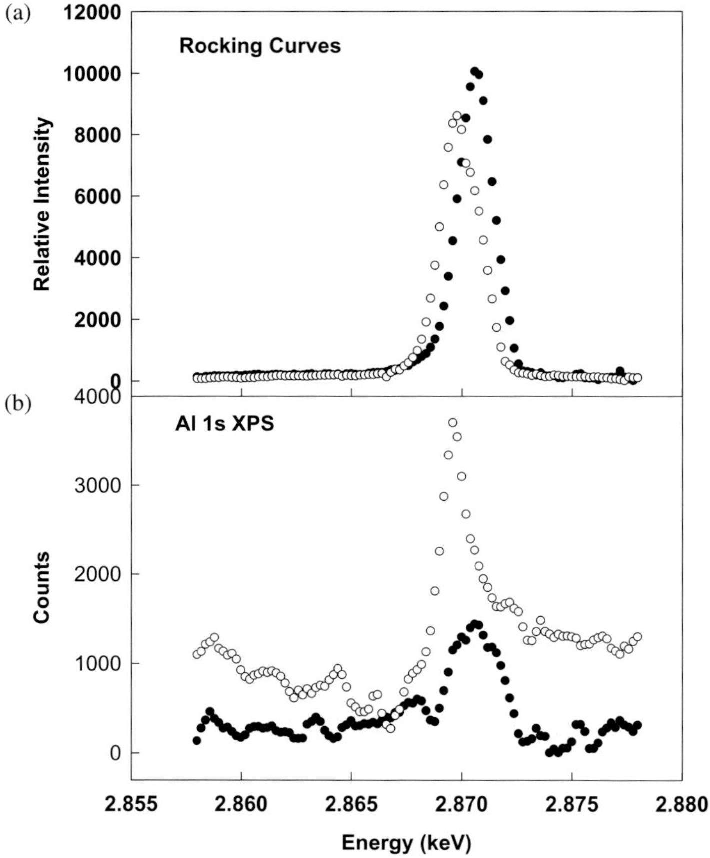

Fig. 15.10. (a) Rocking curves from the (42222$\bar{2}$) reflection from Al-Pd-Mn. The rocking curve as a function of the beam energy was first obtained after sputtering the surface (•), then repeated after the sample was annealed (○); (b) the integrated Al 1s X-ray photoemission intensity simultaneously obtained as a function of beam energy after the sample was sputtered (•), and after annealing (○).

15.7. Conclusions and Remarks

The existence of quasicrystals, materials that do not have translational symmetries but still show sharp X-ray diffraction peaks, is a remarkable phenomenon. Almost equally surprising is the fact that they display the effects of dynamical diffraction, including the Borrmann effect and XSW. XSW, almost uniquely, allow us to investigate element-specific characteristics of the quasicrystals, including elemental distributions, symmetries, and ordering. This is not possible from diffraction peaks, where

$I(\mathbf{q}) = I(-\mathbf{q})$ even when the susceptibilities $\chi_q \neq \chi_{\bar{q}}$ are not equivalent. It is because we have not lost the phase information of the reflection that we are able to see these extra effects.

We have summarized here XSW investigations of the icosahedral quasicrystal Al-Pd-Mn. Although the XSW associated with a specific Bragg reflection do not give individual atomic positions, coherent positions and coherent fractions are observed in X-ray fluorescence and X-ray photoelectron emission. We have demonstrated that the coherent fraction is a function of the correlation between the standing wave and the elemental atomic distribution, and that the coherent position is a measure of the displacement of the centroid of the elemental atomic distribution relative to the phase of the diffracting Fourier component of electron charge. These quantities are readily calculated from atomic models of quasicrystals and give unique element-specific grounds for comparison with real materials. A comparison here of positions from a centrosymmetric model differs significantly from observed atomic distributions.

This field of research is still in its early days, with many possibilities for further inquiry. For example, XSW have not yet been applied to the study of decagonal quasicrystals, which show *2D* tiling behavior of atoms in planes that are periodically aligned in the remaining direction. Many of the most interesting and potentially useful properties of quasicrystals are associated with their surfaces, and the surface science of these materials has much room for future investigation.[44] Surface XSW generated from bulk quasicrystals could be used to determine adsorbate heights and positions, as has been done so successfully for crystalline substrates.[37] One of the great pleasures of working with quasicrystals has been the highly exploratory nature of the results. For every well-established observation in the dynamical diffraction of X-rays from crystals, there is an equivalent phenomenon to be observed in quasicrystalline structures.

References

1. D. Shechtman, I. Blech, D. Gratias and J. W. Cahn, *Phys. Rev. Lett.* **53** (1984) 1951.
2. M. Duneau and A. Katz, *Phys. Rev. Lett.* **54** (1985) 2688.
3. L. Pauling, *Nature* **317** (1985) 512.
4. J. W. Cahn, D. Shechtman and D. Gratias, *J. Mater. Res.* **1** (1985) 13.
5. H. A. Bohr, *Almost Periodic Functions* (Chelsea, New York, 1947).
6. V. Elser, *Acta Crystallogr. A* **42** (1985) 36.
7. L. Bendersky, *Phys. Rev. Lett.* **55** (1985) 1462.

8. N. Wang, H. Chen and K. H. Kuo, *Phys. Rev. Lett.* **59** (1987) 1010.
9. T. Ishimasa, H.-U. Nissen and Y. Fukano, *Phys. Rev. Lett.* **55** (1985) 511.
10. R. K. P. Zia and W. J. Dallas, *J. Phys. A* **18** (1985) L341.
11. A. I. Goldman and K. F. Kelton, *Rev. Mod. Phys.* **65** (1993) 213.
12. D. Levine and P. J. Steinhardt, *Phys. Rev. Lett.* **53** (1984) 2477.
13. M. Gardner, *Sci. Am.* **236** (1977) 110.
14. A. L. Mackay, *Physica A* **114** (1982) 609.
15. A. Yamamoto, *Phys. Rev. B* **45** (1992) 5217.
16. R. Berenson and J. L. Birman, *Phys. Rev. B* **34** (1986) 8926.
17. J.-S. Chung and S. M. Durbin, *Phys. Rev. B* **51** (1995) 14976.
18. E. Prince, *Acta Crystallogr. A* **43** (1987) 393.
19. B. W. Batterman and H. Cole, *Rev. Mod. Phys.* **36** (1964) 681.
20. P. Bak, *Phys. Rev. Lett.* **54** (1985) 1517.
21. M. de Boissieu, P. Stephens, M. Boudard and C. Janot, *J. Phys.: Condens. Matter* **6** (1994) 363.
22. T. Jach, *Mater. Sci. Eng.* **294–296** (2000) 315.
23. M. de Boissieu, M. Durand-Charre, P. Bastie, A. Carabelli, M. Boudard, M. Bessiere, S. Lefebvre, C. Janot and M. Audier, *Philos. Mag. Lett.* **65** (1992) 147.
24. M. Widom, in *Introduction to Quasicrystals*, ed. M. V. Jaric (Academic Press, NY, 1988).
25. M. Widom, K. J. Strandberg and R. H. Swendsen, *Phys. Rev. Lett.* **58** (1987) 706.
26. M. Cornier-Quiquandon, R. Belissent, Y. Calvayrac, J. W. Cahn, D. Gratias and B. Mozer, *J. Non-Cryst. Solids* **153** (1993) 10.
27. M. Boudard, M. de Boissieu, C. Janot, J. M. Dubois and C. Dong, *Philos. Mag. Lett.* **64** (1991) 197.
28. M. Saito, M. Tanaka, A. P. Tsai, A. Inoue and T. Masumoto, *Jpn. J. Appl. Phys.* **31** (1992) L109.
29. H. Lee, R. Colella and Q. Shen, *Phys. Rev. B* **54** (1996) 214.
30. M. Boudard, M. de Boissieu, C. Janot, G. Heger, C. Beeli, H.-U. Nissen, H. Vincent, R. Ibberson, M. Audier and J. M. Dubois, *J. Phys.: Condens. Matter* **4** (1992) 10149.
31. A. Yamamoto, A. Sato, K. Kato, A. P. Tsai and T. Masumoto, *Mater Sci. Forum* **150** (1994) 211.
32. D. W. Delaney, T. E. Bloomer and T. A. Lograsso, in *New Horizons in Quasicrystals: Research and Applications*, eds. A. I. Goldman, D. J. Sordelet, P. A. Thiel and J. M. DuBois (World Scientific, Singapore, 1997), pp. 45.
33. S. W. Kycia, A. I. Goldman, T. A. Lograsso, D. W. Delaney, D. Black, M. Sutton, E. Dufresne, R. Brüning and B. Rodricks, *Phys. Rev. B* **48** (1993) 3544.
34. T. Jach, S. M. Thurgate, Y. Zhang, R. Colella, A. I. Goldman, S. Kycia, M. de Boissieu, and M. Boudard, in *Proc. 6th International Conference on Quasicrystals*, eds. S. Takeuchi and T. Fujiwara (World Scientific, Singapore, 1998) pp. 43.
35. K. Kohra and T. Matsushita, *Z. Naturforsch.* **27A** (1972) 484.

36. A. Caticha and S. A. Caticha-Ellis, *Phys. Rev. B* **25** (1982) 971.
37. D. P. Woodruff, D. L. Seymour, C. F. McConville, C. E. Riley, M. D. Crapper, N. P. Prince and R. G. Jones, *Surf. Sci.* **195** (1988) 237.
38. T. Jach, Y. Zhang, R. Colella, M. de Boissieu, M. Boudard, A. I. Goldman, T. A. Lograsso, D. W. Delaney and S. Kycia, *Phys. Rev. Lett.* **82** (1999) 2904.
39. K. Edagawa, *Philos. Mag. Lett.* **61** (1990) 107.
40. T. Jach, G. Cappello, Y. Zhang and R. Colella (unpublished).
41. N. Tamura, L. Robertson, S. Moss (unpublished) private communication.
42. F. Schmithüsen, G. Cappello, and J. Chevrier, *Ferroelectrics* **250** (2001) 289.
43. G. Cappello, A. Dechelette, F. Schmithüsen, S. Decossas, J. Chevrier, F. Comin, V. Formoso, M. de Boissieu, T. Jach, R. Colella, T. A. Lograsso, C. Jenks and D. Delaney, *Mater. Sci. Eng.* **294–296** (2000) 863.
44. P. A. Thiel, A. I. Goldman and C. J. Jenks, in *Physical Properties of Quasicrystals*, ed. Z. M. Stadnick, Springer Series in Solid-State Sciences, Vol. 126 (Springer, Berlin, 1999).

Chapter 16

X-RAY STANDING WAVES IN THIN CRYSTALS: PROBING THE POLARITY OF THIN EPITAXIAL FILMS

ALEXANDER KAZIMIROV

Cornell High Energy Synchrotron Source (CHESS), Cornell University, Ithaca, NY 14853, USA

JÖRG ZEGENHAGEN and TIEN-LIN LEE

European Synchrotron Radiation Facility (ESRF), BP 220, F-38043, Grenoble, France

MICHAEL BEDZYK

Northwestern University, Evanston, IL 60208, USA and Argonne National Laboratory, Argonne, IL 60439, USA

In thin crystals, an X-ray standing wave field is generated by the interference between the strong incident wave and the weak Bragg diffracted X-ray wave. Though the spatial modulation of the interference field is greatly reduced, the XSW method retains the same structural sensitivity as in thick perfect crystals diffracting dynamically. Because the diffraction curve from a thin crystal is broad, the XSW technique is much less sensitive to lattice imperfections. As examples, we demonstrate in this chapter how this technique can be applied to determine the "as-grown" and "externally switched" polarity in epitaxial thin films. With modern SR sources, the standing wave in an ultra-thin film of only 25 unit cells can be generated and utilized for precise structural investigation.

16.1. Introduction

The ultimate goal of any diffraction analysis is to obtain information about the crystal structure that can then be linked to physical or chemical properties. One example would be the polarity of non-centrosymmetric crystals such as GaAs(111), where the (111) Ga-terminated, the (111)A surface and the $(\bar{1}\bar{1}\bar{1})$ As-terminated, the (111)B surface exhibit very different electronic (surface states, core level shifts), chemical (reactivity, etching), structural (range of surface reconstructions, adsorption) and growth properties. Determining polarity in a nondestructive way with X-rays has been a challenging problem since the early days of X-ray diffraction.[1] It is well known[2] that in absorbing crystals $|F_{hkl}| \neq |F_{\bar{h}\bar{k}\bar{l}}|$ and that Friedel's law breaks down providing the basis for determining polarity by measuring integral intensities of the reflections (hkl) and $(\bar{h}\bar{k}\bar{l})$ from either face of a polar crystal. The sensitivity of the technique is greatly enhanced if the measurements are performed across[3] or close to[4] an absorption edge of one of the atoms constituting the polar crystal.

Early in the development of the XSW method it was recognized that the inward movement of the standing wave pattern through the atomic layers of the unit cell could be used for determining crystal polarity. The first such experiment utilizing the XSW formed during dynamical diffraction was performed on a GaP crystal by observing fluorescence[5] and then later photoelectrons.[6] This was followed by experiments on GaAs,[7–9] CdS,[10] and CdTe.[11] In Ref. 9, an energy-dependent shift of the non-centrosymmetric diffraction planes below and above Ga and As absorption edges was studied in detail. Note that the diffraction planes do not necessarily coincide with the Miller planes, but depend on the scattering strengths of the atoms. The advantage of the XSW method for determining polarity is that it does not require anomalous scattering conditions and the polarity can be determined unambiguously by a measurement on only one face of the crystal. The limitation is the requirement of highly perfect dynamically diffracting crystals.

With technological advances, new complex materials with tailored physical properties have become available and/or are used in the form of thin films grown by different growth techniques on a variety of substrates. In this context, determining the polarity of thin films emerged again as an important technical issue. Many of these materials do not exist as bulk crystals and, when grown as crystalline thin films, they are often imperfect, with a high defect density. Traditional anomalous X-ray diffraction techniques cannot be used as they require measuring $|F_{hkl}|$ and $|F_{\bar{h}\bar{k}\bar{l}}|$, i.e., reflections from both polar faces, and for epitaxial films only one face is accessible. This is the motivation for applying the XSW method to determine the polarity of thin films.

X-ray standing waves were first discovered in perfect bulk crystals (*cf.* Chapter 1) and described in the frame of the dynamical theory of X-ray diffraction for semi-infinite crystals (*cf.* Chapter 2). During the first decades of the development of the XSW method, it was successfully applied to study the atomic structure of impurities in and on substrates such as Si, Ge, GaAs, and other perfect crystals, leaving the wrong impression in the X-ray research community that large perfect crystals are a necessary prerequisite for the XSW technique. However, the phase of the diffracted beam changes by π radian across the Bragg peak (*cf.* Chapters 1 and 2) whether the diffracted beam is strong or weak.

For a thin crystal with a thickness t much smaller than the extinction lengths ($t \ll L_{ex}$), the maximum reflectivity R decreases as $R \sim t^2$. The visibility $v = (I_{\max} - I_{\min})/(I_{\max} + I_{\min})$ of the XSW interference fringes is $v \sim \sqrt{R} \sim t$. Thus, even for a thin film with a maximum reflectivity $R = 10^{-4}$, $v = 1\%$ and the modulation of the photo-absorption, which can be up to $4 \times v$ can be observed and reliably measured. It is an important advantage that the width $\Delta\omega$ of the rocking curve for a thin crystals increases $\Delta\omega \sim 1/t$ making the XSW technique much less sensitive to imperfections. For thin films with $t < L_{ex}$ the effect of the mosaic spread can be taken into account by a simple convolution with a Gaussian function.

The first experiment in which the XSW method was applied to a multi-component thin epitaxial film was performed on a $In_{0.5}Ga_{0.5}P$ film grown by liquid phase epitaxy on a GaAs(111) surface.[12] The polarity of this film was predetermined by the polarity of the substrate, but as will be shown later, the relationship between the polarity of an epitaxial film and the properties of the substrate may not be self-evident. Later, the XSW generated in thin films was utilized to determine the distribution of Bi^{3+} ions over dodecahedral lattice sites in $Y_{3-x}Bi_xFe_5O_{12}$ thin film[13] and to study the lattice location of rare-earth atoms in thin $R$$Ba_2Cu_3O_{7-\delta}$ high-temperature superconducting films.[14]

In the following, applications of the thin film XSW technique for studying the polarity of GaN and ferroelectric thin films will be reviewed. For all experimental results presented here the dynamical theory of X-ray diffraction in epitaxial thin crystals[15] (*cf.* Chapter 3) was used for analyzing the data.

16.2. GaN Thin Films

The III-V compounds GaN, AlN and InGaN possess the wurtzite crystal structure. As epitaxial films, they typically grow along the polar axis with

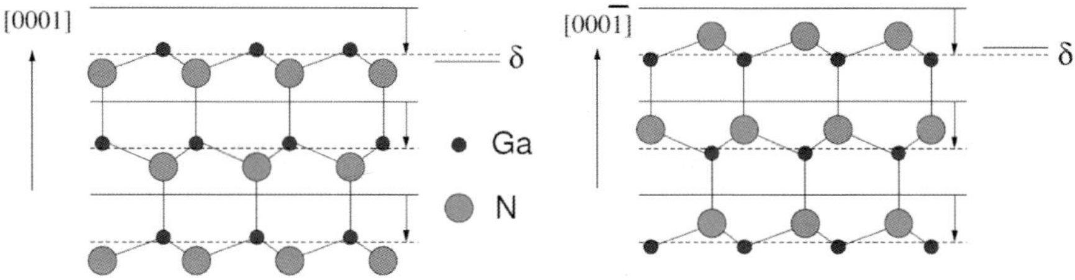

Fig. 16.1. Crystal structure of GaN for Ga- (left) and N- (right) faces. The positions of the antinodal XSW planes at the low angle side of the GaN(0002) rocking curve are shown as solid lines and at the high angle side as dashed lines. The downward pointing arrows indicate the inward shift of the antinodal planes by half of d_{0002} while crossing the GaN(0002) Bragg peak. Due to the difference in the atomic scattering factors for Ga and N atoms, the diffraction planes are shifted by δ from the center of the double layer towards heavier Ga atoms.

either [0001] orientation (Ga-face), or [000$\bar{1}$] orientation (N-face) normal to the surface (Fig. 16.1). The c-axis orientation is largely favored even if the substrate is amorphous.[16] Many techniques have been used to probe the GaN polarity, including XPS and Auger analysis, TEM and convergent beam electron diffraction, ion channeling, and chemical etching (see for a review, Ref. 16). The results obtained by these different analysis methods even investigating films grown by the same technique on the same substrate material were often contradictory and lead to confusing conclusions.

The basis of using the XSW technique for determining the polarity of GaN films is illustrated in Fig. 16.1. Approaching the Bragg reflection from the low angle side, the XSW field is formed with its maxima, i.e., antinodal planes, positioned between the GaN bilayers (horizontal solid lines). By tuning the crystal through the diffraction region, the maxima of the XSW field move inward by one-half of the XSW period ($d_{\text{xsw}} = d_{0002}$). At the high angle side, the antinodes approach the diffraction planes, which are positioned inside the bilayer between the Ga and N planes (horizontal dashed lines). The movement of the antinodal planes is shown by arrows in Fig. 16.1. For reflection from the N-face (right panel) the maxima of the XSW field never reach the Ga plane, while for reflection from the Ga-face film (left panel) the antinodes cross the Ga layer at the high angle side thus resulting in a higher Ga fluorescence yield.

The experiments on GaN films grown by different methods on different substrate materials were performed at the X15A beamline of the National Synchrotron Light Source, BNL. In the first experiment, a 1-μm-thick

GaN film grown by plasma induced molecular beam epitaxy (MBE) on an α-Al_2O_3 single crystal was studied.[17] The energy of the beam was tuned by the double crystals (111) monochromator to 10.5 keV, slightly above the Ga K edge of 10.372 keV. An energy-dispersive Si(Li) solid state detector was used to record the Ga-K fluorescence from the film. The (0002) XSW experimental curve is shown in Fig. 16.2(a). With the maximum reflectivity of only $R = 0.45\%$ the modulations of the XSW yield is within 1%.

To determine the polarity of the film, the following fitting procedure was applied. First, the theoretical reflectivity curve was convoluted with a Gaussian function to account for the broadening of the experimental curve due to the mosaicity. The best fit yielded $\sigma = 0.05°$. This value was fixed and used for the convolution of the fluorescence yield curve. Then, the static Debye–Waller (DW) factor e^{-W} was used as the only fitting parameter for two models of N- and Ga-polarity of the film. The results of the fit are shown in Fig. 16.2(a) and the χ^2 values as a function of the DW factor for the two models are plotted in Fig. 16.2(b). For both χ^2 curves, the minimum is at $e^{-W} = 0.36$, with the χ^2 value for the N-face polarity being four times lower than for the Ga-face polarity. Assuming a mixture of both types of polarity, i.e., the presence of two domains in the films, worsens the fit.

This XSW measurement unambiguously proved that films grown by MBE on sapphire exhibit N-face polarity which is in contrast with earlier reports in which the Ga-face polarity was found for GaN on sapphire, however, grown by metal organic chemical vapor deposition (MOCVD).[18] This XSW result had important practical applications. It was shown[19,20] that the strong confinement and high sheet carrier concentration of two-dimensional (2D) electron gas in AlGaN/GaN/AlGaN and GaN/AlGaN/GaN heterostructures, are the consequences of interfaces charges induced by piezoelectric and spontaneous polarization effects and are the specific properties of the N- or Ga-polarity of the films. Combining the MBE and MOCVD growth techniques, the interface properties of the heterostructure field-effect transistors can be controlled by choosing the appropriate polarity of the growing film.

GaN films grown by hydride vapor phase epitaxy (HVPE) on the Si-face of a 6H-SiC(0001) substrate were studied in 2001.[21] The polytypes of the hexagonal SiC exhibit a structure similar to that of GaN[22] and the two 0001 surfaces are polar, i.e., they are either Si- or C-terminated. The polarity of the growing GaN film depends on the polarity of the substrate, whether it offers an Si or C surface. An X-ray photoelectron spectroscopy

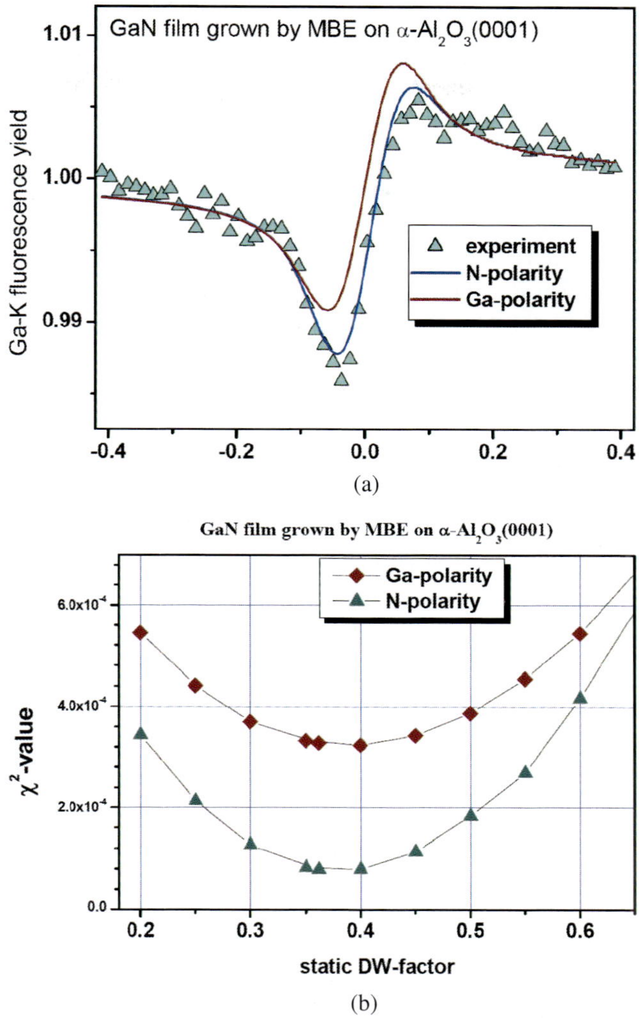

Fig. 16.2. (a) Experimental (0002) Ga K fluorescence XSW yield curve from a 1-μm-thick GaN film grown by molecular beam epitaxy on a sapphire (0001) surface. The horizontal scale is in degree. The best fit (blue line) corresponds to the N-polarity of the film and a static Debye–Waller factor of $e^{-W} = 0.36$. The best fit for the Ga-polarity is shown also for comparison as the red line. (b) The χ^2 values as a function of the static Debye–Waller factor.

study of GaN films grown by MOVPE on SiC showed that GaN epitaxial films on Si- and C-faces of SiC are terminated by nitrogen and gallium atomic layers, respectively.[23] A high-resolution electron microscopy study of the AlN/SiC interface gave the opposite result,[24] i.e., the Al-face AlN was found on Si-face of the SiC. The latter result is in agreement with *ab initio* total energy calculations.[25]

A typical experimental XSW data set is shown in Fig. 16.3(a). The χ^2-plots for the Ga-face and N-face polarity of the film are shown in Fig. 16.3(b). For all experimental data, taken from different parts of the sample, the XSW analysis gives four to six times lower χ^2-values for the Ga-face polarity of the film. The best fit for the Ga-face film is shown in Fig. 16.3(a) by the red line. The theoretical curve for the N-face film is also shown for comparison. Thus, our analysis unambiguously shows that the GaN film grows with the Ga-face on a Si-face SiC substrate. This result is in agreement with high-resolution electron microscopy data[24] and *ab initio* total energy calculations[25] and supports the "standard framework" proposed in Ref. 16.

The broadening of the X-ray rocking curves is primarily due to the mosaic spread. The values of the FWHM of the X-ray curves measured from different regions of the sample varied from 62 to 95 arc sec while dynamical diffraction theory gives a FWHM of 39 arc sec. The corresponding σ-values of the Gaussian function varied from 20 to 36 arc sec. The fact that the X-ray standing wave is generated by the "coherent" part of the lattice gives us the unique opportunity to look at the correlation between the mosaic spread and the perfection of the lattice within the mosaic blocks. The static DW factor derived from the XSW data as a function of the mosaic spread is shown in Fig. 16.3(c) for a set of different regions on the sample. This analysis yields the interesting conclusion that a region with a larger mosaic spread exhibits a more perfect lattice within the individual mosaic blocks. This can be explained by the annihilation of point defects by dislocations. The dislocation network serves as a sink for point defects, generated during the epitaxial growth. The higher degree of mosaic spread means that a higher density of dislocations is present in the boundaries between the individual mosaic blocks. The denser dislocation structure is more effective in "sweeping" the lattice of the adjacent blocks from point defects leading to the higher values of the static DW factor. Recently, this effect was studied in more detail[26] and a point defects diffusion length in thin layers of about $0.5\,\mu$m was determined.

It should be noted that from a dynamical diffraction point of view the investigated GaN films were highly imperfect. Even in case of the best quality films, the dislocation density was as high as 10^7–$10^8\,\mathrm{cm}^{-2}$. Because of the "size broadening" of the rockingcurve of thin films, the XSW method can be applied to materials of such moderate crystalline quality.

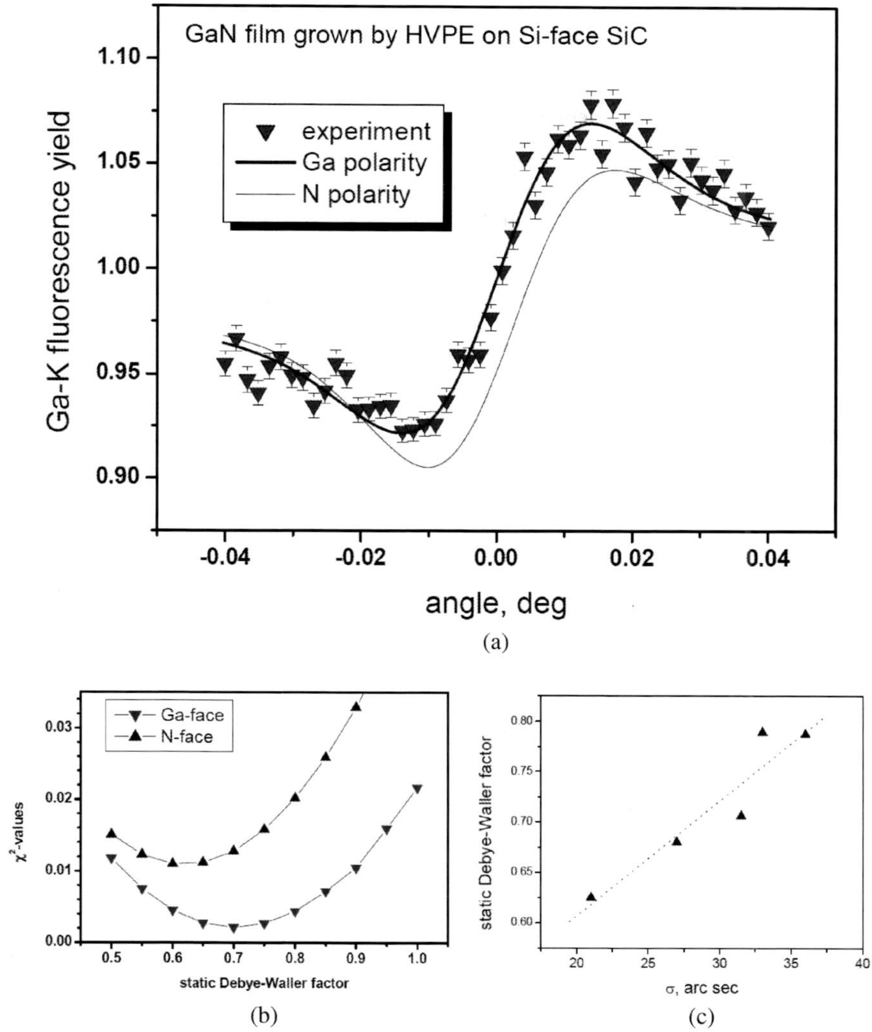

Fig. 16.3. (a) Experimental (0002) Ga K fluorescence yield curve from a 300-nm-thick GaN film grown by hydride vapor phase epitaxy on the Si-face SiC substrate. The red line is the best fit for the Ga-polarity yielding a static Debye–Waller factor of $e^{-W} = 0.71$. The best fit for the N-polarity is shown also for comparison. (b) The χ^2 values as a function of the static Debye–Waller factor. The Ga-face model gives five times lower χ^2 values than the N-face model. (c) Correlation between the static Debye–Waller factor of the GaN lattice and the broadening of the X-ray diffraction curve due to mosaic spread.

The normal incidence XSW technique (*cf.* Chapter 4) also allows using crystals of moderate crystal quality, but requires fixed X-ray energies. The thin film XSW technique is not fixed in energy, at the cost of a very weak intensity modulation of the XSW interference field.

16.3. PTO and PZT Ferroelectric Thin Films

In ferroelectric crystals, electric polarization is caused by a shift of anion and cation sublattices relative to each other along a certain crystallographic direction, which results in a net dipole moment.[27] The "up" and "down" unit cells of a typical ferroelectric, the perovskite $PbTiO_3$ (PTO) or $PbZr_xTi_{1-x}O_3$ (PZT), are shown in Fig. 16.4. The positions of the diffraction planes are shown for the (001) and (002) reflections as the solid and dashed lines, respectively. The inward movement of the antinodal XSW planes for these two reflections are indicated for both polarization states as the solid and dashed arrows. The path difference between "up" and "down" unit cells clearly illustrates the phase difference in the XSW field intensities measured at the location of Pb and Ti atoms in these two polarization states.

The polarization can be switched by applying an external electric field, thus providing the basis for nonvolatile random access memory.[28] Bulk ferroelectric crystals usually consist of regions (domains) of many unit cells polarized in the same direction. In thin PTO films of thickness below the critical value of 50–70 nm grown by the MOCVD on $SrTiO_3$ substrates,

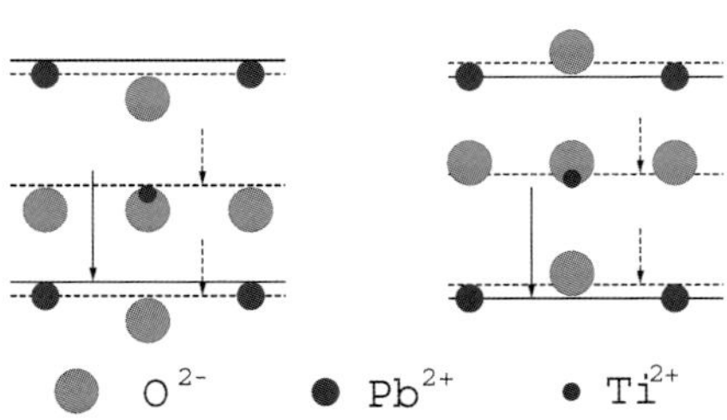

Fig. 16.4. The a-axis projection of the $PbTiO_3$ perovskite "up" and "down" unit cells. The polarization is the result of the relative displacement of the three O^{2-} ions by $-0.112 \cdot c$ from the face-centered positions and the Ti^{4+} ion by $-0.04 \cdot c$ from the body-centered position, where c is the c-axis length. In $PbZr_xTi_{1-x}O_3$ a fraction x of Ti^{4+} ions are replaced by Zr^{4+}. The solid and dashed horizontal lines indicate the position of the (001) and (002) diffraction planes. The solid and dashed arrows show the inward shift of the XSW maxima while crossing the (001) and (002) Bragg reflections, respectively.

the formation of a-domains with in-plane polarization is suppressed and coherently strained single-crystal c-domain films can be grown.[29] A remaining question is whether the c-films are polarized "up" or "down", or whether there is a mixture of the two-type domains.

To answer this question, the XSW technique was applied to PTO films grown by MOCVD on $SrTiO_3$ substrates.[30] The film thicknesses were 100, 200, and 400 Å. The experiments were performed at the 5ID-D and 12ID-C undulator beamlines at the Advanced Photon Source. The Pb L fluorescence yield was measured at an incident energy of 13.5 keV. Though the XSW data from the Pb sublattice is sufficient for polarity determinations, the data from the Ti sublattice can be used as a stringent test of the proposed structural model. The Ti K fluorescence yield was measured at an incident beam energy of 8 keV. To suppress the contribution from the Ti atoms within the underlying $SrTiO_3$ substrate the measurements were performed by using the evanescent-wave emission technique[31] at a take-off angles in the range of 0° to 0.6°, which is below the 0.65° critical angle for the 4.5 keV Ti K_α X-rays escaping from the PTO. The incident beam was typically 0.4 mm and 0.2 mm in width and height, respectively. Several lateral positions on each sample were examined.

The experimental results are presented in Fig. 16.5. An overview of the experimental data reveals a remarkable counter-phase relationship between the (001) Pb and Ti modulations, which is due to the approximately $d_{001}/2$ separation between the Ti and Pb atomic planes, illustrating the unique structural sensitivity of the XSW technique. Another interesting observation is that the amplitude of the XSW-induced modulation of the Ti K fluorescence yield is nearly twice as strong as that for the Pb L yield. This feature is the consequence of the evanescent-wave technique applied to the Ti signal and the enhanced visibility of the XSW interference fringes at the surface of the film.

This data was analyzed with the same approach previously described for the analysis of GaN films. As a first approximation, the film was assumed to have a single, either "up" or "down", polarity and the experimental Pb L yield curves were fitted using the static DW factor as the only fitting parameter. The fit for the 40-nm-film revealed 6.3 times lower χ^2-values for the "up" unit cell yielding $e^{-W} = 0.75$ for both (001) and (002) data. The best fits for the "up" and "down" polarity are shown in Figs. 16.5(a) and 16.5(b). The Ti K fluorescence data was fitted similarly, except that the depth of the probed layer L_{yi} was used as an additional fitting parameter. Again, the fit unambiguously reveals the "up" polarity

References

1. D. Coster, K. S. Knol and J. A. Prinz *Z. Physi.* **63** (1958) 345.
2. W. H. Zachariasen, *Theory of X-ray Diffraction in Crystals* (Dover, New York, 1994).
3. H. Cole and N. R. Stemple, *J. Appl. Phys.* **33** (1962) 2227.
4. R. L. Barns, E. T. Keve and S. S. Abrahams *J. Appl. Crystallogr.* **3** (1970) 27.
5. P. Trucano, *Phys. Rev. B.* **13** (1976) 2524.
6. T. Takahashi and S. Kikuta, *J. Phys. Soc. Jpn.* **47** (1979) 620.
7. J. R. Patel, *Phys. Rev. Lett.* **50** (1983) 1858.
8. M. Bedzyk, G. Materlik and M. Kovalchuk *Phys. Rev. B.* **30** (1984) 2453.
9. M. Bedzyk and G. Materlik, *Phys. Rev. B.* **32** (1985) 6456.
10. S. Annaka, T. Takahashi and S. Kikuta *Jpn. J. Appl. Phys.* **23** (1984) 1637.
11. B. G. Zakharov, A. Y. Kazimirov, V. G. Kohn, E. A. Sozontov and A. N. Sosphenov *Sov. Tech. Phys. Lett.* **15** (1989) 885.
12. A. Yu. Kazimirov, M. V. Kovalchuk and V. G. Kohn, *Sov. Tech. Phys. Lett.* **14** (1988) 587.
13. A. Yu. Kazimirov, M. V. Kovalchuk, A. N. Sosphenov, V. G. Kohn, J. Kub, P. Novak and M. Nerviva, *Acta Crystallogr.* **B48** (1992) 577.
14. A. Kazimirov, T. Haage, L. Ortega, A. Stierle, F. Comin and J. Zegenhagen, *Solid State Commun.* **104** (1997) 347.
15. V. G. Kohn, *Phys. Status Solidi B* **231** (2002) 132.
16. E. S. Hellman, *MRS Internet J. Nitride Semicond. Res.* **3** (1998) Article 11.
17. A. Kazimirov, G. Scherb, J. Zegenhagen, T. -L. Lee, M. Bedzyk, M. K. Kelly, H. Angerer and O. Ambacher, *J. Appl. Phys.* **84** (1998) 1703.
18. B. Daudin, J. L. Rouviere and M. Arley, *Appl. Phys. Lett.* **69** (1996) 2480; M. Seelmann-Eggebert, J. L. Weyher, H. Obloh, H. Zimmermann, A. Rar and S. Porowski, *Appl. Phys. Lett.* **71** (1997) 2635.
19. O. Ambacher *et al.*, *J. Appl. Phys.* **85** (1999) 3222.
20. O. Ambacher *et al.*, *J. Appl. Phys.* **87** (2000) 334.
21. A. Kazimirov, N. Faleev, H. Temkin, M. J. Bedzyk, V. Dmitriev and Yu. Melnik, *J. Appl. Phys.* **89** (2001) 6092.
22. P. Käckell, B. Wenzien and F. Bechstedt, *Phys. Rev. B* **50** (1994) 17037.
23. T. Sasaki and T. Matsuoka, *J. Appl. Phys.* **64** (1988) 4531.
24. F. A. Ponce, C. G. Van de Walle and J. E. Northrup, *Phys. Rev. B* **53** (1996) 7473.
25. R. Di Felice, J. E. Northrup and J. Neugebauer, *Phys. Rev. B* **54** (1996) R17351.
26. N. Faleev, H. Temkin, I. Ahmad, M. Holtz and Yu. Melnik, *J. Appl. Phys.* **98** (2005) 123508.
27. F. Jona and G. Shirane, *Ferroelectric Crystals* (Dover, New York, 1993).
28. O. Auciello, J. F. Scott and R. Ramesh, *Phys. Today* **51** (1998) 22.
29. C. M. Foster, Z. Li, M. Buckett, D. Miller, P. M. Baldo, L. E. Rehn, G. R. Bai, D. Guo, H. You and K. L. Merkle, *J. Appl. Phys.* **78** (1995) 2607.

30. M. J. Bedzyk, A. Kazimirov, D. L. Marasco, T.-L. Lee, C. M. Foster, G.-R. Bai, P. F. Lyman and D. T. Keane, *Phys. Rev. B* **61** (2000) R7873.
31. S. Becker, J. A. Golovchenko and J. R. Patel, *Phys. Rev. Lett.* **50** (1983) 153.
32. C. Thompson, C. M. Foster, J. A. Eastman and G. B. Stephenson, *Appl. Phys. Lett.* **71** (1997) 3516.
33. D. L. Marasco, A. Kazimirov, M. J. Bedzyk, T.-L. Lee, S. K. Streiffer, O. Auciello and G.-R. Bai, *Appl. Phys. Lett.* **79** (2001) 515.
34. M. Drakopoulos, J. Zegenhagen, T.-L. Lee, A. Snigirev, I. Snigireva, V. Cimalla and O. Ambacher, *J. Phys. D: Appl. Phys.* **36** (2003) A214.

Chapter 17

ISOTOPIC EFFECT ON THE LATTICE CONSTANT OF GERMANIUM AND SILICON

ALEXANDER KAZIMIROV

Cornell High Energy Synchrotron Source (CHESS), Cornell University, Ithaca, NY 14853, USA

JÖRG ZEGENHAGEN

European Synchrotron Radiation Facility (ESRF), BP 220, F-39043, Grenoble, France

EVGENY SOZONTOV

Institute of Crystallography (Kaluga Branch), RAS, Akademicheskaya 8, Kaluga, 248640, Russia

VICTOR KOHN

Russian Research Center "Kurchatov Institute", 123182, Moscow, Russia

MANUEL CARDONA

Max-Planck-Institut für Festkörperforschung, Heisenbergstr. 1, 70569 Stuttgart, Germany

Angular dependences of intensity of photoelectron or fluorescence yield excited by the XSW field from the topmost layer of a sufficiently thick epitaxial film are extremely sensitive to the lattice constant difference between substrate and the film. This sensitivity has been used to study the influence of the isotopic mass on the lattice constant of Ge and Si. The samples were homoepitaxial films on a substrate with different isotopic composition. For samples of moderate crystalline quality the measurements were performed at the near backscattering geometry. This approach requires only miniscule amounts of isotopic pure materials and can be applied to materials that cannot be grown as high-quality single crystals.

17.1. Introduction

Crystals of the same chemical composition but built from different isotopes exhibit different lattice constants. This phenomenon is due to the effect of nuclear mass on the zero-point vibrations and the related anharmonicity of interatomic potentials. The effect is largest at low temperature, when the total energy is dominated by the zero-point motion, and vanishes when approaching the Debye temperature. It is stronger for lighter isotopes due to the higher zero-point energy. The lattice constant of a crystal consisting of the heavier isotope is smaller than that built from the lighter one. London, in one of the first theoretical papers on the subject,[1] explained this fact in the following way. Suppose we replaced the atoms in the crystal by their lighter isotope, then their vibration frequency, $\nu \sim \frac{1}{\sqrt{m}}$ for monatomic solids, will be higher. This will also usually happen when the temperature of the crystal is increased. This consideration clearly emphasizes the intrinsic relation of the isotopic effect to crystal thermal expansion (anharmonicity).

The interest in this effect has its origin in the 1930s in the discussions on the condensation of liquid[2,3] He a phenomenon in which the zero-point energy plays a critical role. Substituting isotopes in a crystal lattice was considered to be a straightforward way to change the total energy of crystal without changing its potential energy. The first measurements were performed for the lattice parameters of ordinary and “heavy” ice by using the oscillation technique[4] and for LiH and LiD by using powder diffraction,[5] followed by experiments on ^{6}Li and ^{7}Li fluorides[6] and metallic *Li*.[7] Ten years later, experiments on oxides ^{24}MgO and ^{26}MgO, ^{40}CaO, and ^{48}CaO, and ^{58}NiO and ^{64}NiO were performed at room and liquid N_2 temperatures.[8] Comparison of the experimental data published at that time revealed a remarkable difference of the isotopic effect between crystal with either van der Waals or ionic bonding. Measurements on lithium hydrides by using powder diffraction were reported later for a wide temperature range.[9] Improvement in accuracy of X-ray diffraction techniques led to precise measurements of the lattice constant difference of single crystal natural Ge and isotopically enriched ^{74}Ge at room and liquid nitrogen temperatures by using a double-beam triple-axis spectrometer.[10] Measurements on synthetic

In the ideal case of $L_{yi} \ll t_f \ll L_{ex}$, the XSW data can be treated by using the standard XSW yield equation for perfect crystals (*cf.* Chapter 1). For real samples, the effect of the layer on the local phase of the XSW field cannot be neglected and the theory of the XSW in layered crystals developed in Refs. 24–27 and presented in Chapter 3, is needed for the accurate data analysis.

17.3. Experiment

Samples with isotopically highly enriched epitaxial films have been grown at the MPI-FKF institute in Stuttgart by using molecular beam epitaxy (Ge) and liquid phase epitaxy (Si). Since the total surface shift is proportional to the thickness of the film, special care has been taken for accurate calibration of the growth rate. The final thickness has been measured by using SIMS and AFM techniques. The XSW measurements have been performed over a wide temperature range by using a laboratory X-ray setup (^{nat}Ge/^{76}Ge) at the MPI-FKF and the RÖMO beamline at HASYLAB (*cf.* Appendix 5 by G. Materlik).

17.3.1. *Lattice constant measurements for germanium: ^{nat}Ge/^{76}Ge and ^{70}Ge/^{76}Ge*

The samples were grown in an MBE system with a base pressure below 10^{-10} mbar. The Ge(111) substrate was prepared in UHV by degassing and sputter-annealing to produce a sharp $c(2 \times 8)$ reconstruction. The cleanliness of the surface was controlled by Auger electron spectroscopy. The source material, about 100 mg of ^{76}Ge, was kept at 1450°C in the Knudsen cell, equipped with a liquid nitrogen shroud, leading to a constant growth rate of 0.92 nm/min during the total deposition time of about 25 h. In order to obtain a smooth Ge film, the substrate temperature was slowly ramped from 250°C to 500°C during the first minutes of growth. To ensure the absence of any residual impurities in the film, the whole MBE chamber had been carefully cleaned before the start of the project. As a proof of cleanliness, we obtained a zero-phase shift in the XSW yield for a ^{nat}Ge/^{nat}Ge control samples prepared under the same conditions. Two samples were grown: (1) an isotopically enriched ^{76}Ge film with a thickness of 1.36 ± 0.01 μm and an average isotope mass of $\bar{M} = 75.72$ on a perfect ^{nat}Ge (111) substrate resulting in a mass difference of $\Delta M = 3.05$, and (2) an isotopically enriched ^{76}Ge film with a thickness of 0.56 ± 0.01 μm and

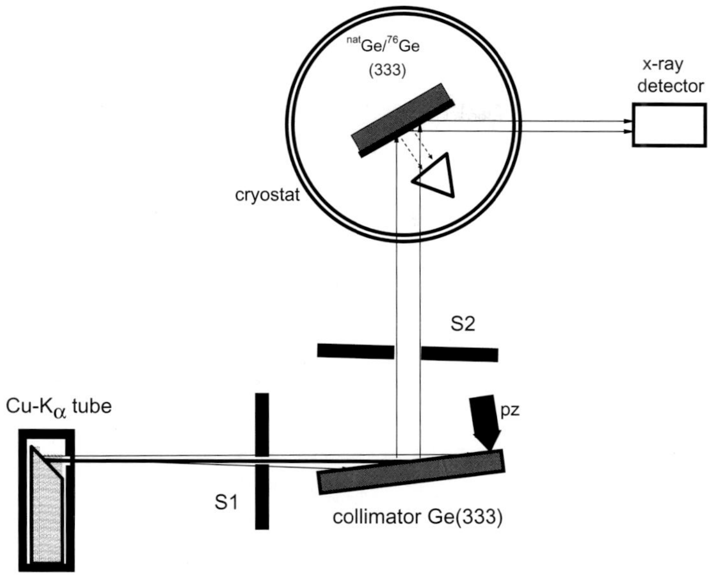

Fig. 17.2. Laboratory XSW setup. The X-ray beam is collimated by an asymmetrical monochromator crystal. The sample is mounted inside a He-flow-through cryostat. The photoelectron yield is measured by a channeltron and the diffracted beam is monitored by the scintillation detector. Angular scans are performed by scanning the collimator crystals by using a piezocrystal.

the same isotopic composition as for the sample #1, but grown on a single crystal ^{70}Ge(111) substrate with the average isotope mass of $\bar{M} = 70.08$.

Sample #1 was measured by using a home-made laboratory setup based on a double-crystal diffractometer, as shown in Fig. 17.2. The Cu-K$_\alpha$ radiation from a 1.5 kW generator, collimated by an asymmetrical Ge(333) crystal (*cf.* Chapter 13), was incident on the sample mounted inside a He flow-through cryostat. The XSW measurements were performed by scanning the collimator crystal with the help of a piezocrystal. The photoelectrons from the sample, excited by the XSW field, were collected by a channeltron at a grazing angle of 0° to 15° to further reduce the escape depth. The photoelectron yield curves for different temperatures and the Ge(333) rocking curve are shown in Fig. 17.3. The phase change by about π is clearly observed while cooling the sample from room temperature to 54 K, which corresponds to a shift of the crystal surface inward by about $d_{333}^{\text{Ge}}/2 = 54.4$ pm. The experimental curves were fitted with the XSW theory for multilayer crystals[27] (*cf.* Chapter 3) by using

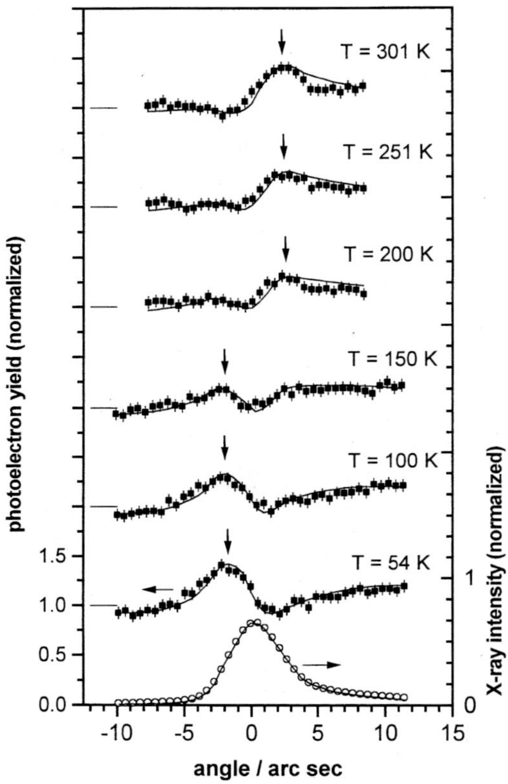

Fig. 17.3. Photoelectron Ge(333) XSW yield curves measured from the $^{nat}Ge/^{76}Ge$ sample at different temperatures. Ge(333) rocking curve is shown at the bottom. The phase shift $\Delta\varphi \approx \pi$ is clearly observed while cooling the sample from 300 K to 54 K (from Ref. 28).

the lattice constant mismatch (difference in Bragg angles of the substrate and the layer) as a fitting parameter. Because of the tetragonal distortion, the values obtained from the fit were then converted to relative lattice constant difference based on the well known elastic constants for Ge. The results scaled to $\Delta M = 1$ are displayed in Fig. 17.4 by the blue solid circles.[28]

The ^{70}Ge single crystal, which served as a substrate for the sample #2, was grown by the Bridgman technique and initially showed excellent X-ray rocking curves, very close to the theoretical ones. However, because of various technical difficulties, it needed three attempts to finally grow the ^{76}Ge epitaxial layer of appropriate thickness. As a result, because of

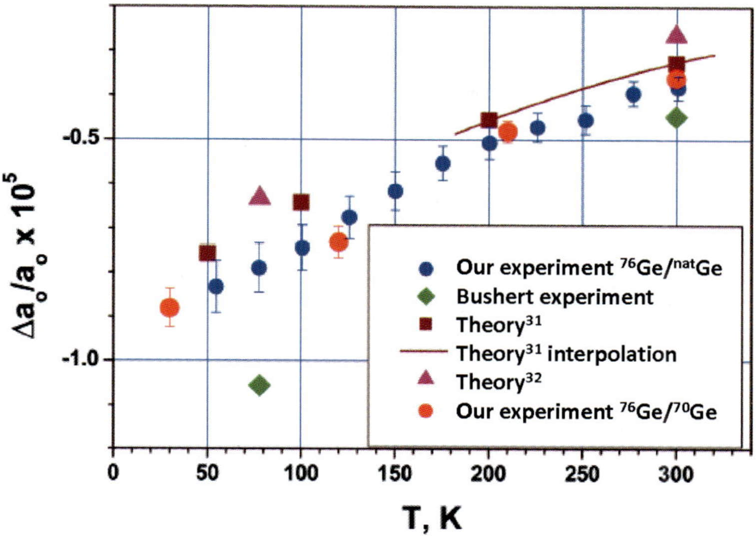

Fig. 17.4. Experimental results from two *Ge* samples scaled to $\Delta M = 1$. For comparison the results from the earlier experiment and from theory are also shown.

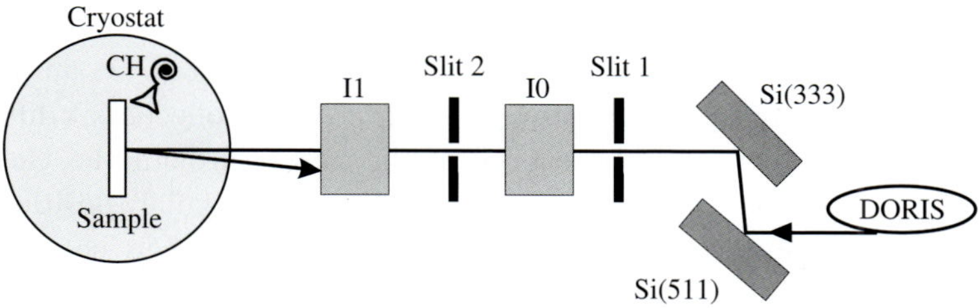

Fig. 17.5. Experimental set-up at the RÖMO beamline at HASYLAB. The X-ray beam is monochromatized by the double-crystal Si(511)/Si(333) monochromator with the energy band-pass of 0.3 eV. The energy of 7.59 keV corresponds to the near backscattering condition for the Ge(444) reflection from the sample. Two ionization chambers (I0, I1) measure the beam intensity. Electrons from the sample are detected by a channeltron (CH) (from Ref. 32).

the thermal treatment and/or the subsequently needed re-polishing, the crystalline quality deteriorated and the sample showed finally a mosaic spread of about 0.3 degree, making it unsuitable for the standard XSW technique. Thus, the sample was measured in the near backscattering geometry[29] which is known to be much less sensitive to mosaicity[30,31] (see Chapter 4). The experimental setup at the RÖMO beamline at HASYLAB is shown in Fig. 17.5.

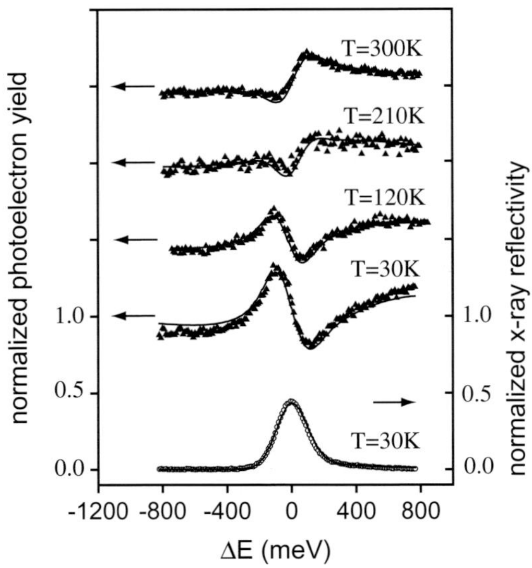

Fig. 17.6. Experimental photoelectron XSW yield curves from the $^{70}Ge/^{76}Ge$ sample measured at near backscattering condition, Ge(444) reflection (from Ref. 32).

A double-crystal Si(511)/Si(333) monochromator was used to produce the X-ray incident beam with an energy band-pass of 0.3 eV at an energy around 7.59 keV. The photoelectron yield curves were collected at different temperatures by scanning the energy of the beam through the Ge(444) reflection.[32] As one can see from Fig. 17.6, the angular positions of maximum and minimum of the XSW yield change when cooling the sample from 300 K to 30 K due to the shift of the surface layers by about $d_{444}^{\mathrm{Ge}}/2 = 40.8\,\mathrm{pm}$.

The experimental values of the isotopic effect scaled to $\Delta M = 1$ for the sample #2 are shown in Fig. 17.4 as red solid circles. In the same graph experimental data from the earlier work[10] are shown along with calculations based on density-functional perturbation theory[33] and path-integral Monte Carlo simulations.[34]

17.3.2. *Lattice constant measurement for silicon:* $^{nat}Si/^{30}Si$

A few years after the start of our project chemically pure silicon isotopes became available in sufficient quantities. An isotopically enriched ^{30}Si film

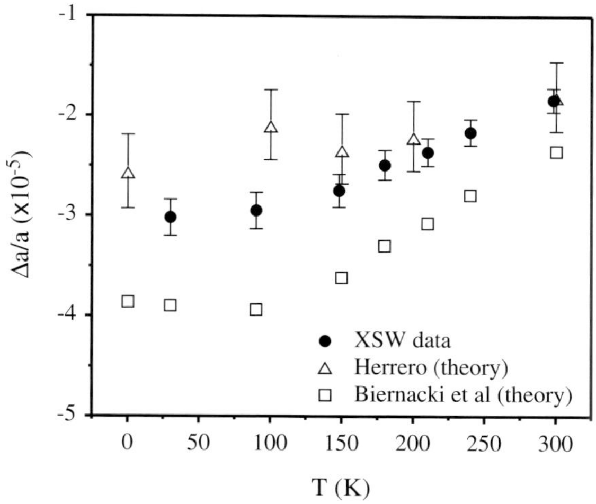

Fig. 17.7. First experimental data (solid circles) from the $^{nat}Si/^{30}Si$ sample measured by using laboratory XSW setup and scaled to $\Delta M = 1$. Results from theory are shown by open triangles[35] and open squares[36] (from Ref. 32).

(60% ^{30}Si, 40% ^{28}Si, $M = 29.20$) with a thickness of 0.92 μm was grown on a perfect single crystal Si(111) substrate with natural isotopic composition (91% ^{28}Si, 4% ^{29}Si, 5% ^{30}Si, $M = 28.14$) by liquid phase epitaxy. The measurements were performed by using the laboratory setup discussed above (*cf.* Fig. 17.2). Experimental data scaled to $\Delta M = 1$ are shown in Fig. 17.7 together with theoretical calculations.[36,37] Note the much stronger effect for lighter Si than for Ge: Fig. 17.7 yields for $T \approx 0\,\mathrm{K}$ $\Delta a_0/a_0 \approx -3 \times 10^{-5}$ whereas in Fig. 17.4 we read $\Delta a_0/a_0 \approx -1 \times 10^{-5}$. It is easy to show[17] that the isotope effect in $\Delta a_0/a_0$ (normalized to $\Delta M = 1$) is proportional to $M^{-3/2}$. The ratio of the $-3/2$ power of the mass of silicon to that of germanium is 3.5, in rather good agreement with the corresponding isotope effects. This scaling law also applies to the corresponding isotope effect measured for diamond: 1.5×10^{-4} for $\Delta M = 1$.[11] The ratio of this effect to that of germanium is 15 whereas $(M_C/M_{\mathrm{Ge}})^{-3/2} = 14.3$. Later, high resolution diffraction measurements at near backscattering conditions on a sample with the same isotopic composition but a film thickness of 15 μm were performed using high order reflections.[37] The results agreed perfectly with our XSW data and revealed anomalous temperature behavior

pump of sarco(endo)plasmic reticulum) is one of the most important energy-dependent calcium-regulating cell components. The protein has the molecular mass of 110 to 115 kDa and the dimensions of $100 \times 80 \times 140$ Å. Various iso-forms of Ca-ATPase consist of 994–1042 amino acid residues, among which are 24 SH-containing cysteinic molecules capable of binding lead ions.

In order to deposit Ca-ATPase on a silicon substrate, we used one of the most efficient methods for immobilizing protein molecules — protein adsorption on a preliminarily formed lipid monolayer carrying a specified electric charge. The general scheme of a deposition technique is as follows: First, a monolayer of dipalmitoylphosphotidylethanolamine (DPPE) is transferred onto a solid substrate by the L-B method. Then, adsorption immobilization of protein molecules on the transferred phospholipid monolayer is performed during the next 24 h. Finally, the substrate with the protein film is taken away through a newly formed Langmuir monolayer of DPPE, thus completing the deposition of a control sample (Fig. 18.2(a)) (for details see Refs. 14 and 15).

Two types of samples have been studied (Figs. 18.2(b) and 18.2(c)). In the sample #1, lead acetate was added into the solution during solubilization and the mixture was incubated for 1 h. Sample #2 was also treated by lead acetate, but upon incubation in the working protein solution, the sample was kept for 3 h in the xydiphone (K, Na — ETHIDRONATE) solution. The goal was to investigate the molecular mechanisms of the protective effect of bisphosphonate drug xydiphon on membrane-bound enzyme damaged by heavy metal (lead in this case).

The TR-XSW measurements have been carried at the station KMC2 (BESSY-II) at the energy of 13.5 keV. The Pb L_α fluorescence angular dependences for two types of samples are presented in Fig. 18.3. One can see a remarkable difference in the curves from the samples #1 and #2: the

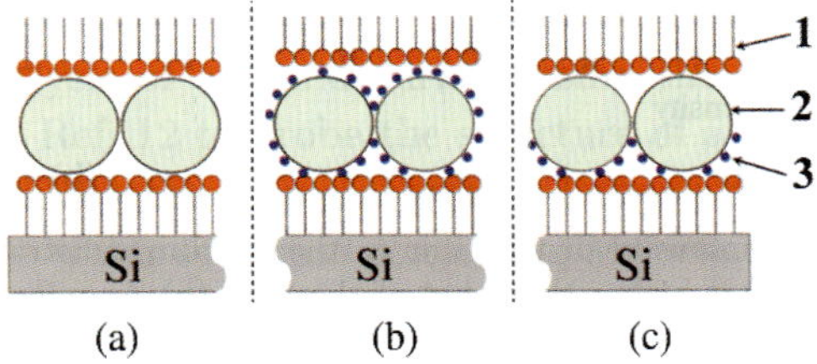

Fig. 18.2. Schematic view of the lipid–protein structure on a solid substrate for three samples: (a) control sample; (b) sample 1; (c) sample 2. (1) Phospholipid layers; (2) protein molecule; (3) lead ions.

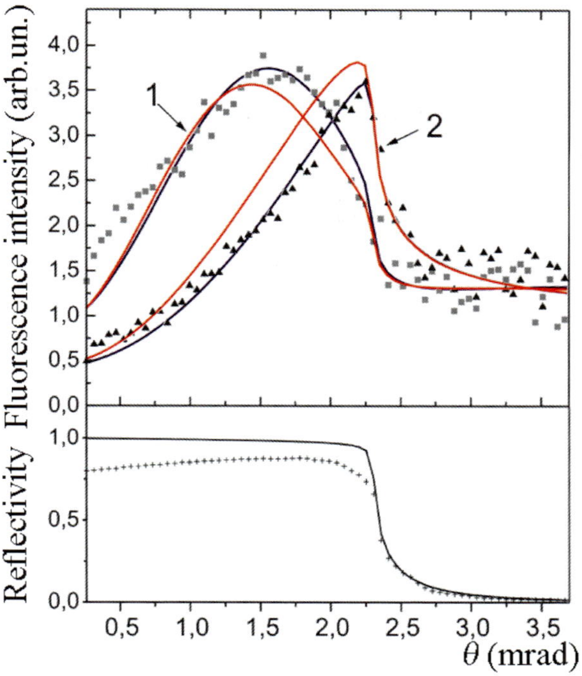

Fig. 18.3. Experimental and calculated angular dependencies of the Pb L_α fluorescence yield from the protein/lipid films of Ca-ATPase damaged by lead ions. Curve 1 represents experimental results for sample #1 depicted in Fig. 18.2(b); blue line — best fit (lead layer thickness of 13 nm), red line — calculations for the lead layer thickness of 15 nm. Curve 2 — experimental results for sample #2 depicted in Fig. 18.2(c); blue line — best fit (lead layer thickness of 1 nm), red line — calculations for the lead layer thickness of 3 nm.

maximum of the fluorescence yield from the sample #1 is shifted toward the small angles, and the fluorescence curve is much broader. This is typical for the situation when the lead ions are distributed over a thicker layer whose thickness is comparable with or exceeds the period of the XSW in the vicinity of θ_c (Fig. 18.1(b), curve 3).

The fluorescence curve from the sample #2 is sharply peaked at $\theta = \theta_c$ which is characteristic for the ions located in a thin layer near the bottom of the film (Fig. 18.1(b), curve 2). We may conclude that treating the lipid–protein film by xydiphone solution resulted in a removal of a large amount of lead ions. The best fit of the Pb L_α fluorescence data yields the thickness of the Pb layer of 13 nm for the sample #1 and only 1 nm for the sample #2. To demonstrate the sensitivity of the XSW measurements to the thickness of the layer containing lead atoms the fluorescence curves calculated for 15 nm thickness and 3 nm thickness are also plotted. Computational algorithm based on a recursive method developed by Parratt[16] has been used to analyze all experimental data presented in this work.

These results demonstrate unique possibility of the technique to probe large organic molecules with the dimensions comparable with the scale of wave field intensity variations and obtain valuable information about conformational, structural and functional properties of the model cell membranes.

18.3. Langmuir Layer on a Liquid Surface

In this section, we will approach a more complicated experimental problem — to study different atomic species of a single monolayer on a liquid surface.

The TR-XSW fluorescence yield is determined by two main factors: the E-field intensity distribution inside a sample and the mean position of atoms with respect to this field. Although the electric field in a thin film on a semi-infinite substrate is determined mainly by the substrate, in the subcritical region $\theta < \theta_c$ a film may change essentially the E-field intensity. Even in the simplest case of a homogeneous layer with the thickness d_f and the refractive index $n_f = 1 - \delta_f - i\beta_f$ there are at least three fitting parameters: the thickness and the electron density of a film determining the electric field inside the film, and the mean position D of the atoms emitting fluorescence.

As it was shown in Ref. 17, the E-field intensity distribution inside the film can be approximated by a linear function of the depth z up to a second order in $k_z z$, where k_z is the z-component of the wave vector in the film, under the following condition:

$$4k_0^2 d_f^2 \delta_f \ll 1 \tag{18.1}$$

where $k_0 = (2\pi/\lambda)$, λ is the wavelength of an incident beam. Theoretical considerations show that at small angles $\theta < \theta_c$ the correction of the E-field intensity caused by such a thin film is completely determined by the value of the surface electron density of the film (the number of electrons per unit area). At larger angles $\theta > \theta_c$ the electric field is practically unaffected by a thin film and its intensity depends only on electron density of a substrate.

For a molecular monolayer, the electron density of the film can be expressed in terms of molecular parameters: N_{mol} — the number of electrons in one molecule and S_{mol} — the area per one molecule. Assuming chemical composition of monolayer is known, i.e., the value of N_{mol} is determined, we obtain that S_{mol} is the only film parameter that is required to describe the wave field inside the film. The advantage of TR studies of molecular monolayers on water surface is the possibility to estimate the S_{mol} value from the pressure–area isotherm. Then, the model can be fitted to experimental fluorescence data by using only one fitting parameter, i.e., the mean position of ions D, thus providing an unambiguous location of ions incorporated in the molecular film.

The objects of our investigation were organic surfactants containing heavy ions and forming well-ordered Langmuir layers on liquid surface: metal-substituted phthalacyanines (Pc), polyorganosiloxanes and phospholipids.[18,19] The Pc molecules are the disk-shaped macrocyclic molecules of high thermal and chemical stability. Among the studied Pc molecules were phthalocyanine derivatives containing Sn-[di(tetradecyloxycarbonyl) phthalocyanine]tin. Metal ion is positioned in the center of a macrocyclic ring and despite the small linear dimensions of these objects (about a nanometer) we aimed to determine the position of phthalocyanine macrocycles with respect to the water surface. The parameters of the studied monolayers were proved to satisfy condition (1), therefore, the approach presented above was used for the analysis of experimental data. A two-box model was chosen: the bottom slab — macrocyclic rings and the top slab — aliphatic tails.

The experimental studies were performed at the station ID10B (ESRF) at the energy of 12.5 keV. The angular curve of the Sn L_α fluorescence yield from the Sn(Pc) monolayer is shown in Fig. 18.4. The experimental curve has a well-pronounced maximum at $\theta \sim \theta_c$ (the node of the XSW lies in the vicinity of the air–water interface) and it was fitted by using only one variable parameter D, the distance between Sn ions and the water surface. The area per molecule determined from the compression isotherm was $S_{mol} = 90\,\text{Å}^2$. The best fit (solid line in Fig. 18.4) was found for $D = 7\,\text{Å}$ (for comparison, dashed line shows calculations for $D = 2.5\,\text{Å}$).

The diameter of the macrocyclic ring of a phthalocyanine molecule obtained from the computer simulation is about 13 Å (see inset in Fig. 18.4). Therefore, the distance between the metal ions and the water surface ($D = 7\,\text{Å}$) determined experimentally allows one to conclude that the macrocycles of phthalocyanine molecules are oriented perpendicular to the water surface. In spite of the low sensitivity of the TR-XSW to such small objects due to much larger size of the XSW pattern, we were able to obtain

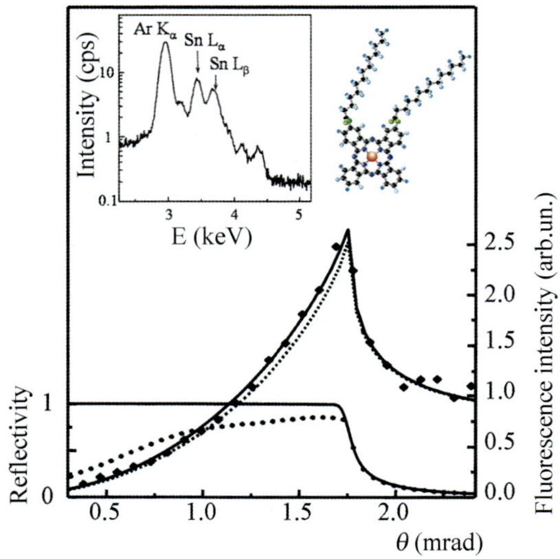

Fig. 18.4. Experimental and calculated angular dependencies of the Sn L_α fluorescence yield from the organic monolayer of Sn(Pc) on a water subphase. Fluorescence curves calculated for different distances D between the Sn ions and the water surface are shown as a solid line ($D = 7$ Å) and dashed line ($D = 2.5$ Å). In the inset, the fluorescence spectrum from the Sn(Pc) monolayer and the schematic of Pc molecule are shown.

quantitative structural information from the individual organic monolayer on liquid surface.

18.4. Molecular Organization in Lipid–Protein Systems on Liquid Surface

Based on the experience and the knowledge gained from the experiments described in the previous sections we want as a next step to apply TR-XSW method to study a self-assembled lipid–protein system on a liquid subphase.[20] The main difficulty is the lack of any prior information about the electron density profile of the structure in which molecules retain their mobility and are free to move over the water subphase.

We studied the alkaline phosphatase — an enzyme catalyzing the hydrolysis of monoesters of the phosphoric acid. The alkaline phosphatase is a zinc-containing enzyme where the zinc ions are necessary for catalytic activity and, probably, for the stabilization of the native structure. It has been shown that phosphatases are mostly dimers. The protein has the molecular mass of about 95 kDa and the dimensions of $90 \times 50 \times 50$ Å. Deposition of a lipid–protein structure on a water subphase was the following. The L-α-phosphatidylinositol ammonium salt was dissolved in water by shaking. To obtain a lipid–protein mixture, the solution of the dispersed phospholipid was added to a water solution of alkaline phosphatase. The lipid–protein mixture was shaken for 10 min and then introduced under the Langmuir monolayer of DPPC (dipalmitoylphosphatidylcholine) preliminarily formed on a water subphase surface. As a water subphase we used the buffer solution of 0.01 M TRIS in a highly purified water (specific resistivity 18 MΩ cm) obtained from a commercial purification system Milipore Corp.

Experimental studies were performed at the station ID10B (ESRF), $E = 13.3$ keV. Figure 18.5 shows the spectrum of the characteristic fluorescence of lipid–protein structure on the water subphase.

The following characteristic fluorescent peaks have been identified: the P K_α peak from the phosphate groups of the heads of phospholipid molecules, DPPC and phosphatidylinositol; the S K_α peak originated from the functional groups of cysteine and methionine residues of protein molecules; the Zn K_α peak from the Zn ions contained in the active centers of the metalloenzyme alkaline phosphatase; the Cl K_α peak from the water subphase (TRIS-solution). Of special interest are the intense Ni K_α and Fe K_α peaks that indicate the presence of these ions in the lipid–protein structure. The surface of protein molecules has many amino acid residues whose functional groups can coordinate metal ions (the carboxyl, ester, amine, alcohol, and imidazole groups); the functional groups of the phospholipids DPPC and phosphatidylinositol can also interact with cations of the subphase.

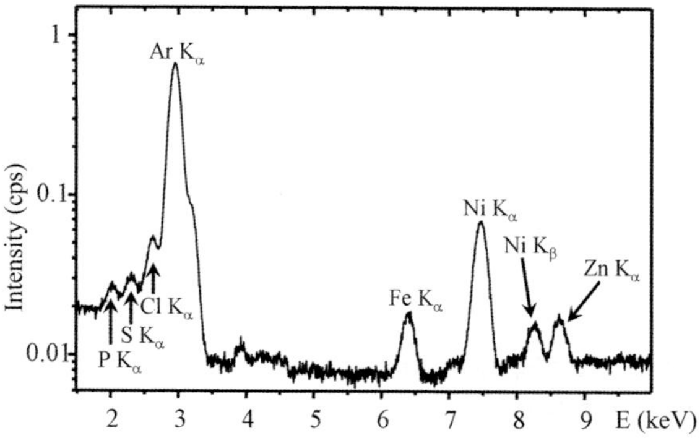

Fig. 18.5. X-ray fluorescence spectrum from the lipid-protein system on the water subphase (cps — counts per second).

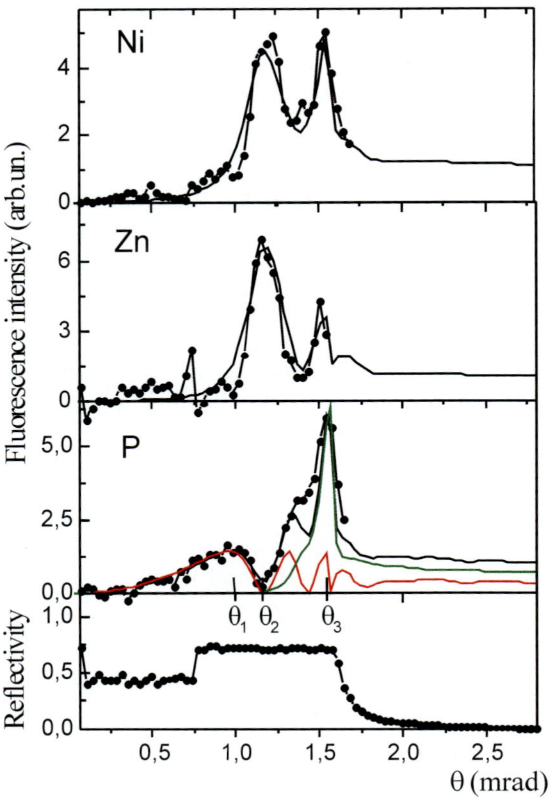

Fig. 18.6. Experimental and calculated angular dependencies of Ni K$_\alpha$, Zn K$_\alpha$, and P K$_\alpha$ fluorescence yield from the lipid–protein structure on the water subphase. Black solid lines — the best fit, dots — experimental results. Red and green lines show the calculated fluorescence yield from the top (layer A) and the bottom (layer D) layers of the proposed model depicted in Fig. 18.7(b). Three angular positions are marked for which E-field intensity distributions are presented in Fig. 18.7(a).

Figure 18.6 shows the angular dependences of the fluorescence yield for the P, Zn, and Ni ions. Even qualitative comparison of these experimental fluorescence curves provides important information on the distribution of different ions in the protein–lipid film. Indeed, the substantial differences between the shapes of P and Zn fluorescence curves indicate that these ions occupy different positions in the protein–lipid film. This observation allows concluding that phospholipid and alkaline phosphatase molecules form separate layers. It should be reminded that phosphorus ions (contained in the phosphate groups of phospholipids) can be considered as markers

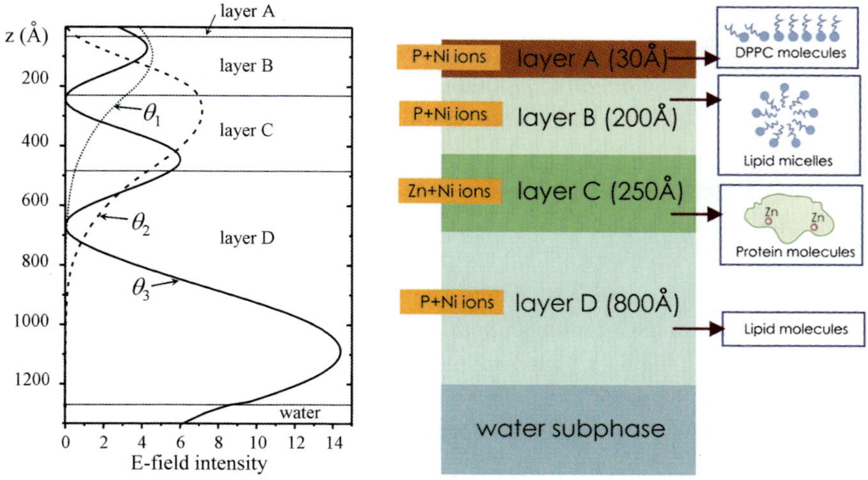

(a) (b)

Fig. 18.7. (a) The normalized E-field intensity distributions over lipid–protein system on the water subphase calculated for the fixed angles θ_1, θ_2, and θ_3 marked on Fig. 18.6. (b) Proposed four-layer model of stratified lipid–protein system formed at the water subphase.

of phospholipids, and zinc ions (contained in the active center of the metalloenzyme alkaline phosphatase) are markers of protein molecules.

A good agreement between the theoretical angular dependences of the fluorescence yield and the experimental data has been obtained in the frame of the four-layer model schematically shown in Fig. 18.7(b).

When fitting the data, it was taken into account that the layers could differ in the amounts of ions of a given species. Thus the resultant theoretical fluorescence curve for each ion species is a linear combination of fluorescence curves for the layers comprising the multilayer model. As an illustration, in Fig. 18.6, the theoretical angular dependences of the fluorescence yield from layer A (red line) and layer D (green line) are plotted. As one can see, the characteristic features of the experimental P K_α fluorescence data can be described by summing these two curves with appropriate weight coefficients.

The following parameters of the layers (thicknesses and electron densities) were determined by fitting the theoretical angular dependencies to all the three, Zn, P, and Ni, experimental fluorescence curves: layer A: thickness 30 Å, density 0.46; layer B: thickness 200 Å, density 0.21;

layer C: thickness 250 Å, density 0.42; and layer D: thickness 800 Å, density 0.68 (the electron densities of the layers are indicated in units relative to the electron density of water ρ_0 taken to be 1). The layer A contains mostly P ions; the layer C, only Zn and Ni ions; and the layer D, mostly Ni and P ions.

The E-field intensity distributions over the lipid–protein structure calculated for three angles of incidence (Fig. 18.7(a)) exhibit a well known "resonance enhancement",[21–23] the phenomena observed in layered systems under the conditions of TR. Thus, the first and the second Ni peaks, the main Zn peak, and the main P peak correspond to the extraordinary high values of the E-field, which are more than four times higher than the intensity of the incident field.

In summary, analysis of the obtained experimental data allowed us to determine the composition and structure of bioorganic nanosystem and to locate trace amounts of metal ions incorporated into the multilayer from the water subphase. The presented results may be explained by the fact that the lipid–protein mixture spontaneously stratified at the water subphase after spreading under a layer of the phospholipid DPPC. Indeed, the mixture of the phosphatidyl inositol and the alkaline phosphatase is thermodynamically unstable disperse system. Most likely, the phosphatidyl inositol molecules formed aggregations, such as micelles or vesicles (layer B in Fig. 18.7(b)), under the DPPC phospholipid layer, whereas molecules of alkaline phosphatase arranged themselves in a self-assembled manner into a pure protein layer containing no phospholipid molecules (layer C in Fig. 18.7(b)). The self-assembling of a complicated lipid–protein system into a layered structure following the model suggested in our study may look rather schematic at first glance. We want to emphasize, however, that according to the liquid mosaic model of the cell membrane discussed earlier, the layered organization of the lipid–protein components constitutes their basic nature. To the best of our knowledge, this is the first TR-XSW characterization of the molecular organization in a single organic monolayer and the self-assembling in lipid–protein systems on the surface of a liquid subphase.

In conclusion, we demonstrated that the TR-XSW technique can be successfully applied not only to study the model biomembranes on a solid substrate, but to a single monolayer and complicated self-assembled lipid–protein system on a liquid subphase as well. In this way, the native conformation of protein molecules remains unaltered and their biological functions preserved, which allows this system to be used as a model for studying various biophysical and biochemical processes in functionally active biological membranes opening up new possibilities for fundamental and applied research in nanobiotechnology, biology, and medicine.

References

1. S. J. Singer and G. L. Nicolson, *Science* **175** (1972) 720.
2. K. B. Blodgett and I. Langmuir, *Phys. Rev.* **51** (1937) 964.
3. R. S. Becker, J. A. Golovchenko and J. R. Patel, *Phys. Rev. Lett.* **50** (1983) 153.
4. J. Wang, M. J. Bedzyk, T. L. Penner and M. Caffrey, *Nature* **354** (1991) 377.
5. M. J. Bedzyk, D. H. Bilderback, G. M. Bommarito, M. Caffrey and J. S. Schildkraut, *Science* **241** (1988) 1788.
6. M. J. Bedzyk, G. M. Bommarito, M. Caffrey and T. L. Penner, *Science* **248** (1990) 52.
7. S. I. Zheludeva, M. V. Kovalchuk, N. N. Novikova, A. N. Sosphenov, M. C. Petty, V. A. Howarth, S. P. Collins and R. I. Cernik, *Mater. Sci. Eng.* **C3** (1995) 211.
8. S. I. Zheludeva, M. V. Kovalchuk, N. N. Novikova, A. N. Sosphenov, V. E. Erochin and L. A. Feigin, *J. Appl. Phys.* **26** (1993) A202.
9. S. I. Zheludeva, M. V. Kovalchuk and N. N. Novikova, *Spectrochim. Acta B* **56** (2001) 2019.
10. S. I. Zheludeva, M. V. Kovalchuk, N. N. Novikova, A. N. Sosphenov, N. E. Malysheva, N. N. Salaschenko, A. D. Akhsakhaljan and Yu. Ja. Platonov, *Thin Solid Films* **232** (1993) 252.
11. J. Wang, C. J. A. Wallace, I. Clark-Lewis and M. Caffrey, *J. Mol. Biol.* **237(1)** (1994) 1.
12. J. M. Bloch, M. Sansone, F. Rondelez, D. G. Peiffer, P. Pincus, M. W. Kim and P. M. Eisenberger, *Phys. Rev. Lett.* **54** (1985) 1039.
13. W. B. Yun and J. M. Bloch, *J. Appl. Phys.* **68**(1990) 1421.
14. N. N. Novikova, E. A. Yurieva, S. I. Zheludeva, M. V. Kovalchuk, N. D. Stepina, O. V. Konovalov, A. L. Tolstikhina, R. V. Gaynutdinov, D. V. Urusova, T. A. Matkovskaya, A. M. Rubtsov, O. D.Lopina and A. I. Erko, *J. Sur. Invest. X-ray, Synchrotron and Neutron Techniques* **8** (2005) 71 (in Russian).
15. N. N. Novikova, E. A. Yurieva, S. I. Zheludeva, M. V. Kovalchuk, N. D. Stepina, A. L. Tolstikhina, R. V. Gaynutdinov, D. V. Urusova, T. A. Matkovskaya, A. M. Rubtsov, O. D. Lopina, A. I. Erko and O. V. Konovalov, *J. Synchrotron Rad.* **12** (2005) 511.
16. L. G. Parratt, *Phys. Rev.* **95** (1954) 359.
17. S. I. Zheludeva, N. N. Novikova, O. V. Konovalov, M. V. Kovalchuk, N. D. Stepina, E. A. Yur'eva, I. V. Myagkov, Yu. K. Godovskil, N. N. Makarova, A. M. Rubtsov, O. D. Lopina, A. I. Erko, A. L. Tolstikhina, R. V. Gainutdinov, V. V. Lider, E. Yu. Tereschenko and L. G. Yanusova, *Crystallogr. Rep.* **48** (2003) S25.

18. N. N. Novikova, S. I. Zheludeva, O. V. Konovalov, M. V. Kovalchuk, N. D. Stepina, I. V. Myagkov, Y. K. Godovsky, N. N. Makarova, E. Yu. Tereschenko and L. G. Yanusova, *J. Appl. Crystallogr.* **36** (2003) 727.
19. S. I. Zheludeva, N. N. Novikova, O. V. Konovalov, M. V. Kovalchuk, N. D. Stepina and E. Yu. Tereschenko, *Mater Sci. Eng.* **C23** (2003) 567.
20. S. I. Zheludeva, N. N. Novikova, N. D. Stepina, E. A. Yurieva and O. V. Konovalov, *Book of Abstracts: International Conference of Nanoscience and Technology*, ICN+T 2006, Basel/Switzerland, 1016 (2006).
21. J. Wang, M. J. Bedzyk and M. Caffrey, *Science* **258** (1992) 775.
22. B. N. Dev, A. K. Das, S. Dev, D. W. Schubert, M. Stamm and G. Materlik, *Phys. Rev.* **B61** (2000) 8462.
23. F. Pfeiffer, U. Mennicke and T. Salditt, *J. Appl. Crystallogr.* **35** (2002) 163.

Chapter 19

APPLICATIONS OF XSW IN INTERFACIAL GEOCHEMISTRY

PAUL FENTER

Chemical Sciences and Engineering Division,
Argonne National Laboratory,
9700 South Cass Avenue, Argonne IL, 60439, USA

The application of XSW for probing structures and processes relevant to interfacial geochemistry is reviewed. Examples given include the imaging of 3D cation adsorption sites, mineral surface terminations, and cation incorporation in biofilms.

19.1. Introduction

The cycling of elements in Earth's environment is of fundamental interest in the geosciences. The transport of elements, whether nutrients (e.g., Ca^{2+}, K^+, etc.) or contaminants (e.g., heavy metals), through the environment is controlled by their mobility. Elements can be effectively sequestered through adsorption to mineral surfaces or by incorporation into minerals as impurities.[1] The interaction of dissolved cations with minerals is therefore fundamental to the area of low-temperature geochemistry.

There are many significant challenges to understanding these processes. First these processes are best studied *in situ* (e.g., with the mineral in contact with water), to reveal the actual processes unperturbed by changes of environment and so that the processes can be observed in real-time. Second, minerals are often rather complex, with many distinct lattice sites that must be distinguished. Finally, elemental sensitivity is critical since adsorption and/or incorporation typically involves only trace quantities. These requirements necessitate the use of X-ray based probes that have the capability to penetrate macroscopic quantities of water, and have sub-angstrom spatial resolution (derived from the short wavelength of X-rays) and elemental specificity (e.g., obtained through X-ray fluorescence which provides a unique signature for each element). The X-ray standing wave technique encompasses all of these characteristics and has been shown to be particularly powerful for probing both ion adsorption and ion incorporation phenomena. See for example, Refs. 2–13.

19.2. Cation Adsorption at the Mineral-Water Interface

The determination and interpretation of surface adsorption sites by XSW is traditionally guided by XSW triangulation followed by model-dependent fits to the XSW-derived parameters (coherent fractions and coherent positions).[12] This approach is best-suited to probe systems in which there exists a single adsorption site so that there is a one-to-one relationship between coherent positions and the adsorbate locations. When the actual distribution is more complex, e.g., consisting of multiple inequivalent sites, as is common with many natural minerals, the interpretation of XSW data becomes more difficult.[7] As described in Chapter 14 in this volume, a more robust approach is to "image" elemental distributions directly by discrete Fourier synthesis. This approach has now been used to probe impurity site distributions in a solid[14] and at the mineral-water interface.[10,11,15]

The recent determination of cation site distributions at the TiO_2 (110)-electrolyte interface provides an excellent example of the need for such a robust approach when probing the potential complexities of the mineral-water interface.[10,11] Adsorbed ion site distributions (Fig. 19.1) were obtained *in situ* using a thin film cell.[7] In this case, only five symmetry distinct reflections were measured, including the **H** = *110*, *200*, *101*, *211*, and *111*, making use of the known surface symmetry (i.e., with two orthogonal mirror planes). In spite of this relatively small number of Fourier components, these XSW images readily distinguished between the various high symmetry adsorption sites. It is evident that Sr^{2+} adsorbs at the two symmetry-equivalent "tetradentate" sites within the surface unit mesh, bridging between two terminal oxygens (TOs) and two bridging oxygens (BOs). This is the same site that was identified for adsorption of Y^{3+} (also by XSW imaging), as well as for Rb^+ and H_2O as determined by separate crystal truncation rod measurements,[10,11] and for various other cations as observed in recent computational studies,[16] suggesting that this is an important reactive site at the rutile-water interface. In contrast, Zn^{2+} is found to have a different distribution with two symmetry distinct sites,

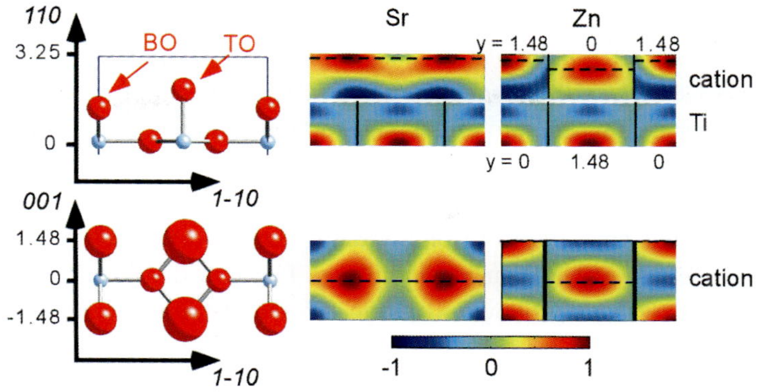

Fig. 19.1. Three-dimensional distributions of Sr^{2+} and Zn^{2+} at the TiO_2 (110)-water interface derived by XSW imaging (Adapted from Ref. 10). Side (top) and top (bottom) views schematics of the rutile surface with bridging (BO) and terminal oxygen (TO) sites. Also shown are 2D cuts though the imaged ion distributions for Sr^{2+} and Zn^{2+} showing the vertical and lateral ion distributions, respectively.

one above the bridging oxygen (BO) and one bridging between terminal oxygen (TO) sites.

Using these preliminary but model-independent images, the adsorption sites were optimized through a quantitative comparison of measured and predicted XSW-derived coherent fractions and positions, revealing the precise cation locations are displaced from the high-symmetry adsorption sites.[11] These results showed excellent consistency with classical molecular dynamics simulation, except for the location of Zn^{2+}. A more in-depth investigation of the interaction of Zn^{2+} with rutile (with X-ray absorption spectroscopy and density functional theory) demonstrated that the adsorption process included a change in the Zn coordination geometry (from six-fold to four-fold coordinated hydration shell) as well as the hydrolysis of a water molecule in the adsorbed Zn hydration shell.[17] Thus, the Bragg XSW results, in conjunction with parallel computation and modeling work, provided a unique window into the interfacial chemistry of the oxide-water interface.

The XSW images are controlled, in part, by the sampling of the element-specific partial structure factor at available Bragg reflections. For instance, the oblong shape of the density lobes is controlled primarily by the experimental resolution, corresponding to $\sim$3 Å along *hkl* $[1\bar{1}0]$, and $\sim$1 Å along *110*. There is, however, some additional broadening (along $[1\bar{1}0]$ and *001* for Sr, and primarily along $[1\bar{1}0]$ for Zn) apparently due to the cation distribution, which was ultimately associated with unresolved lateral displacements of the cations from the high symmetry adsorption sites, confirmed with a quantitative model-dependent analysis. The derived images also show two inequivalent adsorption sites per unit mesh. Since the derived coverage for the two ions is ~0.4 cations per unit mesh, it is clear that this is an artifact of the XSW imaging process in which the full density profile is folded into a single unit cell (*cf.* Chapter 14).

The imaged Zn^{2+} distribution highlights an important issue concerning the completeness of the Fourier sampling. The two distinct Zn sites that were imaged by XSW follow the same lateral distribution as that of Ti within the substrate lattice. These two Ti sites are not distinguished (i.e., they were in phase) for most of the reflections that were used in the XSW measurement (i.e., $\mathbf{H} = 110, 200, 101, 211$). They were distinguished only by the XSW measurements at $\mathbf{H} = 111$ (a so-called "oxygen only" reflection) where the two Ti sites are exactly out of phase. In the absence of the *111* Fourier component, therefore, the images would show both of these two sites as being occupied regardless of the actual site-specific occupation factors. These "oxygen only" reflections are more challenging to measure due to their intrinsically narrow Darwin widths and weak structure factors, but inclusion of the *111* Fourier component provides critical insight by demonstrating that both of the two inequivalent sites are occupied by Zn. This example highlights the importance of obtaining a complete sampling of available Bragg reflections (at least to some maximum $|H|$) to avoid unintentional artifacts in XSW imaging of complex systems (*cf.* Chapter 14).

19.3. Imaging Mineral Surface Terminations with XSW

While the termination of many mineral surfaces often is known from simple crystallographic considerations and knowledge of sample preparation, in many cases minerals may have multiple potential terminations and the actual terminating plane is not known *a priori*. There have been numerous reports of mineral surfaces that have partial layers or coexisting inequivalent terminations.[15,18,19] This can substantially complicate interfacial structural analyses. Consequently, it is useful to have a method that can readily image the termination(s) of a mineral.

Multiple mineral surface terminations can be imaged through the use of a sorbing agent that adsorbs equivalently on the different terminations, and whose distribution can be imaged directly by XSW. This was recently demonstrated for the hematite (012) aqueous interface.[15] In this case, there are two possible terminations determined by whether the outermost

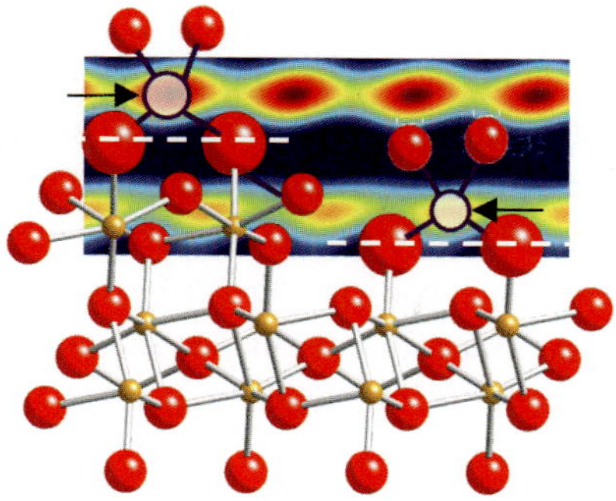

Fig. 19.2. Side view schematic of the hematite (012) surface with two distinct terminations (the half- and full unit cell terminations, noted by the horizontal dashed lines) and defined by the terminal oxygen sites (enlarged red spheres). This leads to two distinct adsorption arsenic adsorption heights for the adsorbed arsenate species (shown as translucent spheres, indicated by black arrows). This schematic is superimposed on the vertical arsenic distribution imaged directly by XSW measurements (indicated by the same color map used in Fig. 19.1) (adapted from Ref. 15).

surface layer consists of a complete- or half-unit cell termination as shown in Fig. 19.2. Each of these terminations, however, expresses a chemically similar surface consisting of terminal oxygen (TO) sites that have similar lateral separations. Arsenate (AsO_4^{3-}) was used as a probe of the mineral termination since it is expected to adsorb in a similar manner on the two terminations. Full 3D XSW imaging showed that As adsorbed in a bidentate configuration, bridging between the TOs.[15] Vertical cuts through the XSW-derived As-distribution, shown in the background of Fig. 19.2, show two distinct heights in the arsenic distribution, vertically separated by 2.6 Å, which is smaller than the height difference for equivalently terminated hematite planes (3.69 Å). Superposition of the XSW-derived distribution with the crystal structure shows that each of these density lobes corresponds to arsenate in a bridging-bidentate adsorption geometry for each of the two terminations. Assuming that the interaction of arsenate with each termination is equivalent, this leads to a conclusion that 75% of the surface was terminated by the half unit-cell termination, with the remainder of the surface terminated by a full-unit cell. These fractions are consistent with that determined by separate specular CTR measurements on the same sample.[15]

19.4. Probing the Reactivity of Biofilm-Coated Minerals

Mineral surfaces are often coated by organic matter that is a potentially significant complication in understanding mineral-surface reactivity in the environment. Organic matter can be adsorbed in a number of forms, ranging from simple aliphatic compounds (e.g., stearate), bio-molecules (e.g., cysteine), natural organic matter (e.g., humic substances derived from the decomposition of plants), and microbial biofilms. Of particular concern is determining how the presence of the organic matter influences the reactivity of the mineral surface since the presence of such organic material may "block" the reactivity of the mineral surface or create additional sites for adsorption increasing the sorption capacity.

Recent work has used total-external reflection XSW (*cf.* Chapter 5) to probe the reactivity of organic-film coated surfaces, including microbial biofilms[13,20] and model polymer films having the chemical functional groups that are present in natural organic matter.[21] The case of microbial biofilms is particularly challenging since it consists of individual microbes as well as their exudates, resulting in a complex, laterally inhomogeneous film that is ~1 μm thick (Fig. 19.3). This work probed the distribution of Pb^{2+} above metal oxide surfaces (Al_2O_3 and Fe_2O_3) that were coated with a biofilm and subsequently exposed to solutions containing Pb^{2+}. The Pb^{2+} distribution was probed with long-period X-ray standing waves generated by X-rays reflecting from the system at incident angles near the

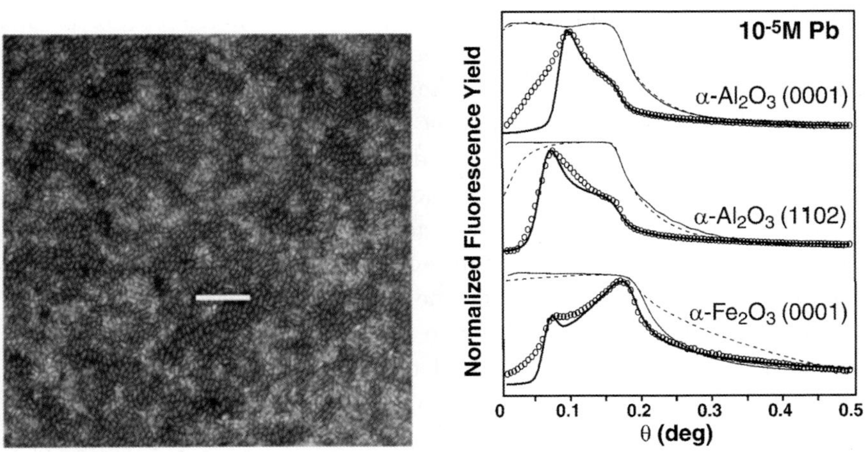

Fig. 19.3. Left: Optical fluorescence image of a microbial biofilm on an Al_2O_3 (0001) surface (scale bar: 10 μm). Right: TR-XSW fluorescence profiles (open circles) of Pb^{2+} interacting with three surfaces with the associated X-ray reflectivity data (dashed lines). Fits to the data are shown as solid lines. (Reproduced with permission from Templeton *et al.*, *Proc Natl Acad Sci USA* **98** (2001), 11897; Copyright (2001) National Academy of Sciences, USA.)

critical angle. The XSW period in this geometry decreases continuously with increasing angles of incidence as described in a separate chapter on the TR-XSW technique. In this sample geometry, the first antinode of the X-ray standing wave intersects each interface (e.g., biofilm-air, mineral-biofilm) when the incident angle equals the critical angle of that interface. Consequently enhancement of the fluorescent yield at the biofilm-air critical angle indicates incorporation within the biofilm, while a maximum near the metal oxide-biofilm interface critical angle indicates accumulation of the cation at the oxide-biofilm interface. This conceptual simplification makes it possible to probe the relative sorption capacity of the metal-oxide surface versus the biofilm.

Systematic measurements of TR-XSW as a function of substrate orientation (e.g., [0001] versus [$1\bar{1}02$] crystallographic surfaces), composition(Al_2O_3 versus Fe_2O_3), and Pb^{2+} solution concentration showed some surprising results. At low Pb^{2+} concentrations, maxima in the fluorescent yield at the substrate critical angle indicated in all cases that the bio-film did not "block" adsorption to the oxide. Surprisingly, comparison of results for Al_2O_3 [0001] versus [$1\bar{1}02$] showed that Pb^{2+} incorporation into the biofilm depended upon the oxide crystal orientation, associated with the different cation adsorption affinities of the two substrates. At higher Pb^{2+} concentrations incorporation into the biofilm was observed for all three surfaces showing that the presence of the biofilm substantially increased the sorption capacity of the mineral surface at high cation concentrations. These results were quantified using a simple box model for adsorption in which the reactivity of the biofilm was assumed to be uniform. Within this framework, the surface versus biofilm reactivity was determined for these two surfaces at a range of cation concentration. A more complete analysis of these spectra, as described by Chapter 5 of this volume on TR-XSW, would allow the full cation distribution to be determined within the biofilm, at least to the spatial resolution of these measurements.

19.5. Conclusions

The explicit phase and elemental sensitivity of XSW is a powerful capability to probe processes at complex mineral surfaces and mineral-water interfaces. The robustness of this approach allows for direct observations through *in situ* measurements in the environment of interest, especially when using the XSW imaging[14] capability to visualize the inherent complexity of these natural systems.

Chapter 20

COMPLEX SURFACE PHASES OF Sb ON Si(113): COMBINING XSW AND DENSITY FUNCTIONAL THEORY

M. SIEBERT, TH. SCHMIDT, J. I. FLEGE and J. FALTA

Institute of Solid State Physics, University of Bremen, P. O. Box 330440, D-28334 Bremen, Germany

We report on recent results regarding the surface structure obtained by adsorption of Sb onto Si(113). This system was identified to show adsorption at multiple surface sites. It yields an example for a structure determination that could only be solved by the combination of X-ray standing wave measurements with an analysis of possible adsorption sites by means of density functional theory.

20.1. Introduction

The X-ray standing waves (XSW) technique has successfully been applied to determine the adsorption site and bonding geometry for a large variety of surface adsorbate systems (see, e.g., Ref. 1 and references therein). An XSW analysis consists of a number of measurements of different Fourier components of the distribution function of the atoms under investigation. With $\mathbf{H} = (hkl)$ describing the corresponding Bragg condition, each measurement yields both the phase Φ_c (coherent position) and the amplitude f_c (coherent fraction) of the (hkl) Fourier component of the atomic spatial distribution (*cf.* Chapter 1):

$$f_c^{\mathbf{H}} \exp(2\pi i \Phi_c^{\mathbf{H}}) = \frac{1}{N} \sum_{n=1}^{N} D_n^{\mathbf{H}} \exp(i2\pi \mathbf{H} \cdot \mathbf{r}_n). \tag{20.1}$$

In this equation, the Debye–Waller factor $D_n^{\mathbf{H}}$ accounts for thermal vibrations of the nth atom along the direction $\mathbf{H}$. In most cases, $D_n^{\mathbf{H}}$ is very close to unity.

For "simple" systems with a single adsorption (or incorporation) site, measurements at three independent Bragg conditions allow to determine this site directly, by three-dimensional geometric triangulation.[2] For more complex structures, however, with multiple adsorption sites, the coherent fractions f_c can become much smaller than unity, and a straightforward interpretation of the coherent positions Φ_c is no longer possible.

In principle, the measurement of a large number of Fourier components, in the literature referred to as XSW imaging (*cf.* Chapter 14 in this volume), can still provide enough information for a direct structure determination. For many cases, however, the measurement of a large number of Fourier coefficients is not possible due to experimental restrictions such as low excitation cross-sections or the presence of strong indirect excitation channels of the secondary signal. Moreover, the number of Fourier components obtainable by XSW can be limited due to insufficient quality of the substrate crystals.[3] In such cases, the combination of XSW experiments with theoretical modeling of the system is an alternative approach to solve the structure under investigation. Competing structural models can be tested by comparing the Fourier components of the model configurations with the experimental XSW results. The reliability of this approach crucially depends on the quality of the theoretical models. Over the last decade, density functional theory (DFT)[4,5] has been established as a state of the art technique for surface structure calculation, especially in conjunction with scanning tunneling microscopy (STM).[6–9] In the present contribution, we demonstrate that a combination of XSW with DFT is also very suitable because both methods are sensitive to the atomic positions on a picometer scale, thus a direct comparison between theory and experiment is possible. Moreover, extending XSW experiments by DFT calculations also allows to investigate the relaxation of the upper atomic layers of the substrate, which in general is not possible with XSW alone.

In order to illustrate this approach, we address the adsorption of Sb on Si(113) in the following. A major technological interest in this system arises from the fact that group V elements are promising candidates for the surfactant mediated epitaxy of germanium on silicon.[10] On Si(001), which is the most common substrate for semiconductor applications, the Stranski–Krastanov growth mode[11] of Ge can be suppressed and smooth Ge films can be grown by pre-adsorption of one monolayer (ML) of arsenic. Nevertheless, these Ge films show many defects.[12] High-quality Ge films can be grown when Sb is applied as surfactant on Si(111).[13] However, Si(111) is of less technological importance, mainly due to the moderate quality of the oxide layers. Si(113) can be considered a promising alternative, because it consists of Si(001)-like and Si(111)-like surface atoms, and excellent oxide films have been achieved on Si(113).[14]

20.2. Experimental and Computational Details

The experiments were carried out at the undulator beamline BW1 at HASYLAB (*cf.* Appendix 5 by G. Materlik). A standard nondispersive monochromator setup with pairs of symmetrically and asymmetrically cut crystals (*cf.* Chapter 13.4.1) was used for XSW measurements in (111) and (113) Bragg reflection. Moreover, XSW measurements were performed dispersively[15] in (202) Bragg reflection with the Si(113) monochromator setup.

For the XSW measurements, the incident photon energy was tuned across the Bragg condition, while X-ray photoelectron spectra (XPS) were recorded with a hemispherical electron energy analyzer, from which the Sb $2p_{3/2}$ photoelectron yield was extracted as secondary signal. Details of the data evaluation can be found elsewhere.[1,16,17]

All samples were investigated under ultra high vaccuum (UHV) conditions with a base pressure of 1×10^{-10} mbar. After RCA-cleaning[18] and subsequent degassing at 600 °C in the UHV chamber for at least 12 hours, the protective oxide layer was removed by heating to 870°C for 5 minutes. Following this procedure, the (3×2) reconstruction[19] of a well-ordered Si(113) surface was verified with low-energy electron diffraction (LEED). Sb was deposited from a Knudsen cell at elevated substrate temperatures. XPS spectra of Si $1s$ and Sb $2p_{3/2}$ photoelectrons were used to determine the relative Sb coverage. The absolute values given in the following were obtained by calibrating the respective signal ratios to corresponding data obtained from a Sb-saturated Si(111) sample prepared at 600°C, which is known to results in a coverage of 1 ML_{111} Sb.[20]

DFT calculations were performed within the local-density approximation.[21,22] using a plane-wave basis set and employing *ab initio* norm conserving pseudopotentials[23,24] as incorporated into the package PWscf by Baroni and coworkers.[25] A repeated-slab geometry with 12 Si layers and a vacuum region of 15 Å thickness was constructed for all calculations. In order to simulate the XSW data of the resulting adsorbate geometries, the bottom bulk-like silicon bilayer was chosen as a reference to compute the coherent positions, since the XSW positions are determined by the bulk lattice planes. Convergence with respect to both the total energy and the geometric properties was obtained by chosing a $4 \times 4 \times 1\,k$ point grid[26] in the surface Brillouin zone and by raising the high energy plane-wave cut-off level to 25 Ry (1 Ry = 13.605 eV). With respect to convergence of the total energy, reliable surface structures may already be obtained at cut-off energies of around 14 Ry. However, with further increased cut-off energies, subtle relaxation effects in the near-surface region were still observed. These small changes in the coordinates of the Sb and Si atoms add up in the calculation of the Fourier components in the XSW calculations since the reference for the XSW signal is the Si bulk, i.e., the bottom Si bilayer atoms held fixed at their corresponding bulk positions. Hence, in this case, the convergence of the XSW Fourier components with respect to the employed high energy cut-off level poses a sharper criterion for convergence than the standard convergence test for the cut-off energy.

20.3. Results and Discussion

Up to now, Sb-covered Si(113) surfaces were investigated with STM in combination with DFT,[6,27] as well as with XPS and LEED.[28] Depending on deposition temperature and Sb coverage, different surface reconstructions have been observed.[28,29] For temperatures below 400°C, a (3×1) LEED pattern is observed for Sb saturation coverage, which changes to a (1×1) pattern around 400°C. At about 500°C a (1×2) LEED pattern evolves with weak and diffuse ($\frac{1}{2}n$) spots which become more pronounced with increasing temperature up to 650°C. Since (2×2) LEED patterns have been reported for this temperature range as well, it is most likely that a (2×2) reconstruction whith a high density of antiphase domain boundaries in $[1\bar{1}0]$ direction is formed.[29] Upon further increase of the temperature, the superstructure spots start to split along the $[33\bar{2}]$ direction. This can be explained by Sb vacancy rows periodically incorporated into (2×2) reconstructed areas, which leads to predominant periodicities of (1×7) and (1×5) at 690°C and 725°C, respectively. At even higher temperatures, a (1×2) pattern appears. At about 790°C, hardly any Sb sticks on the surface, and a (1×1) reconstruction is observed, coexisting with bare (3×2)-Si(113).

A comparable low-coverage Sb:Si(113) surface can also be obtained by submonolayer Sb deposits at 500–600°C. For a coverage of $\Theta_{\text{Sb}} = 0.03$ ML,[a]

[a]Here and in the following 1 ML = 4.1×10^{14} cm^{-2}.

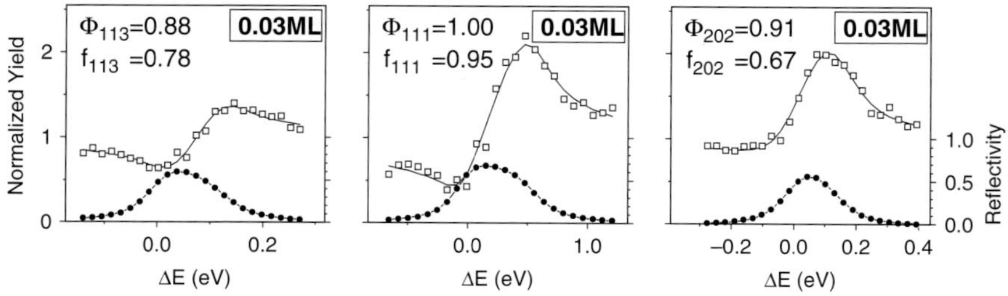

Fig. 20.1. XSW data and fit (solid lines) of reflectivities (•) and Sb $2p_{3/2}$ photoelectron yield (□) at a photon energy of $E = 4.5$ keV in (113), (111), and (202) reflection geometry (from left to right), for an Sb coverage of $\Theta_{\rm Sb} = 0.03$ ML.

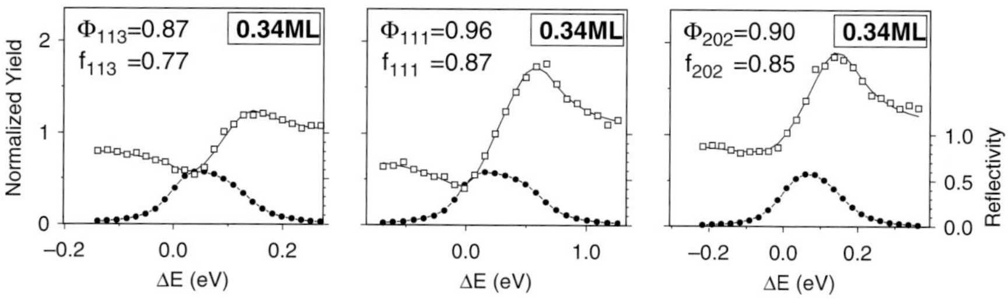

Fig. 20.2. XSW data and fit (solid lines) of reflectivities (•) and Sb $2p_{3/2}$ photoelectron yield (□) at a photon energy of E=4.5 keV in (113), (111), and (202) reflection geometry (from left to right), for $\Theta_{\rm Sb} = 0.34$ ML.

a (1×1) LEED pattern is obtained with faint contributions of a (3×2) reconstruction. The XSW data of such a sample are shown in Fig. 20.1. For all Bragg reflections used here, the coherent fractions f_c are high. The deviation of the coherent fraction $f_c^{111} = 0.95 \pm 0.02$ from unity is so small that it could be fully assigned to the influence of the Debye–Waller factor.[b] The values for $f_c^{113} = 0.78 \pm 0.02$ and $f_c^{202} = 0.67$ show that there is one dominating adsorption site, and only marginal contributions of additional sites have to be considered.

For $\Theta_{\rm Sb} = 0.34$, a (1×2) LEED pattern with weak and diffuse superstructure spots is observed (not shown here). The related XSW data is shown in Fig. 20.2. The coherent positions are virtually identical, and the coherent fractions are only slightly decreased with respect to the results for $\Theta_{\rm Sb} = 0.03$ ML. Hence, this surface can also be described by a single

[b] According to neutron diffraction data obtained for bulk Sb,[30] Debye–Waller factors of $D^{111} = 0.96$–0.98 and $D^{113} = 0.90$–0.94 can be estimated.

Table 20.1. Measured (XSW) and calculated (DFT) coherent positions Φ_c^{hkl} and fractions f_c^{hkl}.

Experiment	Sb coverage	Φ_c^{113}	Φ_c^{111}	Φ_c^{202}	f_c^{113}	f_c^{111}	f_c^{202}
XSW	0.03 ML	0.88	1.00	0.91	0.78	0.95	0.67
XSW	0.34 ML	0.87	0.96	0.90	0.77	0.87	0.85
	low coverage average	0.88	0.98	0.91	0.78	0.91	0.76
XSW	1.26 ML	0.93	0.97	0.91	0.58	0.67	0.71
XSW	1.59 ML	1.01	0.95	0.89	0.38	0.41	0.55
Theory	**Unit mesh**	Φ_c^{113}	Φ_c^{111}	Φ_c^{202}	f_c^{113}	f_c^{111}	f_c^{202}
DFT Sb adatom	(1×1)	0.85	0.96	0.91	1.00	1.00	1.00
DFT Si–Si dimer	(2×2)	0.94	0.96	0.95	0.92	0.92	1.00
DFT Sb–Si dimer	(2×2)	0.96	0.94	0.94	0.86	0.55	1.00
DFT Sb–Sb dimer	(2×2)	0.98	0.89	0.91	0.78	0.31	0.87
DFT Sb adatom–tetramer	(2×2)	1.13	1.02	1.08	0.36	0.52	0.56
DFT Sb tetramer–tetramer	(2×2)	1.23	1.11	0.18	0.84	0.43	0.84

Note: The error bar $\Delta\Phi_c$ is in the order of ±0.02. Note that the values for f_c^{202} are much less reliable as compared to f_c^{113} and f_c^{111}; for details see text.

adsorption site, and it is reasonable to evaluate the averaged coherent positions of the investigated coverages, as shown in Table 20.1, in order to determine the main site by triangulation directly. This procedure yields an adsorption site[31] which we denote as the *antimony adatom* site, in accordance with Wolff *et al.*,[6,27] who proposed this bonding configuration to occur at anti-phase domain boundaries between (2×2) domains.

The adatom site has also been calculated with DFT (see Table 20.1). The results are in excellent agreement with the experiment, since the differences between the averaged measured and the calculated Φ_c of a (1×1) adatom reconstruction are within the accuracies of measurements and calculation. Thus, the adatom site is confirmed. An overview of the calculated Sb adatom surface configuration and the bond lengths is given in Fig. 20.3(a). The bond lengths between the Sb and Si surface atoms were calculated to 2.64 Å for bonds to Si(001)-like atoms and to 2.88 Å for bonds to Si(111)-like atoms. For Sb/Si(111), an Sb–Si bond length of 2.66 Å was determined with surface extended X-ray absorption fine structure (SEXAFS) measurements.[32] Hence, the bonds of the Sb adatoms to the Si(111)-like atoms on the Si(113) surface are significantly stretched and high tensile strain is induced along the $[33\bar{2}]$ direction. The bonds between the Si(111)-like and Si(001)-like surface atoms have a length of 2.55 Å in comparison to the Si–Si bulk value of 2.35 Å. Therefore, they

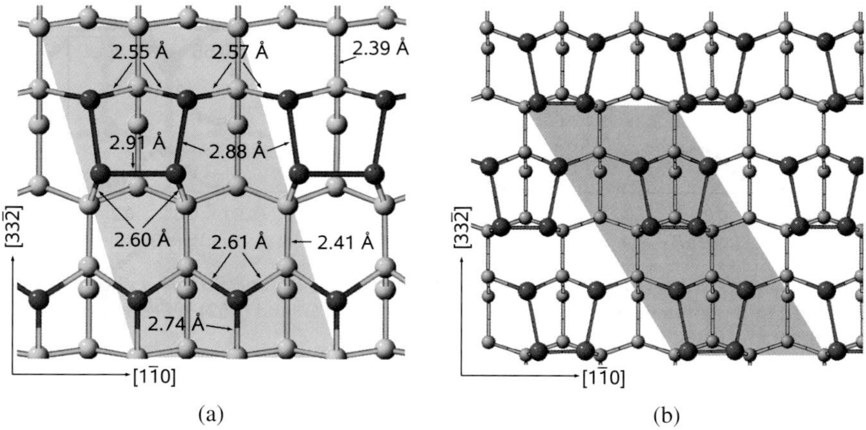

Fig. 20.5. (a) (2×2) Sb adatom–tetramer reconstruction as resulting from DFT. Sb atoms are displayed in dark and Si atoms in light shade. Bond lengths are indicated. (b) (2×2) Sb tetramer-tetramer reconstruction. Unit meshes are shaded.

A comparison between the calculated and experimentally determined Fourier components, which are shown in Table 20.1, yields that for $\Theta_{\mathrm{Sb}}>1$ ML the XSW data cannot be explained by the presence of any of the (2×2) reconstructions alone. Therefore, we computed "mixed" Fourier components, i.e., Fourier components for coexisting surface reconstructions, based on the Fourier components of the pure phases as determined by DFT. Using the relative contributions of the individual phases as free parameters, these mixed components were fitted to the XSW data. For this least-square fitting, Debye–Waller factors of 0.90 and 0.95 were incorporated for the calculated (111) and (113) components, respectively. The coherent fraction f_c^{202} was neglected in this procedure, because the (202) measurements were performed with a dispersive setup.[c]

One result of this procedure is that the contribution of the Sb tetramer–tetramer phase was always computed below 0.02 for the surfaces investigated here. Therefore, this reconstruction is very unlikely to occur. This is not very surprising when considering the large differences between the calculated Fourier components for this structure and the experimentally determined values, as listed in Table 20.1.

[c]For the evaluation of XSW measurements with dispersive setups, additional parameters are required,[15] especially when using X-ray sources with significant divergence. These source parameters turned out to strongly affect the deduced coherent fraction f_c^{202}, which is therefore considered much less accurate as compared to f_c^{111} and f_c^{113}.

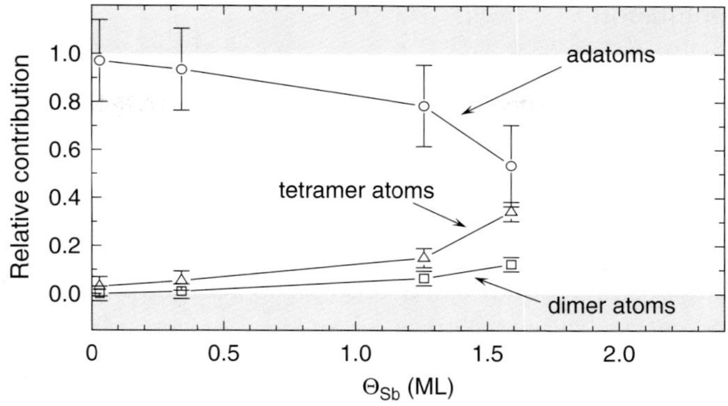

Fig. 20.6. Contribution of Sb adsorption sites to the surface coverage as resulting from least square fit calculations for mixtures of different reconstruction patterns.

In order to evaluate the course of occupation of different adsorption sites, the calculated surface mixtures are evaluated site-specifically, as shown in Fig. 20.6. For low Sb coverage, the adsorbate structure is dominated by adatom sites, and the contribution of all other sites can be neglected within the accuracy of the approach. At higher coverages the relative contribution of Sb adatom sites decreases, especially in benefit of Sb atoms in tetramers. At a coverage of 1.26 ML, the contribution of Sb on dimer sites increases only slightly to 0.07, but the contribution of tetramer sites increased to 0.15. This trend continues at a coverage of 1.59 ML, where Sb atoms in dimers have a contribution of 0.12 and tetramer atoms of 0.34.

The successive occupation of the different surface sites can be understood in terms of chemical surface passivation: For very low coverages, Sb atoms reside in adatom sites only, since in this site, each Sb atom bonds to three substrate atoms, thus the number of dangling bonds is reduced as efficiently as possible, even though a high surface strain is imposed. If more Sb is available on the surface, this strain can be reduced by the insertion of interstitial dimers partly containing Sb atoms, even if the number of dangling bonds saturated per Sb atom (which amounts two for a Sb atom in a dimer) is reduced. In average, the Sb atoms in the tetramer configuration bond to 1.5 substrate atoms only. Hence, this building block can energetically only be favored under Sb-rich conditions. Therefore, it might be predicted that an even higher total Sb coverage is necessary to stabilize domains of a tetramer–tetramer reconstruction which was not observed here, for $\Theta_{\mathrm{Sb}} \leq 1.59\,\mathrm{ML}$.

20.4. Conclusion

We have investigated the adsorption of Sb on the Si(113) surface by an approach combining X-ray standing waves and density-functional theory. XSW measurements were performed in (113), (111) and (202) Bragg reflection geometry for a wide range of Sb coverages from 0.03 ML to 1.59 ML. The results were compared to the Fourier components of the surface structures calculated by DFT. Both XSW and DFT show that the Sb adatom site plays an important role as a structural element for the Sb induced surface reconstructions on Si(113). At very low coverages this site is almost exclusively occupied. Since the adatom site imposes a highly tensile surface strain, the formation of interstitial dimers consisting of Si or Sb atoms is favored, as already observed at a coverage of 0.34 ML. At higher coverages Sb progressively occupies additional surface sites, leading to a complex mixture of Sb adatoms, dimers and tetramers, which tends toward a (2×2) Sb adatom–tetramer reconstruction.

The results shown here demonstrate that the combination of XSW measurements and DFT calculations allow to quantitatively determine the contribution of several adsorption sites in complex mixtures of surface phases. This approach is also very promising for the structural investigation of pure phases, e.g., of complex organic molecules. It might be of particular interest in cases where the number of Fourier coefficients one can determine by XSW experiments is limited, e.g., for substrates with large twist mosaicity, for which reflections with components parallel to the surface are broadened and can hardly be used for XSW.

References

1. J. Zegenhagen, *Surf. Sci. Rep.* **18** (1993) 202.
2. D. P. Woodruff, *Prog. Surf. Sci.* **57** (1998) 1.
3. M. Siebert, T. Schmidt, J. I. Flege, S. Einfeldt, S. Figge, D. Hommel and J. Falta, *Phys. Status Solidi C* **3** (2006) 1729.
4. P. Hohenberg and W. Kohn, *Phys. Rev.* **136** (1964) B864.
5. W. Kohn and L. J. Sham, *Phys. Rev.* **140** (1965) A1133.
6. G. Wolff, H.-J. Müssig, J. Dabrowski, W. Arabczyk and S. Hinrich, *Surf. Sci.* **357/358** (1996) 667.
7. H. Brune, *Surf. Sci. Rep.* **31** (1998) 121.
8. B. Engels, P. Richard, K. Schroeder, S. Blügel, P. Ebert and K. Urban, *Phys. Rev. Lett.* **58** (1998) 7799.
9. R. L. Rosa, J. Neugebauer, J. E. Northrup, C.-D. Lee and R. M. Feenstra, *Phys. Rev. Lett.* **58** (1998) 7799.

10. M. Copel, M. C. Reuter, E. Kaxiras and R. M. Tromp, *Phys. Rev. Lett.* **63** (1989) 632.
11. P. M. J. Maree, K. Nakagawa, F. M. Mulders, J. F. van der Veen and K. L. Kavanagh, *Surf. Sci.* **191** (1987) 305.
12. F. K. LeGoues, M. Copel and R. M. Tromp, *Phys. Rev. B.* **42** (1990) 11690.
13. M. Horn-von Hoegen, M. Copel, J. C. Tsang, M. C. Reuter and R. M. Tromp, *Phys. Rev. B.* **50** (1994) 10811.
14. H.-J. Müssig, J. Dabrowski, K.-E. Ehwald, P. Gaworzewski, A. Huber and U. Lambert, *Microelectron. Eng.* **56** (2001) 195.
15. T. Gog, A. Hille, D. Bahr and G. Materlik, *Rev. Sci. Instrum.* **66** (1995) 1522.
16. B. W. Batterman and H. Cole, *Rev. Mod. Phys.* **36** (1964) 681.
17. J. Zegenhagen, G. Materlik and W. Uelhoff, *J. X-Ray Sci. Technol.* **2** (1990) 214.
18. W. Kern, *Semicond. Int.* **7** (1984) 94.
19. W. Ranke, *Phys. Rev. B.* **41** (1990) 5243.
20. S. Andrieu, *J. Appl. Phys.* **69** (1991) 1366.
21. D. M. Ceperley and B. J. Alder, *Phys. Rev. Lett.* **45** (1980) 566.
22. J. P. Perdew and A. Zunger, *Phys. Rev. B.* **23** (1981) 5048.
23. G. B. Bachelet, D. R. Hamann and M. Schlüter, *Phys. Rev. B.* **26** (1982) 4199.
24. X. Gonze, R. Stumpf and M. Scheffler, *Phys. Rev. B.* **44** (1991) 8503.
25. S. Baroni, A. D. Corso, S. de Gironcoli and P. Giannozzi, Available at: http://www.pwscf.org.
26. H. J. Monkhorst and J. D. Pack, *Phys. Rev. B.* **13** (1976) 5188.
27. J. Dabrowski, H.-J. Müssig, G. Wolff and S. Hinrich, *Surf. Sci.* **411** (1998) 54.
28. K. S. An, C. C. Hwang, C.-Y. Park and A. Kakizaki, *Jpn. J. Appl. Phys., Part I.* **39** (2000) 2771.
29. M. Siebert, T. Schmidt, J. I. Flege and J. Falta, *Phys. Rev. B.* **72** (2005) 045323.
30. P. Fischer, I. Sosnowska and M. Szymanski, *J Phys. C: Solid State Phy.* **11** (1978) 1043.
31. M. Siebert, J. I. Flege, T. Schmidt and J. Falta, *Physica B.* **357** (2005) 115.
32. J. C. Woicik, T. Kendelewicz, K. E. Miyano, C. E. Bouldin, P. L. Meissner, P. Pianetta and W. E. Spicer, *Phys. Rev. B.* **43** (1991) 4331.
33. A. Hirnet, K. Schroeder, S. Blügel, X. Torrelles, M. Albrecht, B. Jenichen, M. Gierer and W. Moritz, *Phys. Rev. Lett.* **88** (2002) 226102.
34. L. E. Sutton, *Table of Interatomic Distances and Configuration in Molecules and Ions* (Chemical Society, London, 1965). Supplement 1956–1959, Special publication No. 18.

Chapter 21

X-RAY STANDING WAVE ANALYSIS OF NON-COMMENSURATE ADSORBATE STRUCTURES PRODUCED BY Ga ADSORPTION ON Ge(111)

JÖRG ZEGENHAGEN

European Synchrotron Radiation Facility (ESRF),
BP 220, F-39043, Grenoble, France

Adsorption of some metals such as Ga, Cu, or Al leads to very unusual reconstructions on the (111) faces of silicon and germanium. The surfaces are tiled by a non-periodic superstructure of two-dimensional domains, each containing many tens of atoms. The determination of the metal adsorption site by the XSW technique allows the understanding the driving force of these complex, non-commensurate surface reconstructions. Stress introduced by the substitutional metal adsorbate, and rehybridization of the tetravalent semiconductor surface atoms renders the interior lattice of the domains incommensurate with the substrate. The strained interior of the domains leads to a non-periodic network of two-dimensional dislocations, called discommensurations, which form the observed irregular superstructure of two-dimensional domains. Two of these discommensurate reconstructions introduced by Ga adsorption on Ge(111) will be discussed in this chapter.

21.1. Introduction

The most common techniques to determine atomic structures with high spatial resolution are diffraction methods. However, the diffraction methods are really powerful only when applied to crystalline materials, since with a diffraction experiment, we probe the autocorrelation of atomic arrangements. Crystalline order or autocorrelation is, in principle, not necessary for the successful application of the XSW technique. In contrast to diffraction, in an XSW experiment, the correlation of the atom(s) with the X-ray interference field is probed. The difference between the XSW and diffraction techniques becomes very clear when we consider that, in principle, the spatial coordinates of a single atom can be obtained by using the XSW technique, which is not conceivable with a diffraction experiment. Localizing a single atom is only a *Gedankenexperiment*, but it helps to highlight the conceptual difference between both techniques. This difference has strong bearing on the application of the XSW technique in surface science. There are in fact several cases in which the atomic arrangements on surfaces exhibit very poor autocorrelation yet a pronounced correlation with the substrate bulk lattice. The reason for the correlation is simple: bond lengths can only vary within a reasonable range, and this constraint imposes strong limitations on surface-adsorbate distances. Setting up the XSW in a substrate crystal, the (mean) surface-adsorbate distance can easily be determined by an XSW experiment using diffraction planes parallel to the surface. The thus determined coherent position $P^{\perp}$ will be a measure of this distance, and the coherent fraction $F^{\perp}$ will reflect the degree of correlation. It was such one XSW measurement, the first one on a surface-adsorbate, which contributed strongly to the sudden rise in popularity of the XSW technique in the 1980.[1]

An adsorbate, which is interacting with the substrate, will be influenced by the (periodic) substrate potential in all three dimensions. This leads in the case of strong chemisorption mostly to commensurate superstructures. In case the interaction is weaker, the overlayer can be incommensurate or can even be characterized as a confined liquid or gas. Yet, as long as the adsorbate is bound to the surface, it will reflect the surface potential and thus be correlated with the substrate lattice. In 1991, Giuseppe La Rocca and me published a paper demonstrating that an XSW measurement is rather sensitive to the structural modulation of an adsorbate imposed by the correlation with the corrugated substrate surface potential.[2] Evidence for the structural correlation of the adsorbate with the substrate is immediately revealed by coherent fractions larger than zero.

The termination of a bulk crystal by a free surface leads to the occurrence of unsaturated chemical bonds for the surface atoms, the so-called dangling bonds for the case of predominantly covalent materials. For a diamond structure semiconductor surface this is shown schematically in Fig. 21.1. The frustrated chemistry of the clean surface is the driving force for reconstructions. In the case of adsorption of foreign atoms on the surface of a semiconductor, the equilibrium structure is likewise predominantly dictated by the attempt to achieve chemical passivation. Owing to the

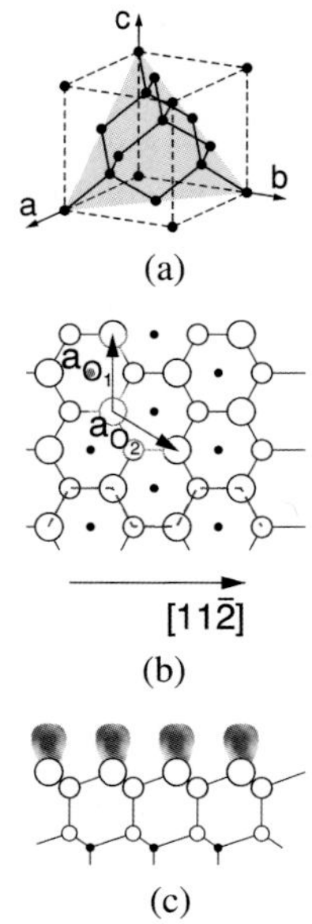

Fig. 21.1. (a) The cubic unit cell of the diamond structure, which is the crystalline form of tetravalent carbon, silicon, germanium, and alpha-tin. (b) The diamond structure (111) surface in top view and (c) in side view along the cut (along the [11–1] direction) indicated by the dashed line in (b). The dangling bonds of the surface atoms (atoms of the upper "double layer") are indicated.

strong and highly directional bonds of covalent semiconductors, the ground state in energy of the resulting surface reconstruction is usually well ordered, periodic, and commensurate with the bulk of the substrate. While the majority of adsorbate-induced reconstructions fit into this scheme, adsorption of some elements gives rise to reconstructions which are astonishingly different.

21.2. Discommensurate Reconstructions

Already in 1964, Lander and Morrison[3] observed rather complex low-energy electron diffraction (LEED) patterns subsequent to the deposition of aluminum (covalent radius 0.126 nm) and indium (covalent radius 0.144 nm) on the surface (Fig. 21.1) of silicon (covalent radius = 0.117 nm). Lacking direct proof, they speculated that the trivalent metals might substitute for tetravalent Si atoms of the top double layer and thus terminate and passivate the surface. They also suggested that "... periodically something is also done to relieve the strain ...” and that this something produced the complex LEED pattern.

Similarly complicated patterns were observed by reflection high-energy electron diffraction much later in 1985 from annealed Si(111) surfaces after Ga (covalent radius = 0.126 nm) deposition in the monolayer (ML) range.[4] An ML is here defined as the coverage corresponding to one atom per substrate surface atom (= $7.84 \times 10^{14}\,\mathrm{cm}^{-2}$ for Si(111)). A likewise complex reconstruction of the Si(111) surface is induced by copper adsorption (metallic radius 0.128 nm) and was studied by LEED, but in the first report published in 1970, the reconstruction was assigned to be commensurate.[5]

Since germanium (covalent radius 0.123 nm) is/was technologically much less exploited than silicon, its surfaces attracted much less attention. However, reconstructions with features very similar to what Lander and Morrison had observed on Si(111) were later also reported for adsorption on (111) Ge surfaces such as Ge(111):In in 1981,[6] Ge(111):Cu in 1989[7] and Ge(111):Ga in 1992.[8]

Resolving the structural properties of reconstructions like these was not an easy task. It required suitably powerful surface analysis tools, which only became available in the 1980s, notably the scanning tunneling microscope (STM) and the XSW technique. First XSW measurements were performed on the Si(111):Ga surface. The results supported the clever speculations of Lander and Morrison.[3] In the ML range, Ga is simply substituting for the Si surface atoms, i.e., the trivalent Ga replaces the tetravalent Si atoms of the outermost double layer[9] (Fig. 21.1), thus passivating the surface. However, the first STM images revealed a surprising strange surface structure[10,11]: the surface was found to be tiled by domains, between six and seven Si lattice units in size, and in a non-periodic way. The interior of the domains exhibits a hexagonal structure with a lattice constant about 7% larger[10] than the 0.384-nm lattice constant of the Si(111) surface. Since the STM is not element specific, the atomic composition of the interior of the domains could not be resolved. The internal domain structure became only clear with the help of the results of XSW measurements performed on this Ga-induced reconstruction.

The STM was obviously imaging the empty states of the trivalent Ga atoms within the domains, which had substituted for the tetravalent Si surface atoms. This assignment was supported by first principle total energy calculations,[9] which also shed light on the reconstruction mechanism. Upon Ga substitution, a change of hybridization ($sp^3 \rightarrow sp^2 + p_z$) creates a flat Si-Ga surface layer, much like graphene, i.e. a single sheet of graphite.

The larger Ga covalent radius in conjunction with an inward relaxation of the Ga atoms produces forceful stress in the surface. Thus, the Si-Ga layer expands and becomes locally, within the domains, incommensurate. The domains are separated by a two-dimensional network of dislocations, called domain walls or discommensurations, which are characteristic of this strange two-dimensional lattice.

To understand the occurrence of such discommensurate reconstructions, one should consider that the structure of an adsorbate at $T = 0\,\mathrm{K}$ is essentially dictated by two competing energies: the adsorbate–adsorbate interaction energy E_{AA} and the adsorbate–substrate interaction energy E_{AS}. If $E_{AA} \gg E_{AS}$, the adsorbate will be incommensurate, i.e., exhibit its own lattice constant. If $E_{AA} \ll E_{AS}$, the adsorbate will be commensurate, i.e., the adsorbate lattice constant will be dictated by the substrate. However, if $E_{AA} \approx E_{AS}$, the resulting structure can be discommensurate, as shown schematically in Fig. 21.2. The particularities of mismatched adsorbate and epitaxial systems were realized already in 1938 by Frenkel and Kontorova.[12] A decade later, the problem was analyzed in the famous paper of Frank and van der Merwe in 1949.[13] Despite extensively simplification of the problem, the principal behavior of mismatched adsorbates and epitaxial systems is surprisingly well described by their theory.

The term discommensurate means that the structure is neither commensurate (coherent with the substrate lattice) nor incommensurate (incoherent with the substrate lattice). The discommensurate phase consists of domains, in which the adsorbate is close to being in registry with the substrate, i.e., is weakly incommensurate, but strongly stressed since

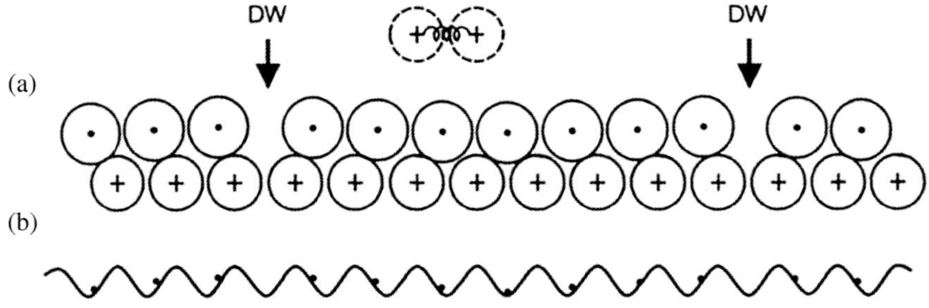

Fig. 21.2. The Frenkel–Kontorova/Frank–van der Merwe model of a discommensurate adsorbate. Competition between the substrate potential and the adsorbate–adsorbate "spring constant" creates a discommensurate reconstruction. (a) Adsorbate on substrate with domain walls (DW). (b) Schematic of the location of adsorbate atoms, represented by black dots, in the harmonic substrate potential.

it is prevented to exhibit its own lattice constant. The domains are separated by boundaries (discommensurations) where the stress, built up in the domains, is released. In fact, the resulting strain changes sign in these boundaries. Depending on the sign of the mismatch with the substrate, the adsorbate will be compressed or stretched in the domains, and thus stretched or compressed, respectively in the domain walls, leading to light or heavy walls, respectively. It is an important property of such discommensurate structures on semiconductor surfaces that the discommensurations not only passively facilitate strain release but due to the covalent, directional nature of semiconductor bonds the domain walls can adopt different bonding topologies and thus actively influence the total energy of the discommensurate structure. This was first realized for the case of Ge(111):Ga.[14,15] The stability of the so-called β-phase (see further below) is largely due to a chemically more passivated bonding situation at the domain boundaries as compared to the γ-phase (see further below), which appears at slightly lower Ga coverage.

The combination of XSW and STM measurements in conjunction with *ab initio* calculations lead to a very thorough understanding of the rich variety of the Ga induced phases on Ge(111) in the ML regime.[16] We will describe in the following as particular examples the XSW analysis of two Ga-induced discommensurate phases on Ge(111), observed in the coverage range of about 0.2 to 0.9 ML. There are other phases at Ga coverage < 0.2 ML and > 0.9 ML, none of which are commensurate (see e.g., Refs. 16, 17). For a comprehensive overview of discommensurate phases induced by Cu, Ga, and In adsorption on (111) surfaces of Si and Ge, a review published in 1997 can be referred.[18]

21.3. XSW and STM Investigations of the Ge(111):Ga γ- and β-phase

The two Ga-induced discommensurate phases on Ge(111) that are treated in the following are observed in a coverage range of about 0.5 to 0.9 ML. A define monolayer, i.e., one atom per Ge(111) surface atom, corresponds to an areal density of 8.16×10^{14} atoms cm^{-2}. STM images obtained for annealed ($\approx$800 K) Ge(111):Ga are shown in Fig. 21.3. Further details of the sample preparation and other experimental conditions can be found in Refs. 14, 18 and 19. With a Ga coverage of about 0.7 ML, the Ge(111) surface is fully covered by a non-periodic superlattice of domains with an average spacing of around $7.4 \cdot a_{110}$ ($a_{110} = 0.40$ nm for Ge). A phase

Fig. 21.3. Empty-state STM images of two discommensurate phases of Ge(111) introduced by Ga at sub-monolayer coverage. The surface is tiled with a non-periodic superlattice of domains about 7 to 8 and 14 to 16 Ge surface lattice constants ($a_{110} = 0.40\,\text{nm}$) in size for the γ (a) and β-phase (c), respectively. The interior of the domains (b, d) exhibits a hexagonal lattice with $a \approx 10\%$ increase in lattice constant comparcd to the Ge surface lattice constant $a_{110} = 0.40\,\text{nm}$. The β-phase exhibits two types of domains in which the atomic rows are shifted relative to each other as the line drawn in (d) shows. This is a clear indication of a stacking fault in the surface layer.

transition occurs above 0.7 ML and at ≈ 0.8 ML the surface is fully covered with much larger domains (14 to 16 a_{110}), again tiling the surface in a non-periodic way. For both, the so-called γ-phase at 0.7 ML and the β-phase at 0.8 ML, the interior of the domains exhibits a hexagonal lattice with $\approx 0.44\,\text{nm}$ lattice constant if empty states are imaged by the STM. Filled state images[14] (not shown here) look dramatically different: The interior of the domains appears almost featureless, but the domain boundaries exhibit irregular atomic protrusions in the case of the γ-phase, whereas for the β-phase the domain boundaries are as featureless as the interior of the domains.

The results of XSW measurements on such Ga covered, annealed Ge(111) surfaces are shown in Fig. 21.4. The XSW measurements were performed recording the Ga K_α X-ray fluorescence. The date were analyzed by fitting the equation (*cf.* Chapter 1.9)

$$I_H = I_0[1 + R + 2\sqrt{R}F_A^H \cos(\upsilon - 2\pi P_A^H)] \tag{21.1}$$

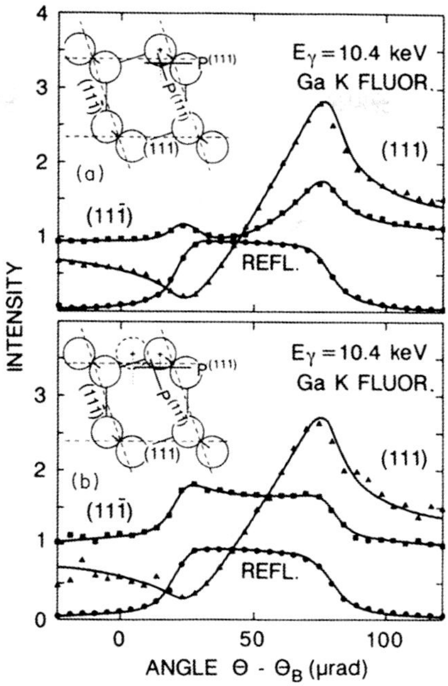

Fig. 21.4. XSW results for the (a) Ge(111):Ga γ-phase (top) and the β-phase (bottom) obtained by employing (111) and (11–1) substrate reflections. Shown are the reflectivity curves and the Ga K fluorescence yields. Symbols are experimentally obtained data and lines are fits to the data. With the fits to the fluorescence yields, the structure parameters P^H and F^H are obtained, yielding $P^{111} = 0.95 \pm 0.005$, $F^{111} = 0.95 \pm 0.02$, $P^{11-1} = 0.83 \pm 0.005$, and $F^{11-1} = 0.48 \pm 0.01$ for the γ-phase and $P^{111} = 0.95 \pm 0.005$, $F^{111} = 0.87 \pm 0.02$, $P^{11-1} = 0.74 \pm 0.01$, and $F^{11-1} = 0.13 \pm 0.01$, for the β-phase. Note that $P^H = 0$ is chosen coinciding with the $\langle 111 \rangle$ planes. In the insets, the P^{111} and P^{11-1} results are indicated schematically with the (111) surface in side view. The dashed circle in (b) indicates the position of a surface atom with a stacking fault.

to the data with the three fitting parameter off-Bragg-yield I_0, coherent fraction F^H, and coherent position P^H. The XSW measurements using the (111) diffraction plane are probing the surface distance of the Ga atoms, whereas the $(11\bar{1})$ measurements are sensitive to the in-plane registry. (Note the equivalence of the nomenclatures $(11\bar{1})$ and (11–1) for denoting the diffraction planes inclined with the surface.) For both phases, the XSW (111) results are almost identical in terms of coherent position P^{111} and coherent fraction F^{111}. The P^{111} values are indicative of an inward relaxed substitutional position of the Ga. However, the P^{11-1} and F^{11-1} values for the γ- and β-phases are different. The coherent positions and fractions are listed in the caption of Fig. 21.4 and the P^H-values are sketched in the insets.

15. E. Artacho and J. Zegenhagen, *Phys. Rev. B* **52** (1995) 16373.
16. P. Molinàs-Mata, M. Böhringer, E. Artacho, J. Zegenhagen, L. Seehofer, T. Buslaps, R. L. Johnson, E. Findeisen, R. Feidenhans'l and M. Nielsen, *Phys. Status Solidi A* **148** (1995) 191.
17. M. Böhringer, P. Molinàs-Mata, J. Zegenhagen, G. Falkenberg, L. Seehofer, L. Lottermoser, R. L. Johnson and R. Feidenhans'l, *Phys. Rev. B* **52** (1995) 1948.
18. J. Zegenhagen, P. F. Lyman, M. Böhringer and M. J. Bedzyk, *Phys. Status Solidi B* **204** (1997) 587.
19. E. Artacho, P. Molinàs-Mata, M. Böhringer, J. Zegenhagen, G. E. Franklin and J. R. Patel, *Phys. Rev. B* **51** (1995) 9952.
20. J. A. Golovchenko, J. R. Patel, D. R. Kaplan, P. L. Cowan and M. J. Bedzyk, *Phys. Rev. Lett.* **49** (1982) 560.
21. J. Zegenhagen, *Phys. Scripta* **T39** (1991) 328.
22. J. Zegenhagen, J. R. Patel and E. Fontes, *Appl. Surf. Sci.* **60/61** (1992) 505.

Chapter 22

PHOTON STIMULATED DESORPTION

JAN INGO FLEGE*, THOMAS SCHMIDT and JENS FALTA

Institute of Solid State Physics,
University of Bremen, Bremen, Germany
**flege@ifp.uni-bremen.de*

ALEXANDER HILLE

Hamburger Synchrotronstrahlungslabor am Deutschen Elektronensynchrotron (HASYLAB/DESY), Hamburg, Germany

GERHARD MATERLIK

Diamond Light Source Limited, Diamond House, Chilton, Didcot, Oxfordshire OX11 0DE, UK

The combination of the X-ray standing wave (XSW) technique and photon stimulated desorption (PSD) offers the possibility to investigate both the atomic structure as well as the mechanisms leading to the X-ray induced desorption of ions from the sample surface. Here, we present a review of surface systems studied by the XSW-PSD technique along with a brief description of the basic differences in XSW data evaluation when desorbing ions are used as secondary signal. We show how direct and indirect desorption processes can be identified and that even site-specific desorption cross-sections may be determined with this method.

22.1. Introduction

The X-ray standing wave (XSW) technique using electrons and fluorescence photons as detected inelastic signals (often referred to as secondary signals, *cf.* Chapter 1) has become a standard technique to determine adsorbate structures. However, when a sample is irradiated by intense

22.3. Experimental Procedure

All the measurements presented here have been performed under ultra-high vacuum conditions at the undulator beamline BW1 at the Hamburg Synchrotron Radiation Laboratory (HASYLAB, *cf.* Appendix 5 by G. Materlik) using monochromatic X-radiation from a double-crystal monochromator in a standard nondispersive setup. The XSW experiments were carried out by tuning the X-ray photon energy through the Bragg condition of the sample. Simultaneously, desorbing ions, photoelectrons, and X-ray fluorescence spectra were recorded. With the DORIS III storage ring operating in single-bunch or two-bunch mode, desorbing ions were identified by time-of-flight (TOF) spectroscopy utilizing the time structure of the incident radiation, i.e., the extended time interval between two successive bunches corresponding to 964 ns or 482 ns with one or two bunches in the storage ring, respectively. Applying a negative bias of more than 3 kV between the sample and the TOF detector leads to distinct flight times for each ionic species as a function of their characteristic m/q values, where m stands for the mass of the ion and q for its charge, respectively. Hence, the desorption products can be identified by varying the distance between the detector and the sample. Examples for the quality and resolution of the obtained TOF spectra will be given in the following section.

22.4. Results and Discussion

We start our survey with the clean Si(111)-(7×7) surface. Its structure has been explained within the so-called dimer adatom stacking-fault (DAS) model as originally proposed by Takayanagi and co-workers.[6] In order to follow the subsequent discussion of the XSW ion data, it is sufficient to note that the (7×7) unit cell partially consists of a local (2×2) Si adatom structure in each triangular subunit separated by dimer rows. In the inset of Fig. 22.1 a TOF spectrum is displayed identifying the desorbate as Si^+ ions created by irradiation at an energy of 5.1 keV. The corresponding ion yield measured in (111) Bragg reflection is given in the main part of the figure. The analysis yields a coherent position of $\Phi = 0.16 \pm 0.03$ and a coherent fraction of $f = 0.53 \pm 0.03$ which have to be compared to the calculated coherent position of the Si adatom since desorption of adatoms from the outermost atomic layer is most likely.[a] Taking the theoretically calculated

[a]Within the framework of Eq. (22.1), this assumption corresponds to setting all desorption cross-sections but the cross-section of the adatom site to zero.

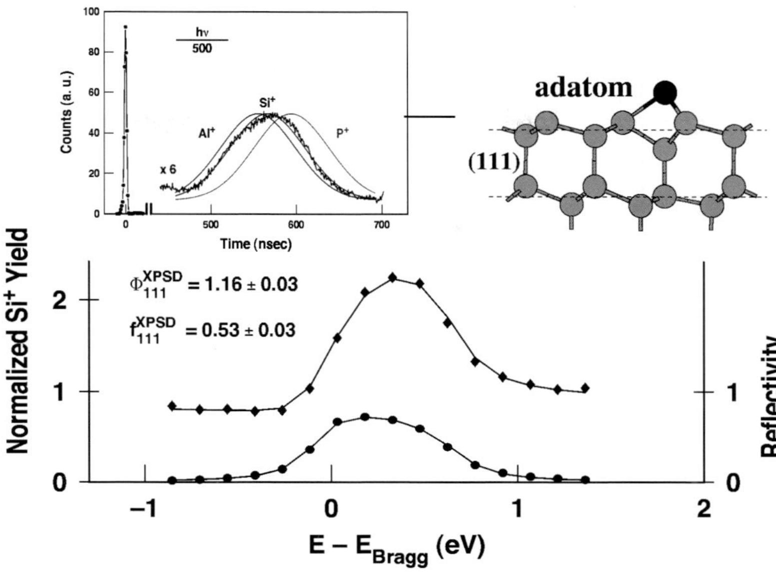

Fig. 22.1. XSW data and fit (solid lines) of the reflectivity (•) and Si^+ (♦) yield as functions of the photon energy E for Si(111)-(7×7) in (111) Bragg reflection. Upper left inset: TOF spectrum identifying the desorbing Si^+ ions. For comparison, the hypothetical peak positions for the neighboring elements Al and P are displayed. Upper right inset: Side view of the adatom configuration on Si(111), which is part of the (7×7) DAS structure.

structure of Qian *et al.*[7] leads to a simulated atomic coherent position of $\Phi = 0.51$ and a coherent fraction of $f = 1.0$. Hence, the predominance of purely direct desorption processes initiated by photo-absorption in the Si 1s level at the adatom can be excluded since both coherent positions should then closely match. Likewise, purely indirect desorption via the XESD mechanism can be ruled out because the coherent position of the Si bulk atoms would then be reproduced (see the case of H/Si(111), which is discussed below). However, the experimental data can be successfully explained by a mixture of a direct mechanism and the presence of indirect processes via photo-excitation of the three bonding partners beneath the adatom. In either case, the core hole predominantly relaxes via Auger decay and subsequently the chemical bonds are destroyed. Assuming a 1 : 3 ratio between direct and nearest-neighbor PSD, the XSW simulation yields a coherent position of $\Phi = 0.14$ and a coherent fraction of $f = 0.54$, reproducing the experimental data within the error bars. In conclusion, we find that using the XSW technique with desorbing ions allows the

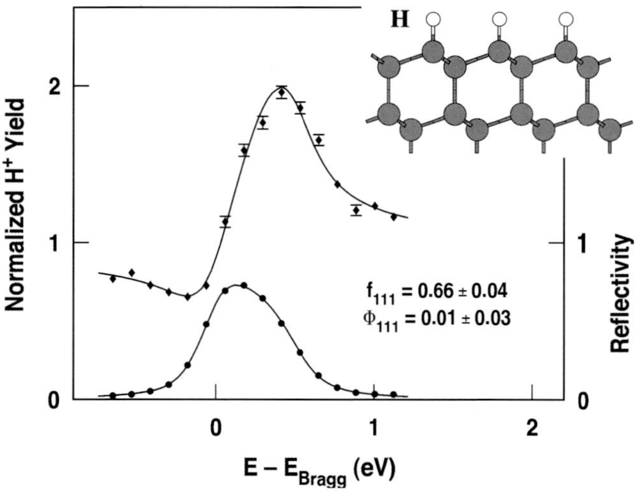

Fig. 22.2. XSW data and fit (solid lines) of the reflectivity (•) and H^+ (♦) yield as functions of the photon energy E for H/Si(111)-(1 × 1) in (111) Bragg reflection. Upper right inset: Side view of the on-top adsorption geometry.

identification of the prevailing stimulated desorption processes for a covalent system in the X-ray regime.

Upon hydrogen adsorption on the clean Si(111) surface, e.g., by wet-chemical etching or exposure at elevated temperatures, the hydrated Si(111) surface gradually transforms into a bulk-like (1×1) adsorbate structure with H occupying the on-top site,[8] as depicted in the upper right inset of Fig. 22.2. This atop geometry with a literature value of 1.51 Å for the Si–H bond length corresponds to an atomic coherent position of 0.61 for the H atom in (111) Bragg geometry. The XSW ion data for the H^+ signal at a Bragg energy of 4.0 keV, however, reveal a coherent position of 0.01 and a coherent fraction of 0.66, bearing the characteristic of XESD processes to which corresponding values of 0.00 and $1/\sqrt{2} \approx 0.71$ can be attributed, respectively. Physically, this result is expected since direct desorption induced by photo-absorption in the Si–H bond is negligible due to the very low associated photo-absorption cross-section at X-ray energies. Therefore, the combined XSW and XPSD experiment with H^+ ions as secondary signal from the saturated Si(111) surface reveals the dominant desorption mechanism, but includes no structural information on the atomic configuration of the adsorbed hydrogen atoms.

Interestingly, this situation changes drastically after deposition of 1.3 bilayers of germanium on the hydrogen-saturated surface at a temperature

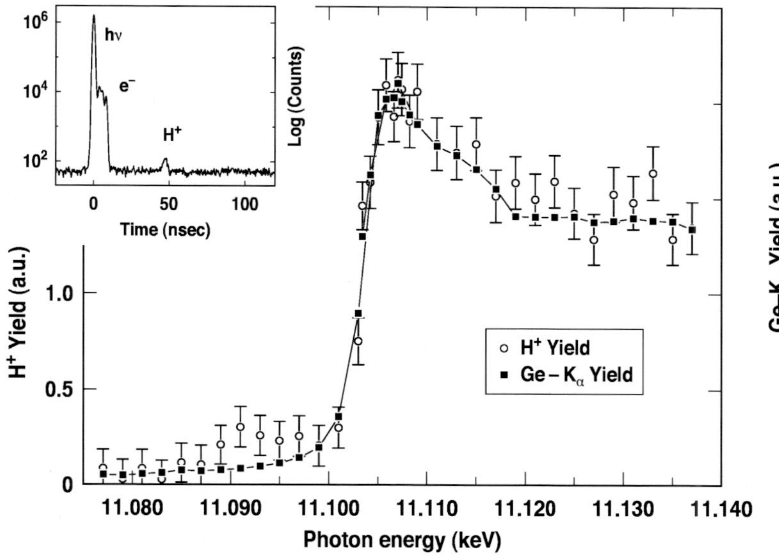

Fig. 22.3. H^+ (○) and Ge K_α (■) yields as functions of the photon energy E at the Ge K absorption edge. Inset: TOF spectrum for H^+ desorption.

of 430°C. This was done in the course of an experiment aimed at studying the heteroepitaxial growth of Ge/Si(111):H. In Fig. 22.3, the H^+ ion yield and the Ge K fluorescence yield are displayed as a function of the photon energy in the vicinity of the Ge K absorption edge. As clearly visible, the ion yield increases by a factor of more than 10. According to Eq. (22.3), this corresponds to $\kappa_{\rm NN}^{\rm Ge\text{-}1s} \geq 0.9$, proving that more than 90% of the desorbing ions are created by photo-excitation of core holes in the Ge 1s level, and that direct desorption processes can be neglected ($\kappa_{\rm direct} \ll \kappa_{\rm NN}$). Since the ion desorption and the fluorescence signal rise at the same energetic threshold, electron-stimulated desorption by Ge 1s photoelectrons can be excluded. Hence, the most probable desorption mechanism involves an interatomic Auger decay following the initial photo-absorption in the Ge K shell. This suggests that H^+ ions can be used as a probe to determine the adsorption sites of their respective germanium bonding partners in an XSW experiment.

The corresponding XSW data for the H^+ ion yield and the Ge K fluorescence signal recorded at a Bragg energy of 11.5 keV are presented in Fig. 22.4. The measured (111) Fourier components are 0.42 ± 0.02 and 0.99 ± 0.02 for the coherent fraction and coherent position of the

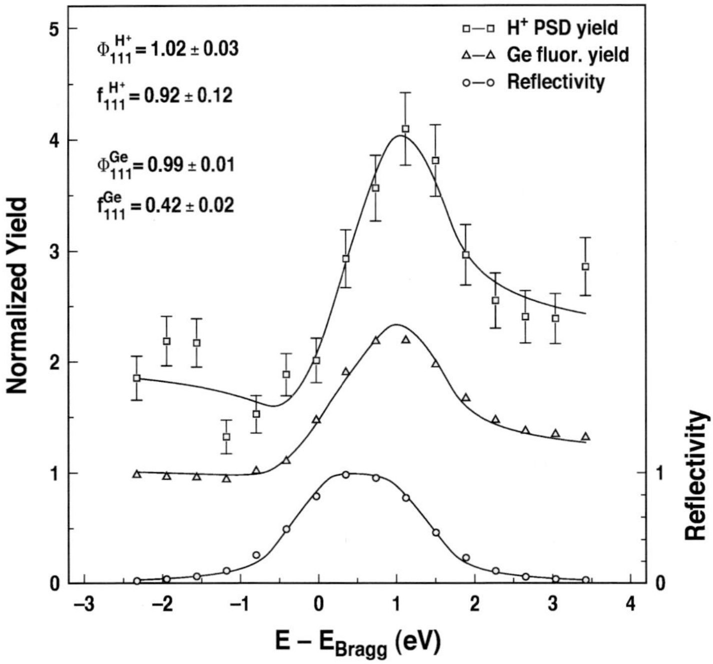

Fig. 22.4. XSW data and fit (solid lines) of the reflectivity (○), Ge K fluorescence (△) and H^+(□) yields as functions of the photon energy E for Ge/Si(111):H in (111) Bragg reflection.

fluorescence yield, respectively. Hence, the thin Ge film is fairly well-ordered. However, the ion signal exhibits a much higher coherent fraction of 0.92 ± 0.12 and a coherent position of 0.02. These values are not compatible with H^+ desorption solely from on-top sites since this would lead to an average coherent position between 0.18 and 0.23 for the Ge atoms on the "accessible" hydrated terraces. Therefore, additional processes need to be taken into account in order to simulate the ion data, such as a more favorable desorption cross section from step edges or point defects. That site-specificity may indeed be pronounced in XPSD experiments will next be demonstrated for the Cl/Si(111) system which is the last example in this review.

The chlorinated Si(111) surface exhibits a geometric structure similar to the previously discussed system H/Si(111). In order to study the influence of photo-absorption in the Cl 1s level on the desorption rate, PSD experiments were performed at the Cl K edge, as reproduced in Fig. 22.5. Both the

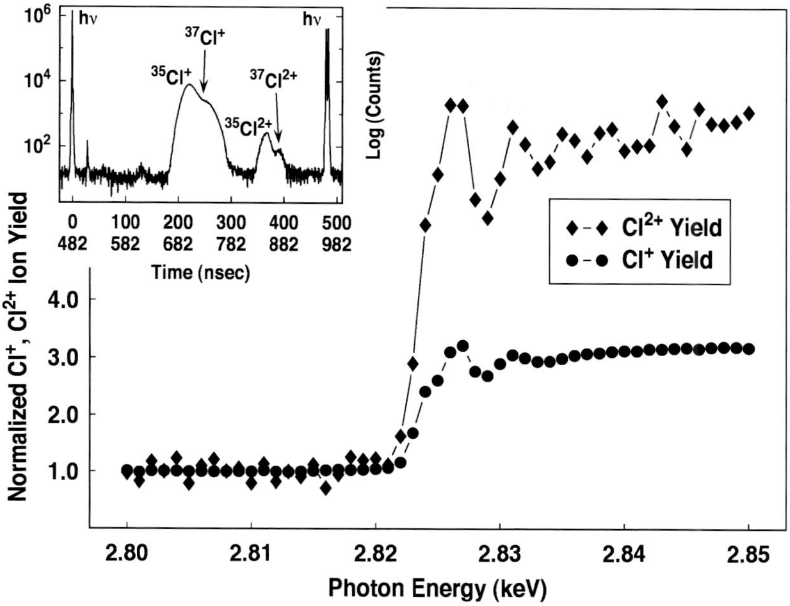

Fig. 22.5. Photon-stimulated desorption and time-of-flight spectra for Cl/Si(111)-(1 × 1). Main figure: Cl^+ (•) and Cl^{2+} (♦) ion yields as functions of the photon energy near the Cl K absorption edge. Inset: Raw TOF spectroscopy data (full spectrum).

Cl^+ and the Cl^{2+} ion yield undergo a significant increase when tuning the photon energy across the absorption edge. The edge jumps $\eta(Cl^+) \approx 3$ and $\eta(Cl^{2+}) \approx 7$ establish above the absorption edge the prevalence of direct desorption processes for both types of charged ions. Consequently, Cl^{2+} ions in particular offer a high Cl elemental specificity if used as a secondary signal in combined XSW and XPSD measurements with Bragg energies above the Cl K edge. The desorption can proceed as follows: After the initial photo-absorption the core hole induces an Auger cascade which leads to the excitation of localized multi-hole states at the Cl atom. Because of the high ionicity of the unperturbed Si–Cl bond, the neighboring Si atom carries a positive charge. Hence, the charge reversal of the highly-excited Cl atom causes a Coulomb repulsion which effectively disrupts the Si–Cl bond, resulting in the emission of a Cl^+ or Cl^{2+} ion.

The result of a typical XSW experiment for Cl/Si(111)-(1 × 1) in (111) Bragg reflection with Cl 1*s* photoelectrons and desorbing ions as secondary signals is shown in Fig. 22.6. While the photoelectron yield nicely reproduces the coordinate of the well-known on-top adsorption geometry,

Chapter 23

DEPTH-PROFILING OF MARKER LAYERS USING X-RAY WAVEGUIDES

AJAY GUPTA

UGC-DAE Consortium for Scientific Research,
Khandwa Road, Indore 452017, India

It is demonstrated that X-ray waveguide structures can be used for precise depth profiling of thin marker layers. The system to be studied forms the cavity of a planar waveguide. Fluorescence from the marker layer as a function of the angle of incidence of X-rays consists of a number of well-defined peaks corresponding to the excitation of various waveguide modes. A detailed analysis of the fluorescence pattern yields the position as well as width of the marker layer with accuracy of the order of 0.1 nm. The technique can be made isotope sensitive by detecting nuclear resonance fluorescence from a specific isotope (Mössbauer active), instead of fluorescence from atomic levels. Depth profiling of isotopes can be used to study self-diffusion of a constituent species in thin films with high precision. Specific examples have been taken in order to demonstrate the capability of the technique in studying impurity diffusion as well as self-diffusion of a constituent species.

23.1. Introduction

Thin films and multilayer form an important class of nanostructured materials. In a multilayer structure, as the thickness of individual layers decreases and becomes comparable to the characteristic length scale of a given physical property (e.g., Fermi wavelength in the case of the electronic properties, exchange length in the case of magnetic properties, electron mean free path in the case of transport properties), that particular property can change drastically. A large variety of new materials have been developed using such multilayer structures. Magnetic Fe/Cr multilayer exhibit large magneto-resistance and are used as magnetic sensors in read-write heads and in memory devices.[1] Multilayer, consisting of alternate layers of a high-Z and low-Z materials such as Pt/C, are used as X-ray mirrors and monochromators.[2] Furthermore, by choice of suitable composition and thickness, the band structure of semiconductor materials can be tailored for various optoelectronic device applications.[3] The interface between two layers in such nanostructures can greatly influence their physical properties.[2,4,5] In this context, the importance of nanoscale diffusion across the interfaces in multilayer has been recognized, as it can modify their properties intentionally or unintentionally.[6,7] Both the characterization of as-prepared multilayer and the study of atomic diffusion in them require depth profiling of various elemental constituents with an accuracy of a fraction of a nanometer (which is typically the width of interfacial region in the multilayer nanostructures). The conventional depth profiling techniques, such as RBS or SIMS, have a typical depth resolution of a few nanometers, which is an order of magnitude less than the required accuracy.

X-ray standing waves formed by the total external reflection (*cf.* Chapter 5) is an excellent tool for elemental depth profiling.[8] In the following, it is demonstrated that the accuracy of the XSW technique in depth profiling of marker layers can be further improved by using X-ray waveguide structures[9,10] (*cf.* Chapter 9). The technique can be made isotope selective if one detects nuclear fluorescence instead of the fluorescence from atomic levels, by tuning the X-ray energy to one of the nuclear transitions (Mössbauer transitions).[11,12]

23.2. Depth Profiling of Thin Marker Layers

The planar X-ray waveguide structure is formed by sandwiching a layer of a low-z element between two layers of a high-z element.[13] With X-rays incident at grazing angle θ_n, transverse electric mode of the nth order can be excited under the condition:

$$\theta_n = (n+1)\lambda/2d, \tag{23.1}$$

where λ is the X-ray wavelength and d is the thickness of the cavity.

For depth profiling, the structure has to be made in the form of the waveguide cavity. Figure 23.1 shows a waveguide structure used for the depth profiling of a thin Ti maker layer in Si. The nominal structure of the multilayer labeled as Ti3 is: Float glass (substrate)/Pt, 70 nm/Si, 16 nm/Ti, 3 nm/Si, 9 nm/Pt, 2 nm. The two Pt layers form the walls of the planar waveguide, while the tri-layer Si/Ti/Si forms the guiding layer. Figure 23.2 shows the X-ray reflectivity and the Ti K_α fluorescence

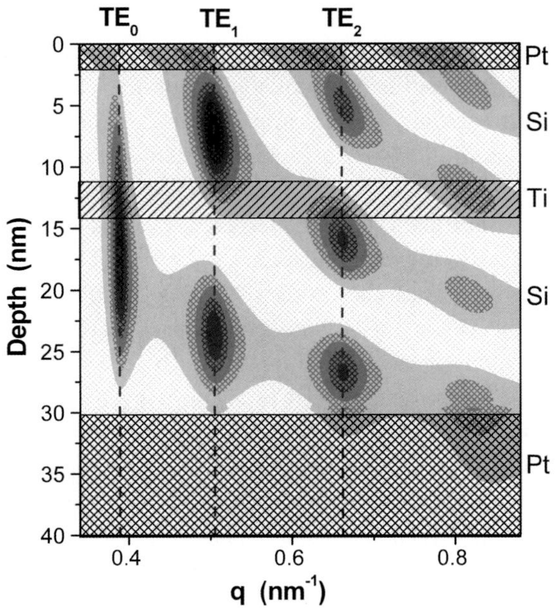

Fig. 23.1. The multilayer Ti3 made of float glass (substrate)/Pt, 70 nm/Si, 16 nm/Ti, 3 nm/Si, 9 nm/Pt, 2 nm, and the contour plot of X-ray field intensity inside as a function of depth and scattering vector q.

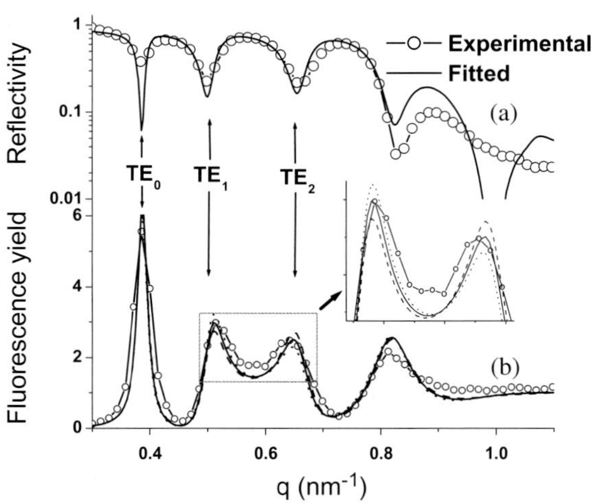

Fig. 23.2. X-ray reflectivity and Ti fluorescence from the multilayer Ti3 as a function of scattering vector q. The continuous curves represent the best fit to the data. In the inset, simulated curves for Ti position shifted by +0.2 nm (- - -) and −0.2 nm (. . .) with respect to the best fit are also shown.

as a function of the scattering vector $q = 4\pi \sin\theta/\lambda$. As expected, the Ti fluorescence exhibits well-defined peaks at q-values corresponding to transverse electric modes (TE_n) TE_0, TE_1, and TE_2 due to the resonance enhancement of X-ray intensity inside the cavity.

The 3-nm-thick Ti marker layer is intentionally placed asymmetrically inside the Si cavity, in order to achieve a higher sensitivity in determining its position.[9,10] To demonstrate this point, Fig. 23.1 also gives the contour plot of the X-ray intensity inside the multilayer structure as a function of q calculated using Parratt's formalism.[14] Well-localized antinodes of the X-ray standing waves corresponding to TE_0, TE_1, and TE_2 modes of the waveguide structure are clearly visible. The position of the marker layer is chosen to lie roughly midway between the antinodes corresponding to TE_1 and TE_2 modes. In this case, even a small variation in the position of the marker layer would result in a significant variation in the intensity ratio of the fluorescence peaks corresponding to the TE_1 and TE_2 modes, because at this depth, the distributions of the field intensity for the TE_1 and TE_2 modes has steep gradients with opposite signs. Thus, fitting the fluorescence peaks of the TE_1 and TE_2 modes provides a sensitive way to determine the position of the marker layer and observe any shift caused by a subsequent sample treatment.

The X-ray reflectivity is very sensitive to the total thickness of the cavity and the position of the top Pt layer, as well as to the interface roughness, but it is not so sensitive to the position and width of the Ti layer. The sensitivity of the fluorescence TE_1 and TE_2 peaks to the position of the Ti marker layer is demonstrated by the inset in Fig. 23.2(b). It shows the simulated patterns corresponding to a shift of the Ti layer by ± 0.2 nm from the value corresponding to the best fit. The obvious mismatch with the experimental data indicates that the accuracy with which the position of the marker layer can be determined is better than 0.2 nm.

The thickness of the Ti layer significantly affects the intensity of the TE_0 peak. This is shown in Fig. 23.3 for a multilayer having the structure similar to that of Ti3 but with the thickness of the Ti marker layer of 4 nm (labeled as Ti4). The intensity of the TE_0 fluorescence peak is significantly reduced compared to the Ti3 peak (*cf.* Fig. 23.2). This is mainly due to the fact that the marker layer itself perturbs the field inside the cavity. Thus, while the position of the marker layer mainly affects the intensities of the TE_1 and TE_2 peaks, its thickness primarily affects the intensity of the peak corresponding to the TE_0 mode. Fitting of the data using several parameters may not always lead to a unique solution. However,

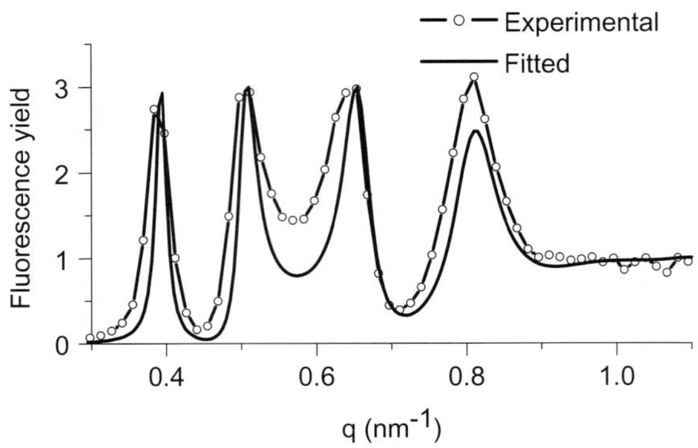

Fig. 23.3. Ti fluorescence from the multilayer Ti4 as a function of scattering vector q. The continuous curves represent the best fit to the data.

since different parameters of the marker layer affect different regions of the fluorescence pattern, an unambiguous determination of the marker structure is possible in most cases.

As an application of the technique, swift heavy ion-induced intermixing of Si with Fe has been studied. Swift heavy ions lose their energy in the target mainly through electronic excitations. In metallic systems, the induced modifications are expected to be small[15] and therefore, the present technique can be usefully applied, providing important clues about the involved mechanisms. The multilayer labeled Fe4, with the structure: Float glass (substrate)/Cr, 20 nm/Au, 70 nm/Si, 17 nm/Fe, 4 nm/Si, 12 nm/Au, 2 nm was studied before and after irradiation with 100 MeV Au ions to a fluence of 1×10^{13} ions/cm^2.

The Fe K_α fluorescence yield measured before and after the irradiation (Fig. 23.4) clearly indicates that the relative intensities of the TE_1 and TE_2 peaks significantly altered. In particular, before irradiation the intensity of the TE_2 peak was higher than that of the TE_1 peak, while after irradiation the situation is reversed. This suggests that the position of the marker layer shifted with respect to the center of the guiding layer as a result of irradiation. A detailed analysis of the reflectivity and the fluorescence data yields that, while in the pristine specimen the position of the Fe marker layer is shifted from the center of the guiding layer by 8.7% of its thickness, after irradiation this shift becomes 9.9%.[9] Thus, the marker layer exhibits an upward movement upon irradiation by >1% of the layer thickness. Furthermore, electron density profile obtained from the fitting shows that, in addition to intermixing between the Fe and Si layers, irradiation results

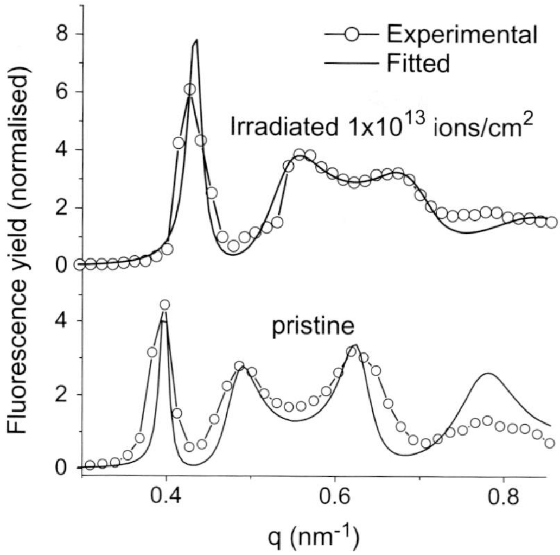

Fig. 23.4. Fe fluorescence from multilayer Fe4 before and after irradiation with 100 MeV Au ions (fluence, 1×10^{13} ions/cm^2).

in a contraction of the guiding Si/Fe/Si layer by about 10%, suggesting that the Si layers are densified upon irradiation. While roughening of the interfaces due to intermixing and densification of Si layer are known effects and have been observed in several studies,[15] the upward movement of the centroid of the Fe layer could be observed only because of the high sensitivity of the present technique. This movement can be understood by an asymmetric mixing at the interfaces.[10]

23.3. Depth Profiling of Isotopic Marker Layers

The high sensitivity to the position and the width of a marker layer can, in principle, also be used for studying thermally induced atomic diffusion. However, in many cases, the quantity that is physically more significant is self-diffusion of one of the constituent species instead of tracer diffusion. For example, in metastable structures like amorphous or nano-crystalline alloys, a structural relaxation as well as a transformation to more stable crystalline phases is governed by self-diffusion of the constituent species.[11,16,17] Self-diffusion is generally measured by depth profiling of one of the isotopes of the constituent species using radioactive tracers or techniques like SIMS, which are capable of differentiating between different isotopes of the same element.[17] The depth profiling technique described in the Sec. 23.2 cannot differentiate between different isotopes of the same element. However, it can be made isotope-sensitive by detecting the nuclear resonance fluorescence when the energy of the incident X-rays (e.g., from a synchrotron radiation source) is tuned to a nuclear transition (Mössbauer transition).[11,12]

In the presence of a nuclear resonance, the total scattering amplitude of an atom must be written as a sum of the electronic scattering and nuclear resonance scattering amplitudes. As a result, the index of refraction of the layer material is written as:

$$n = \sqrt{1 + \frac{\lambda_0^2}{\pi} \sum_i \sigma_i f_i} \approx 1 + \frac{\lambda_0^2}{2\pi} \sum_i \sigma_i (f_i^e + f_i^n). \tag{23.2}$$

Here λ_0 is the wavelength of the X-rays, f_i^e and f_i^n are, respectively, the electronic and nuclear scattering amplitudes, and σ_i is the atomic density of the species i. The nuclear scattering amplitude is given for instance in Ref. 18 and is proportional to the enrichment of the resonant nuclei of species i.

For the Mössbauer transition of ^{57}Fe nuclei (M1 transition), in absence of a magnetic splitting, and for a polycrystalline sample, the nuclear forward scattering amplitude can be written as:

$$f^n = \frac{\lambda_0}{4\pi} \frac{f_{LM}}{1+\alpha} \frac{2j_1 + 1}{2j_0 + 1} \frac{A}{x - i} \tag{23.3}$$

where f_{LM} is the Lamb–Mössbauer factor, j_0 and j_1 are the spin quantum numbers of ground and excited states, respectively, α is the coefficient for internal conversion, $x = (\Delta E - \hbar\omega)/\Gamma_0$, A denotes the inhomogeneous broadening, ΔE is the quadrupole splitting, and Γ_0 the natural line width. The nuclear part of the refractive index shows a resonance behavior and is significant only within a narrow energy range of the order of μeV around the nuclear transition energy. However, in this energy range the Mössbauer isotope presents a strong scattering contrast, and the nuclear scattering amplitude can be more than an order of magnitude larger than the electronic part. The information about the depth distribution of this isotope can be obtained from the q-dependence of the nuclear resonance reflectivity and nuclear fluorescence.

The multilayer used for depth profiling of ^{57}Fe isotope in a compositionally homogeneous film of FeZr had the nominal structure: substrate/Pt, 70 nm/$Fe_{60}Zr_{40}$, 12.5 nm/$^{57}Fe_{60}Zr_{40}$, 5 nm/$Fe_{60}Zr_{40}$, 12.5 nm.[19] In the FeZr layer, a 5-nm-thick part in the center was enriched in ^{57}Fe isotope to about 90%, while the rest of the film had the natural abundance of ^{57}Fe, which is 2.5%. The ID 18 beamline of the ESRF was used for studying the nuclear resonance fluorescence under the standing wave conditions. The storage ring operated in a 16-bunch mode providing short pulses of X-rays with the duration of ~100 ps every 176 ns. The radiation from the undulator source, optimized for the 14,413 eV transition in ^{57}Fe, was filtered by a double crystal Si(111) upstream monochromator followed by a high-resolution nested monochromator providing a bandpass of 4 meV.[20] The scattered radiation as well as fluorescence was detected using fast avalanche photodiodes with the time resolution of ~1 ns. The nuclear and electronic parts of the signal were separated by using a time delay of the nuclear transitions due to a finite life time of the Mössbauer excited states which is 140 ns in the case of ^{57}Fe isotope).[20] Thus, photons detected within a few nanoseconds of the incident X-ray pulse constitute the prompt signal due to electronic scattering, while those detected in an interval of 20–160 ns after the incident X-ray pulse correspond to nuclear resonance scattering. The measurements of the q-dependence of the reflectivity as well as fluorescence were done simultaneously using two different sets of avalanche photodiodes.

Figure 23.5 gives the electronic and nuclear parts of the reflectivity and fluorescence from the as-deposited multilayer. The electronic fluorescence

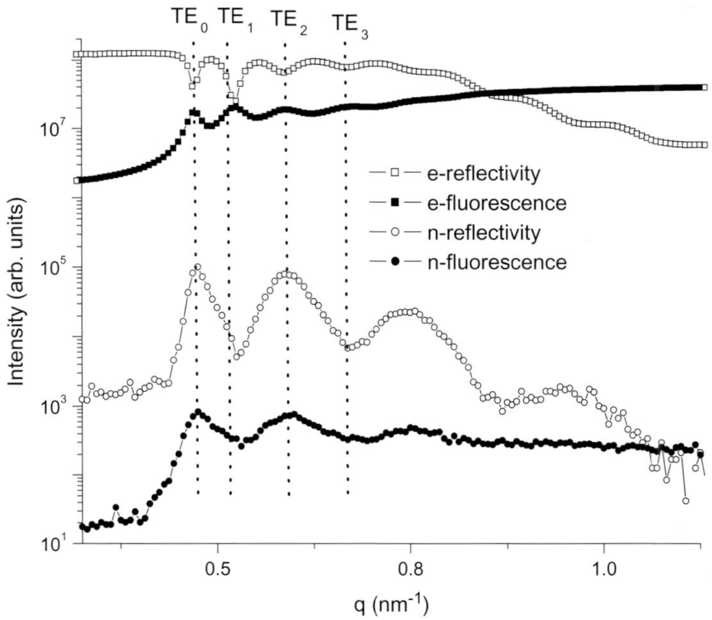

Fig. 23.5. The electronic as well as nuclear reflectivity and fluorescence yield data for the multilayer Pt/FeZr/^{57}FeZr/FeZr.

includes both the K-edge fluorescence from Fe and the L-edge fluorescence from the bottom Pt layer. However, since the fluorescence from the Pt layer is expected to increase monotonically with the angle of incidence up to θ_c, the observed features in the electronic fluorescence originate mainly from the Fe atoms. Since the X-ray scattering contrast at the air/FeZr and FeZr/Pt interfaces is of the same order, the above structure acts as an asymmetric waveguide. Therefore, the field inside the FeZr layer will have resonances whenever the Eq. (23.1) is satisfied. This is confirmed by the peaks in the electronic fluorescence at $\theta = 0.187°$, $0.21°$, $0.235°$, and $0.267°$, corresponding to the TE_0, TE_1, TE_2, and TE_3 modes, respectively.

It may be noted that for θ corresponding to TE_1 and TE_3 modes, although the average field inside FeZr layer is enhanced, there is a node at the center of the layer. Therefore, the nuclear resonance fluorescence from the ^{57}FeZr marker layer, which lies in the center of the FeZr layer, will exhibit a dip at this q value. The experimentally observed nuclear resonance fluorescence does exhibit such behavior: it shows peaks corresponding to TE_0 and TE_2 modes only. Furthermore, there is a one to one correspondence between the nuclear reflectivity and the nuclear fluorescence.

Figure 23.6 gives the nuclear reflectivity of another film with the structure: float glass (substrate)/Pt, 70 nm/FeN, 20 nm/^{57}FeN, 4 nm/FeN, 20 nm, after annealing at different temperatures for 60 min each. Significant changes in the peak intensities have taken place with the thermal annealing, which can be explained by a broadening of the depth profile of the ^{57}Fe marker layer as a result of self-diffusion of Fe. The concentration profile of ^{57}Fe layer as obtained from the fitting of the nuclear reflectivity data is shown in the inset of Fig. 23.6. Typical error bars in the width of concentration profiles are ± 0.2 nm.

In the limit of thin film solution, diffusivity $D(T)$ at a given temperature T can be written as:

$$D(T) = \frac{\sigma_T^2 - \sigma_0^2}{2t}, \tag{23.4}$$

where σ_0 and σ_T are the standard deviation of the concentration profiles in the pristine sample and the sample annealed at temperature T for time t, respectively. Diffusivity in the nano-crystalline FeN film obtained by using Eq. (23.4) are 1.0×10^{-22} m^2/s and 3.7×10^{-22} m^2/s at 200°C and 300°C, respectively. These values are well below the grain boundary diffusion in polycrystalline Fe,[21] however, they are comparable to the diffusivity in nano-crystalline FeZr,[22] where diffusion occurs mainly through the grain

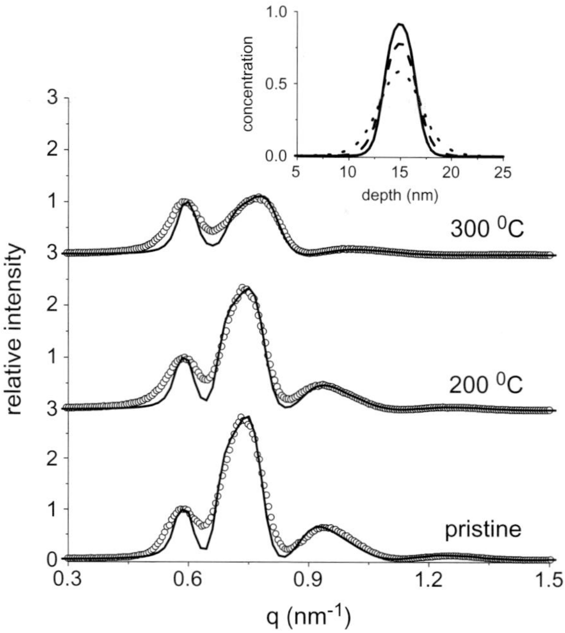

Fig. 23.6. Nuclear reflectivity of multilayer Pt/FeN/^{57}FeN/FeN in the pristine state as well as after annealing at 200°C and 300°C. Curves are shifted vertically with respect to each other for the sake of clarity. Inset shows the concentration profiles of ^{57}Fe layer as obtained from the fitting of nuclear reflectivity data of pristine (——) , 200°C annealed (-----), and 300°C annealed (......) samples.

boundaries, which are amorphous in nature. These results demonstrate that the nuclear resonance reflectivity and fluorescence from isotopic marker layers under XSW conditions can be used to study self-diffusion with an accuracy that is roughly an order of magnitude better than obtainable using conventional techniques like SIMS.

In conclusion, it is shown that the waveguide structures can be used to get depth-resolved information in thin films and multilayer with a sub-nanometer resolution. Depth profiling of a marker layer can be done with an accuracy of the order of 0.1 nm, which is more than an order of magnitude better than that achievable using conventional techniques such as RBS or SIMS. It can be used to study subtle effects associated with nanoscale diffusion or intermixing at interfaces induced by external perturbations as irradiation with heavy ions or laser pulses. The technique can be made isotope selective if one detects nuclear fluorescence instead of the fluorescence from atomic levels, by tuning the X-ray energy to one of the nuclear transitions (Mössbauer transition). The technique can then be used to study self-diffusion in thin films.

References

1. M. N. Baibich *et al.*, *Phys. Rev. Lett.* **61** (1988) 2472.
2. Y. H. Phang *et al.*, *J. Appl. Phys.* **74**, 181 (1993).
3. Y. H. Xie *et al.*, *Appl. Phys. Lett.* **63** (1993) 2263.
4. A. Gupta *et al.*, *J. Phys. Soc. Jpn.* **69** (2000) 2182.
5. W.-S. Kim *et al.*, *Phys. Rev. B* **58** (1998) 6346.
6. Z. Erdelyi *et al.*, *Science* **306** (2004) 1913.
7. G. L. Katona *et al.*, *Phys. Rev. B* **71** (2005) 115432.
8. J. Wang *et al.*, *Science* **258** (1992) 775; S. K. Ghose *et al.*, *Phys. Rev. B* **64** (2001) 33403.
9. A. Gupta *et al.*, *Phys. Rev. B* **72** (2005) 075436.
10. P. Rajput *et al.*, *J. Phys.: Condens. Matter* **19** (2007) 036221.
11. A. Gupta *et al.*, *Phys. Rev. B* **72** (2005) 014207; M. Rennhofer *et al.*, *Phys. Rev. B* **74** (2006) 104301.
12. A. Gupta, *Hyperfine Interact.* **160** (2005) 123.
13. Y. P. Feng *et al.*, *Phys. Rev. Lett.* **71** (1993) 537.
14. L. G. Parratt, *Phys. Rev.* **95** (1954) 359.
15. A. Gupta, *Vacuum* **58** (2000) 16; W. Bolse, *Radiat. Meas.* **36** (2003) 597; J. Marfaing *et al.*, *Appl. Phys. Lett.* **57** (1990) 1739.
16. M. Gupta *et al.*, *Phys. Rev. B* **74** (2006) 104203; M. Gupta *et al.*, *Phys. Rev. B* **70** (2004) 184206.
17. F. Faupel *et al.*, *Rev. Mod. Phys.* **75** (2003) 237.
18. J. P. Hannon *et al.*, *Phys. Rev. B* **32** (1985) 6363.
19. A. Gupta, S. Chakravarty and R. Rüffer, (to be published).
20. R. Rüffer and A. I. Chumakov, *Hyperfine Interact.* **97/98** (1996) 589.
21. H. Tanimoto, P. Farber, R. Würschum, R. Z. Valiev and H.-E. Schaefer, *Nanostruct. Mater.* **12** (1999) 681.
22. A. Gupta, M. Gupta, U. Pietsch, S. Ayachit, S. Rajagopalan, A. K. Balamurgan and A. K. Tyagi, *J. Non-Cryst. Solids* **343** (2004) 39.

Chapter 24

COHERENT DIFFRACTION IMAGING WITH HARD X-RAY WAVEGUIDES

LIBERATO DE CARO and CINZIA GIANNINI

Istituto di Cristallografia — Consiglio, Nazionale delle Ricerche (IC-CNR), via Amendola 122/O, I-70126, Italy

DANIELE PELLICCIA

School of Physics, Monash University
ARC Centre of Excellence for Coherent X-Ray Science
Victoria 3800, Australia

ALESSIA CEDOLA

Istituto di Fotonica e Nanotecnologie — CNR, V. Cineto Romano, 42 00156, Roma

STEFANO LAGOMARSINO

Instituto Processi Chimico Fisici — CNR
c/o Phys. Dept. Universita' Sapienza
P. le A. Moro 2, 00185, Rome, Italy

Coherent X-ray diffraction imaging (CXDI) has been widely applied in the nanoscopic world, offering nanometric-scale imaging of non-crystallographic samples, and permitting the next-generation structural studies on living cells, single virus particles and biomolecules. The use of curved wavefronts in CXDI has caused a tidal wave in the already promising application of this emergent technique. The non-planarity of the wavefront allows to accelerate any iterative phase-retrieval process and to guarantee a reliable and unique solution. Nowadays, successful experiments have been performed with Fresnel zone plates and planar waveguides as optical elements. Here we describe the use of a single planar waveguide as well as two crossed waveguides in the experiments which first showed this optical element a promising tool for producing a line- or point-like coherent source, respectively.

24.1. Introduction

Diffraction experiments can be realized both using planar and curved wave fields. The former is obtained when the source is very far from the sample, the latter for source–sample distances smaller than the Fresnel distance which is defined as a^2/λ, where a is the size of the object and λ is the wavelength. When the detector is placed far from the sample ($\gg a^2/\lambda$) the diffraction experiment is realized in far-field Fraunhofer regime; conversely for sample–detector distances less then the Fresnel distance ($\leq a^2/\lambda$), the diffraction intensity is collected in near-field Fresnel regime. In both cases only the intensity of the complex diffracted wave field is measured, with a loss of the phase contribution. In the far-field regime, as the diffraction pattern is proportional to the square modulus of the Fourier transform of the sample scattering function, the loss of phase information prevents from using the inverse Fourier transform to retrieve the unknown scattering function directly from the diffraction pattern. In the near-field regime, the effect of the wave field curvature can be described by using a wave propagator. Indeed, the final diffraction pattern will be proportional to the convolution of the wave propagator with the object scattering function. For hard X-ray, the object scattering function is related to the mean refractive index, integrated along the beam path, which, in turn, is related to the mean electron density. The real part of the refractive index causes phase shift of the wave field interacting with the sample. Instead, the imaginary part of the refractive index causes an attenuation of the wave field amplitude. The use of curved wave field leads to a "distorted reconstructed object" until the effect of convolution with the wave propagator is properly deconvolved.

Coherent X-ray diffraction imaging (CXDI) is one of the most promising techniques to study the assembly of non-periodic single objects at the nanoscale. The first experiments were performed using planar incident waves.[1,2] It was proven that by illuminating an unknown object with a coherent X-ray beam and recording the diffracted intensity in far-field Fraunhofer regime, it was possible to recover the object shape and size in real space.[1,2] Phase retrieval techniques, classified as oversampling techniques, were applied to directly Fourier transform the diffraction pattern into the object image. Oversampling requirements were clearly indicated to ensure the correct transversal and longitudinal coherence of the beam as well as to record the diffraction pattern in reciprocal space at the right sampling step, finer than the Nyquist frequency $1/2a$, related to twice the size of the sample for each spatial dimension.[1,2]

Just recently, non-planar waves have been used with success, demonstrating how to extend CXDI to the near-field Fresnel regime by the introduction of a "distorted object" for the calculation of the coherent diffraction pattern, to take into account the effects of the wave propagator causing the presence of Fresnel fringes.[3,4] A Fresnel-zone construction is embedded on an original object and then Fourier transformed to form a diffraction image. Simulated numerical examples have indicated that, also for near-field coherent diffraction, suitable Fourier-based iterative phasing algorithms can be realized.[3,4] CXDI experiments with non-planar wave fields have been recently performed in near-field Fresnel regime using either X-ray zone plates[5] and planar waveguides[6,7] as optical elements.

Two striking benefits have emerged with using non-planar wavefronts: a faster convergence of the iterative phase-retrieval algorithms[8] and the possibility to measure easily the diffracted intensity at low-q, which is frequently lost when using planar wave fields, since the presence of a beamstop obscures the signal at small scattering angles. This pattern at low-q contains a two-dimensional in-line magnified hologram of the unknown object,[5] if the incident wave field is emitted by a point-like source. However, a straightforward holographic reconstruction of the unknown object from in-line holograms is usually prevented by the presence of the twin image pattern interfering with the object image pattern. Later-on we will show that for particular geometric conditions this drawback can be overcome.

Another extremely important aspect to consider in CXDI experiments is the degree of coherence of the incident wave. In practice, in most of the experiments, the beam which arrives onto the sample has encountered several optical elements on its path from the source, e.g., monochromators, mirrors, slits, etc., that lead to coherence degradation at the sample position.[9] This lack of coherence can severely disturb the convergence of the phasing process, as recently discussed by Williams *et al.*[9] In this respect, a waveguide is an optical element providing a (mostly) coherent beam. In conjunction with the divergent beam from a point- or line-like source this is extremely important for CXDI experiments.[6,7,10]

The next paragraph gives a first example of CXDI using the beam produced by an X-ray waveguide[6] illuminating a Kevlar fiber by the coherent cylindrical wave field provided by a planar waveguide. In the third paragraph, a two-dimensional image reconstruction of a gold butterfly sample obtained by two crossed planar waveguides will be discussed.[7] Both experiments can be classified as in-line coherent diffraction imaging

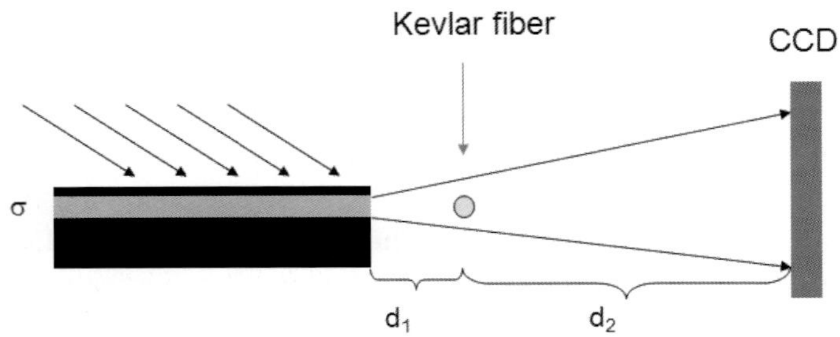

Fig. 24.1. Projection geometry for a line-like coherent source (single planar waveguide).

experiments. An example of an off-axis holographic reconstruction obtained with planar waveguides can be found in Ref. 11.

24.2. One-Dimensional CXDI with Planar Waveguides

Let us consider the projection geometry in Fig. 24.1, where the beam emitted by a line-like coherent and monochromatic source of nanosized dimension σ (planar WG) interacts with a sample at a distance d_1; then it propagates towards an area detector, located at a distance d_2 from the sample, oriented perpendicular to the average direction of the transmitted beam. The radiation emitted by the source is assumed to be monochromatic and coherent.

Applying the paraxial approximation of the Fresnel propagators, the complex field amplitude $A(\mathbf{u})$ at the detector plane can be calculated using the Fresnel–Kirchoff integral[6]:

$$A(\mathbf{u}) \cong A_0 \exp\left\{\frac{i\pi u^2}{d_2\lambda}\right\} \int d\mathbf{r} f_{\text{chirp}}(\mathbf{r}) \exp\left\{-\frac{2i\pi \mathbf{u}\cdot\mathbf{r}}{d_2\lambda}\right\} \tag{24.1}$$

where

$$f_{\text{chirp}}(\mathbf{r}) = A(\mathbf{r}) \exp\left\{\frac{i\pi r^2}{D\lambda}\right\} f(\mathbf{r}) \tag{24.2}$$

$$A(\mathbf{r}) = \int ds \Psi(\mathbf{s}) \exp\left\{\frac{i\pi s^2}{d_1\lambda}\right\} \exp\left\{\frac{-2i\pi \mathbf{r}\cdot\mathbf{s}}{d_1\lambda}\right\} \tag{24.3}$$

and the defocus distance $1/D = 1/d_1 + 1/d_2$. Here, A_0 is a complex constant; $\Psi(\mathbf{s})$ is the complex amplitude emitted by the source and the integral in $d\mathbf{s}$ is performed on its surface S; the vector $\mathbf{r}$ defines a point in the sample plane; λ is the wavelength. The vectors $\mathbf{s}$, $\mathbf{r}$, and $\mathbf{u}$ are referred to origins, lying on the optical axis with their individual reference system; $f(\mathbf{r})$ is the complex transmission function of the object. Thus, in projection geometry the amplitude observed at the detector plane would be the Fourier Transform (FT) of a Fresnel diffracted, phase-chirped, distorted object function,[3,4] modulated in amplitude due to the spatial distribution of the incident complex wave $A(\mathbf{r})$ at the object plane (Eq. (24.3)), related to the complex amplitude emitted by the source $\Psi(\mathbf{s})$. In point- or line-projection geometries[12] one should either experimentally reconstruct the function $\Psi(\mathbf{s})$[5] or evaluate it theoretically, which may be possible when $d_1 \gg \sigma^2/\lambda$. In this case, the incident complex amplitude $A(\mathbf{r})$ at the object plane would be given by the FT of the complex amplitude emitted by the source $\Psi(\mathbf{s})$. For source sizes on the sub-micrometer scale (let say $\sigma \approx 100\,\text{nm}$) and hard X-rays (let say $\lambda \approx 1\,\text{Å}$), the condition $d_1 \gg \sigma^2/\lambda$ is even fully satisfied at source to object distances of the order of 1 mm, giving the possibility to approximate the real $\Psi(\mathbf{s})$ with an effective Gaussian source.[12] For planar (one-dimensional nanosized) WGs this leads to

$$A(\mathbf{r}) = \exp\{-\sigma^2 r^2/d_1^4\lambda^4\}. \tag{24.4}$$

Equation (24.4) enables us to quantify $f_{\text{chirp}}(\mathbf{r})$ as follows:

$$f_{\text{chirp}}(\mathbf{r}) = \exp\left\{\frac{i\pi r^2}{D\lambda} - \frac{\sigma^2 r^2}{d_1^4\lambda^4}\right\} f(\mathbf{r}) \equiv M_{fr}(\mathbf{r})f(\mathbf{r}), \tag{24.5}$$

and all real-space constraints applicable to $f(\mathbf{r})$ in the object plane may be transferred onto $f_{\text{chirp}}(\mathbf{r})$. As usual, the Fourier-space constraint is given by the square root of the measured wave intensity at the detector plane. Thus, one can realize a suitable Fourier-based iterative error-reduction phasing algorithm for retrieving the unknown function $f(\mathbf{r})$, taking into account the complex illumination factors determined by the Fresnel propagators.[4,6,8]

In order to test the possibility to retrieve an unknown transmission function from the measured Fresnel diffraction pattern in projection geometry, a Kevlar fiber, with a circular section of radius $R = 7.2\,\mu\text{m}$, has been used. The measurements were carried out at the second hutch of the optical beamline BM5 of ESRF. The 12 keV photon beam was monochromatized by a Si(111) monochromator delivering a $\Delta E/E$ of 10^{-4}. The beam size was limited by a slit placed at 50 m from the bending magnet source. A planar WG was used in Resonant Beam Coupling mode located 55 m far from the source to condition the beam and to generate a coherent cylindrical beam. The WG had a Be guiding core layer 130-nm thick and Mo claddings as described in Ref. 13. The CCD area detector had a de-magnified pixel size of $0.625\,\mu\text{m}$; 4×4 binning was used in the

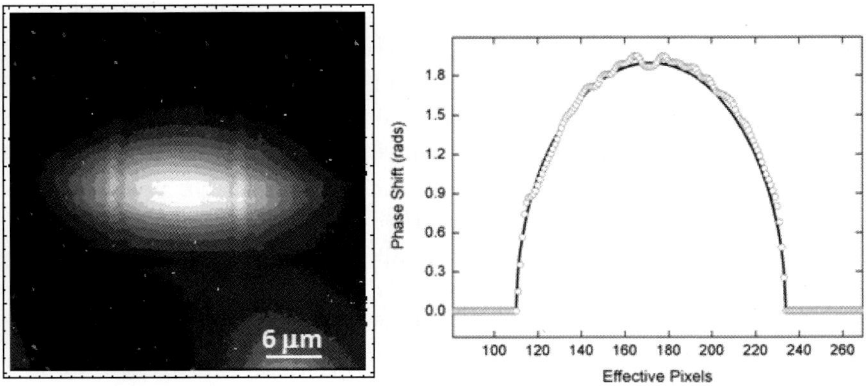

(a) (b)

Fig. 24.2. (a) Square root of the measured diffraction pattern. (b) The symbols represent retrieved values of $\varphi(\mathbf{r})$ in correspondence of the minimum of the FOM; the continuous curve gives the expected values for an ideal Kevlar fiber with a circular section of $7.2\,\mu$m radius.

measurements, i.e., the pixel size Δ was $2.5\,\mu$m. Figure 24.2(a) shows the square root of the diffraction pattern with $d_1 = 0.017\,$m, $d_2 = 0.348\,$m.

As the beam energy was $12\,$keV ($\lambda = 1.033\,$Å), the density of the fiber $1.44\,$g/cm^3, the refraction index $n = 1 - 2.1639 \times 10^{-6} + 1.5237 \times 10^{-9}\, i$, the fiber can be approximated to a pure phase contrast object. Thus, we can put $f(\mathbf{r}) = \exp[i\varphi(\mathbf{r})]$, where $\varphi(\mathbf{r})$ is the phase shift introduced by the fiber, i.e., the unknown function to be retrieved is a complex one.

In fact, one is dealing with a one-dimensional phase-retrieve problem (perpendicular to the fiber axis), since a planar WG generates a Gaussian-like cylindrical wave with its axis parallel to the fiber axis, leading to a line-projection set-up.[12] Thus, in the real space, one can impose that the transmission function is equal to 1 outside of a region of size $2R$, i.e., $f(\mathbf{r}) = 1$ if $\mathbf{r} \notin [-R, R]$, defining a support suitable for a line-projection-geometry diffraction experiment on a pure phase object. However, the size of this space region ($2R$) cannot *a priori* be fixed for an unknown object, but it can be chosen as a parameter to be determined with the phase-retrieve procedure. In our experiment we should expect that the phasing algorithm would converge when R is sufficiently close to the actual fiber radius.

As initial phases one could adopt random phases, but to improve the convergence of the algorithm we have found it useful to start from the Fresnel-term phase values defined in Eq. (24.5), which constitute a partial, *a priori* knowledge about the unknown phase of the transmitted beam reaching the detector plane. We have monitored the phasing retrieval by means of a Figure of Merit (FOM), defined as the sum over all the pixels of the deviation of the retrieved phase from the expected values

$$\mathrm{FOM} = \sum |\varphi(\mathbf{r}) - \varphi_R(r)| / \sum \varphi_R(\mathbf{r}), \qquad (24.6)$$

where $\varphi_R(\mathbf{r})$ is the phase-shift expected for an ideal Kevlar fiber of radius R and $\varphi(\mathbf{r})$ is the retrieved phase-shift. The minimum value for the FOM was reached for $R = 7.2\,\mu$m, i.e., the best convergence of the algorithm was obtained for a support of size $14.4\,\mu$m ($= 2R$). This value is in good agreement with the fiber radius value determined independently both by comparing the measured diffracted intensity transmitted from the fiber with the simulation obtained using the free-space propagation method[14] and by scanning electron microscope measurements.

Figure 24.2(b) shows the results obtained for the retrieved function $\varphi(\mathbf{r})$ by our phasing algorithm for a support of size $14.4\,\mu$m. showing the retrieved values (symbols) and the expected value for an ideal Kevlar fiber with a circular section of 7.2-μm radius (line). In Fig. 24.2 the effective pixel size is given by the real pixel size scaled by the magnifying factor $m = (d_1 + d_2)/d_1$ due to the projection geometry: $\Delta_{\mathrm{eff}} = \Delta/m$.[6,12] For this experimental set-up the effective pixel size was about $0.12\,\mu$m.

24.3. Two-Dimensional CXDI with Two Planar WGs in a Cross Configuration

Fuhse and coworkers[11] using a waveguide-based experimental set-up have performed a hard X-rays off-axis holographic reconstruction of a tungsten tip with a spatial resolution of about 100 nm. They used a two-beam technique with an additional reference beam to probe the phase of the diffracted wave, thus avoiding complications due to the presence of twin images of in-line holographic reconstructions.

De Caro *et al.*[7] have recently proposed a new experimental set-up with two crossed planar waveguides as optical elements to produce a hard X-ray virtual point-like source for a digital in-line holographic experiment (Fig. 24.3).

The experiment was performed at the ID1 beamline of the European Synchrotron Radiation Facility (ESRF), using photons of 11 keV ($\lambda = 0.1127$ nm) monochromatized by a Si(111) double crystal and pre-focused onto the first WG entrance by Be Compound Refractive Lenses (CRLs).

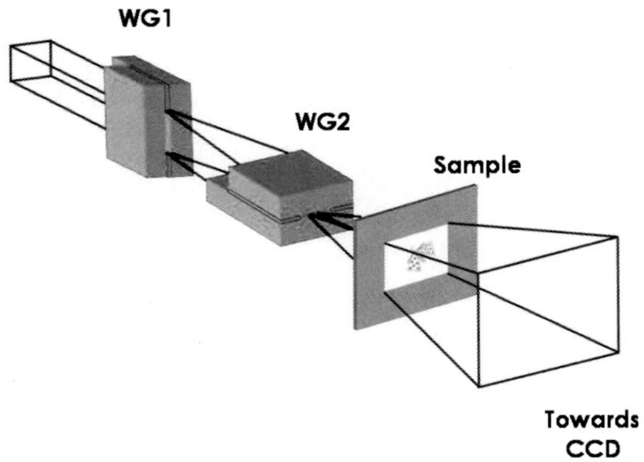

Fig. 24.3. Experimental set-up with two crossed planar waveguides as optical elements to produce a hard X-ray virtual point-like source.

A gold butterfly of maximum size of 4.6 μm and 600 nm in thickness, electrochemically deposited onto a Si_3N_4 membrane, was used as test object. The two planar waveguides (WG1 and WG2) were made of two silicon slabs, each of 500 μm thickness, bonded together via a Microposit® S1800® photoresist layer, acting as a spacer. Each waveguide was fabricated in order to work in asymmetric front coupling scheme.[15] In this case, the incident beam makes a reflection at the very entrance of the waveguide. The standing wave pattern is established at the entrance with a period that depends on the incidence angle.

This method allows the selection of a single mode propagating into the air-filled gap. Moreover, the 500-μm thickness of the cladding layers fully stops the direct beam, naturally avoiding the use of any additional beam stopper. The gaps of WG1 and WG2 were 290 (t_1) and 140 nm (t_2), respectively. The beam focused by beryllium lenses onto the WG1 entrance transforms into a divergent coherent line-like wave field. This wave field propagates as a cylindrical wave onto the second planar waveguide (WG2) producing another line-like wave field perpendicularly to WG1. A WG1–WG2 distance of 10 mm (d_{12}) and a WG2-sample distance of 9 mm (d_{2s}) were chosen in order to approximately satisfy the following condition

$$\frac{d_{12} + d_{2s}}{d_{2s}} \approx \frac{t_1}{t_2}. \tag{24.7}$$

In this way, WG1 was placed at about twice the distance from the sample than WG2, to compensate for the beam divergence of the wave exiting WG1 (proportional to λ/t_1), the latter being about one half the beam divergence at the exit of WG2 (proportional to λ/t_2).

A $W_1 \times W_2 = 10 \times 10\,\mu\text{m}^2$ sized beam coherently illuminates the sample, about twice the size of the object, as required by oversampling requirements. Let us note that the oversampling has to be satisfied only to allow the iterative phase-retrieval; instead the direct holographic reconstruction does not require this condition.[11] A Princeton single photon counting CCD detector with $20 \times 20\,\mu\text{m}^2$ pixel size (Δ) and 1340×1300 imaging array (PI-SX:1300) was put at a distance $L = 3.65\,\text{m}$ Consequently, the magnifying factor of the realized projection geometry was $M = L/d_{2s} \cong 406$.

The combined action of the two waveguides is expected to generate an overall point-like virtual source at the exit of WG2. Figure 24.4(a) shows the pattern collected moving the object out of the beam path, namely the incident beam intensity (I_0). Taking the modulus of the FFT of the diffracted intensity, $|\text{FFT}[I_0]|$, a bi-dimensional wave field image of the autocorrelation of the source at the exit surface of WG2 can be visualized. Figure 24.4(b) is its three-dimensional representation, showing that the source can be described as a point-like one.

The ratio between the intensity of diffraction patterns, measured with and without the sample in the beam path, shown in Fig. 24.5(a), clearly shows an in-line hologram of the test sample. The SEM image of the butterfly test sample, the object which ought to be retrieved, is shown in Fig. 24.5(b).

With a 600-nm-thick gold sample and working with a $\lambda = 0.1127\,\text{nm}$ wavelength, the first-order Born approximation[16] can be adopted to

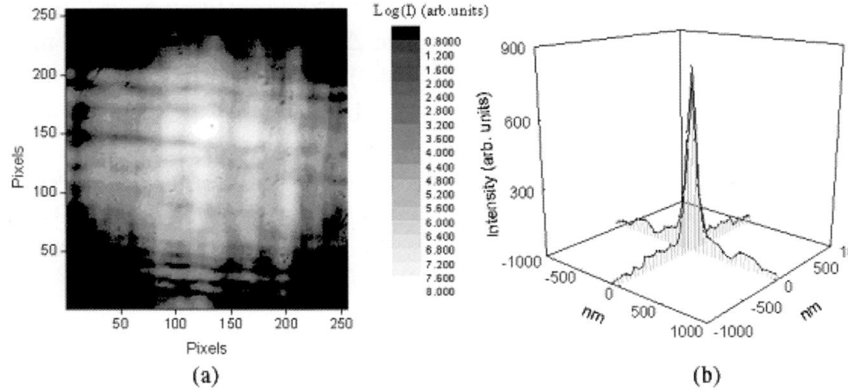

Fig. 24.4. (a) Measured intensity without the object; (b) 3D-representation of the modulus of the FFT of the raw intensity data collected without the butterfly.

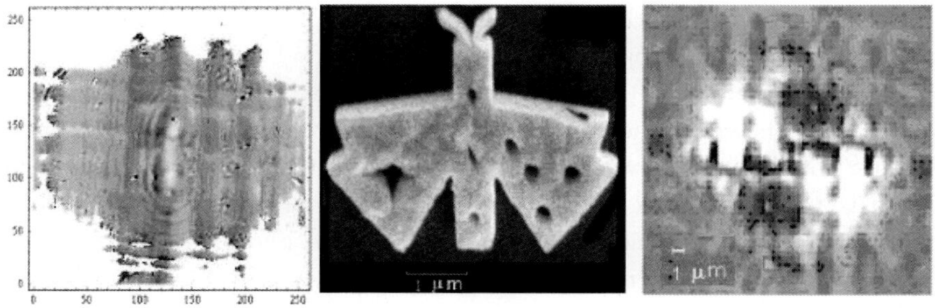

(a) (b) (c)

Fig. 24.5. (a) Ratio between diffracted intensity with and without the butterfly; (b) SEM image of the 4.2-μm test object; (c) digital holographic reconstruction of the object image and its twin image.

calculate the diffraction wave. In this way, the overall wave field is approximately given by the unperturbed incident wave field $A_{\text{ref}}(\mathbf{r})$ plus a small perturbation due to the presence of the object $A_{\text{scat}}(\mathbf{r})$, where $\mathbf{r}$ is the vector indicating a generic point in the detector plane. In fact, at 11 keV the wave field incident on a gold object of 600 nm thickness is modified by a transmission function T such as $\Delta T = 1 - |T| = 0.046 \ll 1$.

From the two measured intensity patterns (with and without the object) one can calculate a contrast image.[7]

$$\begin{aligned} I(\mathbf{r}) &= |A_{\text{ref}}(\mathbf{r}) + A_{\text{scat}}(\mathbf{r})|^2 - |A_{\text{ref}}(\mathbf{r})|^2 \\ &\cong A^*_{\text{ref}}(\mathbf{r})A_{\text{scat}}(\mathbf{r}) + A^*_{\text{scat}}(\mathbf{r})A_{\text{ref}}(\mathbf{r}) + |A_{\text{scat}}(\mathbf{r})|^2 \end{aligned} \tag{24.8}$$

containing a linear term in the scattered wave (holographic term), and the classical diffraction pattern (quadratic term in the scattered wave). Here, A^* indicates the complex conjugate. Since $\Delta T \ll 1, |A_{\text{scat}}| \ll |A_{\text{ref}}|$ and the holographic term dominates in Eq. (24.8).

The holographic reconstruction of the object image can be performed via the Kirchhoff–Helmoltz transform of Eq. (24.8).[17,18] This, in turn, can be implemented via an FFT, and actually gives the autocorrelation function of the contrast intensity pattern of Eq. (24.8):

$$\text{FFT}[I(\mathbf{r})] \approx S(\mathbf{u}) * \bar{O}(\mathbf{u}) + \bar{S}(\mathbf{u}) * O(\mathbf{u}). \tag{24.9}$$

Here, the symbol "$*$" indicates the convolution product; $S(\mathbf{u}) = \text{FFT}[A_{\text{ref}}(\mathbf{r})]$ and $O(\mathbf{u}) = \text{FFT}[A_{\text{scat}}(\mathbf{r})]$ are the source and object complex functions, respectively. For a point-like source ($S(\mathbf{u}) \approx \delta(\mathbf{u})$, with δ denoting the Dirac-delta function, Eq. (24.9) leads to a digital holographic reconstruction of the object image plus its twin image numerically made via FFT. If the object is sufficiently shifted with respect to the optical axis, i.e., $O(\mathbf{u}) \to O(\mathbf{u} - \Delta\mathbf{u})$ with the shift $\Delta\mathbf{u}$ larger than the object size, the object and its twin images are well separated in the holographic reconstruction. This condition occurred in the present experiment, as shown in Fig. 24.4(c). Conversely, if $\Delta\mathbf{u}$ would have been smaller than the object size, the object image would have been overimposed on its twin image and the object not recognizable. The spatial resolution of the digital in-line holographic reconstruction shown in Fig. 24.4(c) is limited by the source size to about 200 nm. This result can be obtained thanks to the point-like nature of the virtual source ($S(\mathbf{u}) \approx \delta(\mathbf{u})$). The secondary maxima of the source cause distortions in the holographic reconstruction of the object shape. Despite of these limitations, this novel experimental set-up has shown for the first time the possibility to perform, with a hard X-ray waveguide-based set-up, a digital in-line holographic reconstruction of a micrometric object with a spatial resolution limited only by the source size.

Let us note that the experimental intensity pattern contains information at higher spatial frequencies than 200 nm corresponding to contributions of radiation scattered outside the incident wave field cone beam. This further source of information can be used for a phase retrieval approach, as argued in Ref. 11, in order to reconstruct the test object to a resolution higher than the holographic one. Error-reduction iterative algorithms have been recently generalized for diffraction intensity data measured with curved coherent wave fields.[4,6–8] The FFT of the square root of the raw intensity data collected without the test sample is a complex function, whose phase is a good approximation of the incident wave field phase, apart of a constant phase term.[7] Starting just from the incident wave field phase so obtained, a partial reconstruction of the butterfly was obtained very quickly, after only 20 iterations, as shown in Fig. 24.6(a).

A comparison with the one-bit version of the butterfly SEM image of Fig. 24.6(b) indicates that many features (enhanced by red circles) are present in the reconstructed image, although only a 40% correlation between the two images is found. The resolution of the partially reconstructed image is close to 50 nm. The incompleteness of the obtained reconstruction could be ascribed to the presence of secondary maxima in the tails of the virtual 2D point-like source (Fig. 24.4(b)). Nevertheless, this work shows that a good estimate of the wave field phase emitted by a point-like source can be obtained directly from its autocorrelation function. Indeed, it is worth to note that for point-like source, only those

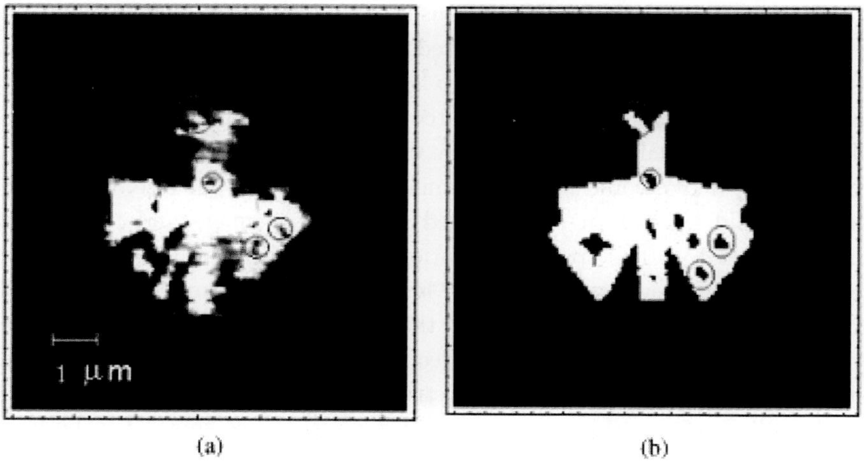

Fig. 24.6. (a) Phased object; (b) one-bit SEM image of the object.

phases associated with the maximum amplitudes, are dominant in the wave field determination. Ideally, for a δ-dirac-like source, only one phase is needed. This constitutes a big advantage for any phase retrieval process using coherent diffraction data obtained with a "nearly-ideal" point-like source.

24.4. Conclusions

This chapter has highlighted one of the most exciting applications of the X-ray waveguides: the CXDI methodology, which we firmly believe will play an important role in the application of fourth-generation synchrotron radiation sources, and will hopefully open the possibility of imaging with wavelength-limited spatial resolution. The optical elements which can produce curved coherent wave fields have proved to retrieve nanometer features even if embedded in extended objects, thus widening the range of applicability of this emerging tool.[19] In addition, it was shown how a partial coherence in the beam can be easily tolerated when working with a curved wavefront.[20]

Acknowledgments

We acknowledge partial financial support from the SPARX project.

References

1. J. Miao, P. Charalambous, J. Kirz and D. Sayre, *Nature (London)* **400** (1999) 342.
2. I. K. Robinson, I. A. Vartanyants, G. J. Williams, M. A. Pfeiffer and J. A. Piteny, *Phys. Rev. Lett.* **87**, 195505 (2001); J. Miao, T. Ishikawa, B. Johnson, E. H. Anderson, B. Lai and K. O. Hodgson, *Phys. Rev. Lett.* **89**, 88303 (2002); J. M Zuo, I. Vartanyants, M. Gao, R. Zhang and L. A. Nagahara, *Science* **300**, 1419 (2003); D. Shapiro, P. Thibault, T. Beetz, V. Elser, M. R. Howells, C. Jacobsen, J. Kirz, Lima, J. Miao, Nieman and D. Sayre, *Proc. Nat. Acad. Sci. USA* **102** (2005) 15343.
3. H. M. Quiney, K. A. Nugent and A. G. Peele, *Opt. Lett.* **30** (2005) 1638.
4. X. Xiao and Q. Shen, *Phys. Rev. B* **72** (2005) 033103.
5. H. M. Quiney, A. G. Peele, Z. Cai, D. Paterson and K. A. Nugent, *Nat. Phys.* **2**, 101 (2006); G. J. Williams, H. M. Quiney, B. B. Dhal, C. Q. Tran, K. A. Nugent, A. G. Peele, D. Paterson, and M. D. de Jonge, *Phys. Rev. Lett.* **97**, 025506 (2006); G. J. Williams, H. M. Quiney, B. B. Dhal, C. Q. Tran, A. G. Peele, K. A. Nugent, M. D. de Jonge and D. Paterson, *Thin Solid Films* **515** (2006) 5553.
6. L. De Caro, C. Giannini, A. Cedola, D. Pelliccia, S. Lagomarsino and W. Jark, *Appl. Phys. Lett.* **90** (2007) 041105.
7. L. De Caro, C. Giannini, D. Pelliccia, C. Mocuta, T. H. Metzger, A. Gagliardi, A. Cedola, I. Burkeeva and S. Lagomarsino, *Phys. Rev. B* **77** (2008) 081408.
8. R. W. Gerchberg and W. O. Saxton, *Optik* **35**, 237 (1972); J. R. Fienup, *Appl. Opt.* **21**, 2758 (1982); V. Elser, *J. Opt. Soc. Am. A* **20** (2003) 40.
9. S. K. Sinha, M. Tolan and A. Gibaud, *Phys.: Rev. B* **57**, 2740 (1998); I. A. Vartanyants and I. K. Robinson, *J. Phys.: Condens. Matter* **13**, 10593 (2001); G. J. Williams, H. M. Quiney, A. G. Peele and K. A. Nugent, *Phys. Rev. B* **75** (2007) 104102.
10. S. Di Fonzo, W. Jark, S. Lagomarsino, C. Giannini, L. De Caro, A. Cedola and M. Mueller, *Nature* **403** (2000) 638.
11. C. Fuhse, C. Ollinger, S. Kalbfleish and T. Salditt, *J. Synchrontron Radiat.* **13** (2006) 69.
12. L. De Caro, C. Giannini, A. Cedola, S. Lagomarsino, I. Burkreeva, *Opt. Commun.* **265** (2006) 18.
13. W. Jark, A. Cedola, S. Di Fonzo, M. Fiordelisi and S. Lagomarsino, N. V. Kovalenko and V. A. Chernov, *Appl. Phys. Lett.* **78** (2001) 1192.
14. S. Lagomarsino, A. Cedola, P. Cloetens, S. Di Fonzo, W. Jark, G. Soullie and C. Riekel, *Appl. Phys. Lett.* **71** (1997) 2557.
15. M. J. Zwanenburg, J. F. Peters, J. H. H. Bongaerts, S. A. de Vries, D. L. Abernathy and J. F. van der Veen, *Phys. Rev. Lett.* **82**, 1696 (1999); D. Pelliccia, I. Bukreeva, M. Ilie, W. Jark, A. Cedola, F. Scarinci and S. Lagomarsino, *Spectrochim. Acta Part B* **62** (2007) 615.
16. J. M. Cowley, *Diffraction Physics*, 2nd edn. (Elsevier Science Publisher, New York, 1990).
17. H. J. Kreuzer and R. A. Pawlitzek, *Europhys. News* **34** (2003) 62.

18. H. J. Kreuzer, K. Nakamura, A. Wierzbicki, H. W. Fink, and H. Schmid, *Ultramicroscopy* **45** (1992) 381.
19. B. Abbey, K. A. Nugent, G. J. Williams, J. N. Clark, A. G. Peele, M. A. Pfeifer, M. D. de Jonge and I. McNulty, *Nature* **4** (2008) 394.
20. L. W. Whitehead, G. J. Williams, H. M. Quiney, K. A. Nugent, A. G. Peele, D. Paterson, M. D. de Jonge and I. McNulty, *Phys. Rev. B* **77** (2008) 104112.

Chapter 25

X-RAY STANDING WAVE FOR CHEMICAL-STATE SPECIFIC SURFACE STRUCTURE DETERMINATION

D. P. WOODRUFF

Physics Department, University of Warwick,
Coventry CV4 7AL, UK

By exploiting the so-called "chemical shifts" in photoelectron binding energies, the use of photoelectron-monitored X-ray standing waves allows one to obtain not only element-specific but also chemical-state specific structural information from surfaces. This approach has been used to study a range of problems, including complex molecular adsorbates, co-adsorbed molecular species, and multiple-site occupation of atoms and molecules at surfaces.

One important feature of all XSW structural studies is that the information one obtains is element specific. By monitoring the X-ray absorption though the photoelectron emission arising directly from the photoionization event, or through the Auger electron or X-ray emission associated with refilling of the resulting core hole, the energy of these emitted electrons or photons provide a direct signature of the atomic species with the absorption occurred. Photoelectron detection, however, offers the possibility of even more specificity, namely the ability to distinguish the photo-absorption at atoms of the same element in different bonding configurations. To achieve this, one exploits the well-established changes in the photoelectron binding energy of specific atomic core levels as a result of changes in the local bonding environment.[1] These changes are widely exploited in the standard surface science technique of X-ray photoelectron spectroscopy (XPS) and indeed are the reason why in the early years of application of this technique it was known as ESCA — electron spectroscopy for *chemical* analysis. The origin of these core-level shifts is complex,[1] resulting from a combination of initial state effects (changes in the one-electron binding energy including those due to charge transfer) and final state effects (changes in the inter-atomic relaxation energy).

From the point of view of exploiting these shifts, however, the key feature is that they provide a spectral fingerprint of the different states of the atoms. Most commonly, these shifts are associated with differences in the chemical bonding of the associated atoms, due to inequivalent sites within an adsorbed molecule, to inequivalent adsorbed species (e.g., different molecular species or molecular and atomic species), or to inequivalence in their bonding to the surface. In such systems, the core-level shifts are commonly referred to as *chemical shifts*. More generally, however, core-level shifts can also be observed in photoelectron emission from substrate atoms due to differences in the coordination or local geometry of surface and sub-surface atoms; in such cases they are commonly referred to as surface core-level shifts. Notice, of course, that one may also find chemical shifts in substrate atom photoelectron binding energies associated with bonding to adsorbates.

The first experiment to exploit these chemical shifts in XSW was presented by Sugiyama *et al.*,[2] but the first demonstrations of this approach to obtain chemical-state specific local adsorption geometries were conducted a few years later by Jackson *et al.*[3,4] To illustrate the way this is achieved, we first consider in detail one of these early examples, specifically a study of the interaction of methanethiol (CH_3SH) with Cu(111).[4] In general, there is still considerable interest in the interaction of the general class of alkanethiols, $CH_3(CH_2)_{n-1}SH$, with the coinage metals Cu, Ag, and Au, and particularly with Au(111) for which this combination of adsorbates and substrate are the archetypal self-assembled monolayer (SAM) systems. The particular case of methanethiol on Cu(111) is a simple model system that shows some interesting complexities. Figure 25.1 shows the results of an investigation of this system using low-energy XPS[5] and shows the S $2p$ photoelectron energy spectra recorded, using a photon energy of 200 eV, from a Cu(111) surface first exposed to methanethiol at low temperature (~124 K) and then heated to successively higher temperatures. A general feature of the way that all the alkanethiols behave on coinage metal surfaces is that the surface interaction causes loss of the thiol hydrogen atom to form an adsorbed thiolate, in the present case methylthiolate, CH_3S-, that bonds to the surface through the S head-group atom. At sufficiently

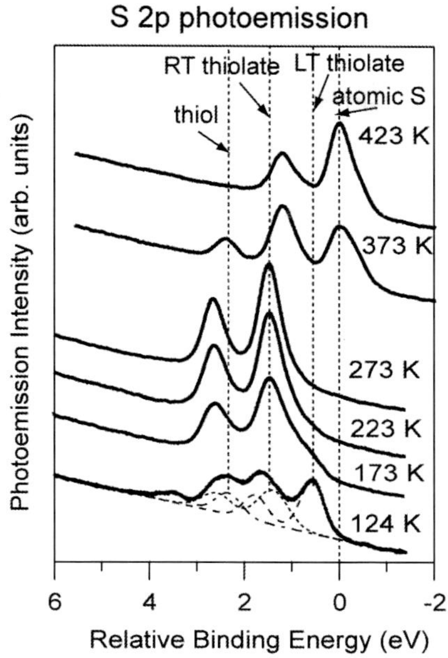

Fig. 25.1. Sulfur $2p$ photoelectron spectra recorded from a Cu(111) surface exposed to methanethiol (CH_3SH) at 124 K and then heated to successively higher temperatures. Photoelectron binding energies are shown relative to the S $2p_{3/2}$ component of the emission from the atomic S species having a value of ~161 eV. After Ref. 5.

high temperatures, decomposition occurs, leaving residual atomic S on the surface. The data of Fig. 25.1 can actually be most easily understood by first considering the spectrum recorded at the highest temperature, after which this decomposition has occurred and no molecular species remain on the surface. This spectrum, recorded after heating to 423 K, shows just two peaks that can be attributed to the two spin-orbit split S $2p_{1/2}$ and $2p_{3/2}$ component peaks with an intensity ratio of 1:2. These two peaks are thus characteristic of a single S species on the surface, and the spectra from all other surface S species represented in the measured spectra must comprise similar doublets with the same energy separation.

Comparing this spectrum with those recorded after heating to 273 K and 373 K, it is then clear that after heating to 273 K the surface must be covered with a different single S species, while heating to 373 K corresponds to a surface in which this species coexists with the atomic S seen as the only species at 423 K. Comparison with the spectrum recorded immediately after adsorption at 124 K shows that at this low temperature, two additional

relative to that of the S $2p$ states (~2472 eV and ~161 eV, respectively) is greater due to lifetime broadening. Far more significant in the present case, however, is the effect of instrumental broadening, in these particular cases being dominated by the incident photon energy spread rather than the electron spectrometer (with relative values of ~1.5 eV and 0.3 eV for the S $1s$ and S $2p$ measurements respectively). Because of this broadening, one might expect that there would be considerable ambiguity in effecting the separation of the component peaks in any one of these photoelectron spectra. However, an important constraint in fitting the S $1s$ spectra associated with the XSW study is that the complete set of such spectra, recorded at photon energy steps of 0.2 eV in an energy range of ~14 eV, must be fitted by a set of three peaks (thiol, LT thiolate, RT thiolate) with the same binding energies and peak widths, the only variables in the fitting being the relative intensities of the three components. This constrains the fitting very severely as these relative intensities change very significantly through the XSW photon energy range. Using this procedure leads to a set of chemical-state specific S XSW absorption profiles for each species, as shown in Fig. 25.3 for measurements of the (111) and (111) XSW conditions at normal incidence to the scatterer planes (normal incidence XSW or NIXSW; see Chapter 4. Figure 25.3 also shows fits to these absorption profiles corresponding to the best-fit values of the associated coherent positions and coherent fractions, leading to identification of the adsorption sites by triangulation (see Fig. 25.4). In particular, the S head-group atom of the adsorbed intact thiol was found to occupy a site atop a surface Cu atom, although for the LT thiolate, an ambiguity in the triangulation means that it is not formally possible to distinguish between occupation of two-fold coordinated bridging sites and occupation of the two inequivalent three-fold coordinated hollow sites in this surface. In fact subsequent experiments by other methods indicate that the LT thiolate actually occupies *both* of these two sets of sites in a complex "honeycomb" ordering.[6,7] A second set of (NI)XSW measurements based on S $1s$ photoelectron spectra recorded from a similar sample heated to 300 K provided information on the local adsorption geometries of the co-adsorbed RT thiolate and atomic S species.

This example serves to illustrate most of the key features of the methodology, namely the use of photoelectron monitoring of the XSW absorption profile, achieved by the measurement of individual photoelectron energy spectra around the peak of interest at each photon step in the XSW range. These individual spectra are then fitted by a discrete set of component peaks at constant relative binding energies, and the component

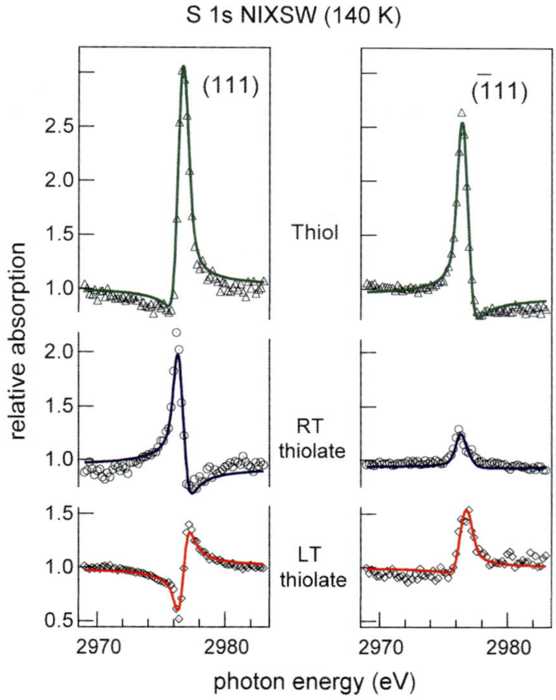

Fig. 25.3. Chemical-state specific S 1*s* photoelectron-monitored NIXSW absorption profiles extracted from the full data set of the experiment summarized in Fig. 25.2, on the co-adsorbed molecular species formed on Cu(111) by exposure to CH_3SH at 140 K. The full lines are theoretical fits to the experimental data points.[4]

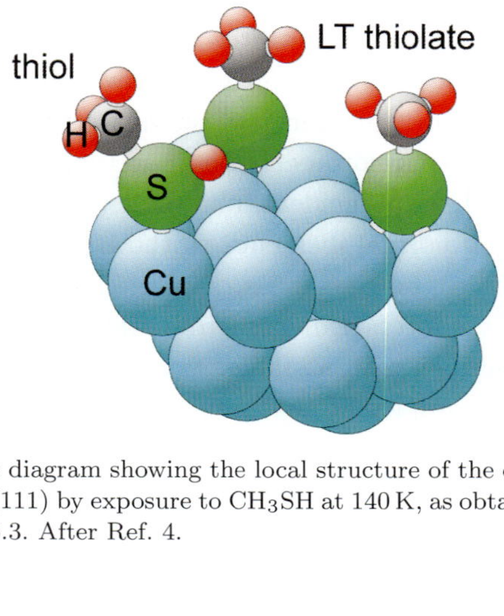

Fig. 25.4. Schematic diagram showing the local structure of the co-adsorbed molecular species formed on Cu(111) by exposure to CH_3SH at 140 K, as obtained from the NIXSW data shown in Fig. 25.3. After Ref. 4.

intensities as a function of photon energy provide the chemical-state specific XSW absorption profiles on which the structure determination is based. One implication of the use of photoelectron monitoring, of course, is that the resulting absorption profiles should be analyzed in such a way as to take account of the angular dependence of the photoelectron emission, which may mean that the photoelectron-monitored XSW profiles are not, truly, absorption profiles. By appropriate choice of scattering geometry (notably using *s*-polarization) to avoid any adverse effects to dipolar angular distributions, and proper corrections for effects due to quadrupolar angular distributions, correct XSW structural parameters may be extracted, as discussed in Chapter 11. A second feature of the technique that is evident from this description is the desirability of the best possible spectral resolution, as this enhances the ability to exploit small chemical shifts. The resolution in the photoelectron spectra is limited by two components, the energy bandwidth of the incident X-radiation, and the resolution of the electron spectrometer. For instrumentation having a constant resolving power ($E/\Delta E$), this clearly means that low photon energies and low electron energies are favored, although the true situation is somewhat more complex. For example, in crystal monochromators the resolving power is independent of photon energy for a given crystal and reflection, but at higher reflection order the X-ray scattering factors are smaller, reducing the rocking-curve width and thus improving the resolving power. Moreover, while the intrinsic resolving power of electron spectrometers is governed only by physical dimensions of the device, the actual resolution for a given photoelectron energy is determined by the pass energy within the analyzer, which may be reduced by pre-retardation. Pre-retardation, however, may lead to reduced detected intensity due to changes in the acceptance angle or effective electron source size "seen" by the analyzer. As a general statement, therefore, it remains true that photoelectron detection favors low XSW photon energies, and typically these experiments have been performed with incident photon energies of ~3 keV or less, most commonly using normal incidence to the Bragg scatterer planes (see Chapter 4).

The example of the interaction of methanethiol with Cu(111), described in detail earlier, together with a similar study on Ni(111),[8] are typical of quite a number of examples of the application of chemical-state specific XSW to study the local adsorption geometry of co-adsorbed species containing a common elemental species. Other similar studies include those of co-adsorbed SO_2, SO_3 and S on both Cu(111)[9] and Ni(111)[10] surfaces, these species being the result of a reaction of the surface with SO_2, and subsequent disproportionation. Another example is a study of co-adsorbed PF_x products on Ni(111) resulting from the effect of the incident synchrotron radiation on a surface initially exposed only to PF_3.[4] This case is a reminder that in studying molecular adsorbates on surfaces, even on those of metals, the intense and often highly focused incident X-ray flux used in the XSW experiments may induce undesired photochemistry. The results of this particular investigation, however, also highlight the way in which the structural information from XSW measurements can help to identify the resultant molecular species. In particular, the P 1*s* photoelectron spectra, recorded in the XSW study of the surface species present after radiation damage at 140 K, clearly showed four distinctly different chemically shifted components. Based on previous P 2*p* spectral fingerprinting,[11] two of these states could be assigned to adsorbed PF_3 and PF_2 (occupying, respectively, one-fold-coordinated atop sites, and two-fold-coordinated bridge sites), but the assignment of the other two states was less clear. The XSW data, however, showed clearly that one of these species was associated with P atoms in three-fold coordinated hollow sites, whereas the layer spacing of the other species could only be reconciled with a site above this hollow-coordinated P atom, leading to the clear implication that these two P atoms were components of a single P_2F_x, attributed, on the basis of consideration of the surface reactions, to P_2F_3. Figure 25.5 shows

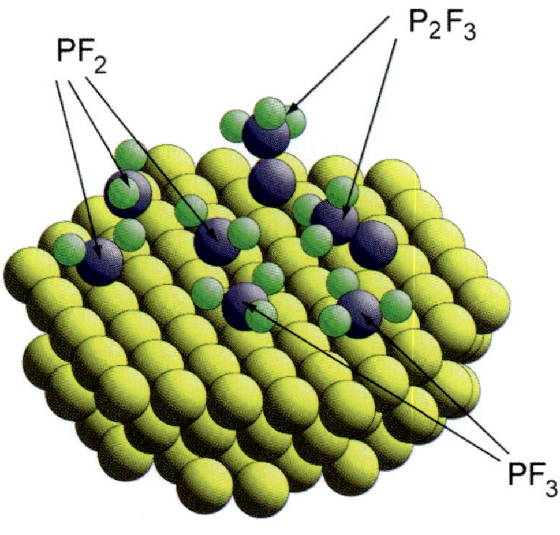

Fig. 25.5. Schematic diagram showing the local structure of the co-adsorbed PF_x species on a Ni(111) surface, resulting from exposure of a PF_3 layer to incident synchrotron radiation at 140 K, as determined by a chemical-state specific NIXSW investigation.[3]

a schematic diagram of the local geometry of the co-adsorbed fragments inferred by this investigation.

Of course, many investigations of molecular adsorbates on surfaces are driven by a desire to understand fundamental issues in heterogeneous catalysis. In this regard, there is certainly interest in the chemistry of S-containing molecules in the context of desulfurization of fossil fuels, but by far the largest class of molecules of relevance to this problem are those containing only low atomic number elements, notably C, N, and O (as well as H, of course). While photoelectron-based XSW studies (mainly NIXSW investigations in a backscattering geometry) do use significantly lower incident X-ray energies (~3 keV) than more general XSW investigations, these energies are still far above the thresholds for photoionization of even the deepest core levels (1*s*) of these light elements, that typically lie in the range 280–530 eV. As such, the absorption in the standing wave field is quite weak, so the experiments are quite challenging; moreover, the resulting photoelectron energies are typically reasonably high (~2500 eV), compounding the problem of performing high-resolution photoelectron detection of the XSW. Nevertheless, such experiments are certainly possible using undulator beamlines installed on modern synchrotron radiation facilities. Figure 25.6 shows a subset of the raw data taken from such a study of CO and atomic oxygen co-adsorption on Ni(111).[12] As in Fig. 25.2, this figure shows a set of photoelectron spectra, in this case from the O 1*s* state, at different photon energies stepped through the normal-incidence $(\bar{1}11)$ Bragg reflection condition from the substrate. There are clearly three distinct O 1*s* component peaks, and these can be separated to obtain XSW absorption profiles and thus local structure determinations for each species. The peak with the highest kinetic energy (lowest associated photoelectron binding energy) is readily assigned to the atomic O surface species, produced in the initial reaction with molecular oxygen. The remaining two peaks are attributed to CO adsorbed in two different local sites on the surface, and the XSW structural analysis identifies these as one-fold coordinated atop sites and three-fold coordinated hollow sites, as labeled in Fig. 25.6.

A rather different example of the use of a chemical-state specific XSW experiment to investigate the adsorption geometry of a molecule comprising only low atomic number elements is an investigation of PTCDA (1,4,5,8-perylene-tetracarboxylicacid-dianhydride)[13] deposited on Ag(111). This molecule, shown in Fig. 25.7, is one of a class of organic molecules being investigated for their potential role in molecular electronics, and while

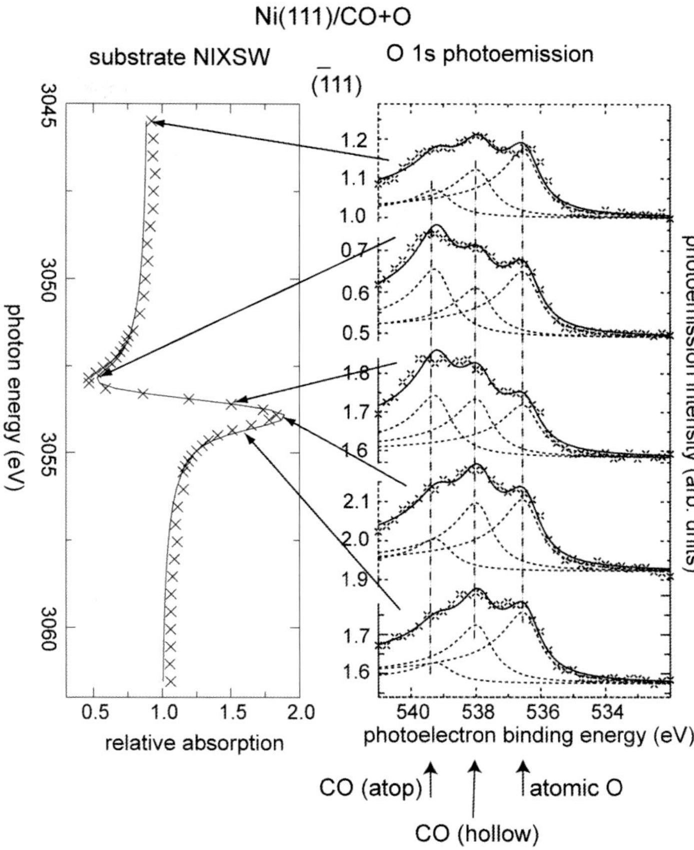

Fig. 25.6. A subset of the NIXSW experimental data in the same format as Fig. 25.2, but from an investigation of CO + O co-adsorption on Ni(111).[12]

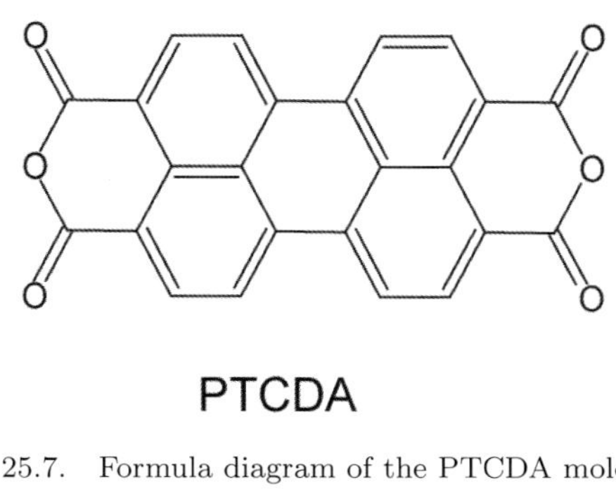

Fig. 25.7. Formula diagram of the PTCDA molecule.

As a final, somewhat different, example of the application of this method to a study of atomic adsorbates, we should mention a recent study of V atoms on the rutile $TiO_2(110)$ surface.[20] The presence of V atoms causes partial reduction of the surface, with chemical shifts in the Ti $2p$ photoelectron spectrum indicative of the presence of some lower oxidation states, while a component of V $2p$ emission indicates the presence of a vanadium oxide on the surface, most probably of local VO_2 character (V^{4+}). In addition, however, the V $2p$ spectra show the presence of a second species with a photoelectron binding energy more nearly consistent with "metallic" V (V^0). Chemical-state selective NIXSW measured on these components provides further insight into the associated local structures. Specifically, the oxidic V atoms are found to occupy either Ti substitutional sites or VO_2 epitaxial sites (equivalent relative to the bulk lattice), while the metallic V species appears to be associated with very small pure V clusters on the surface.

In summary, the use of chemical shifts in core-level photoelectron spectroscopy, monitored in an X-ray standing wave, offers many important advantages in studying the structure of surfaces, particular, but not exclusively, involving molecular adsorbates and co-adsorbates. The need for reasonable spectral resolution in both incident X-rays and photoelectron detection clearly favors the use of low energy Bragg reflections (e.g., photon energies of $\sim$3 keV), and analysis of the data must take proper account of the influence of the angular dependence of the photoelectron emission, but with these provisos, the method offers a route to investigations of increasing complex surfaces.

References

1. D. P. Woodruff and T. A. Delchar, *Modern Techniques of Surface Science*, 2nd edn. (Cambridge University Press, Cambridge, 1994).
2. M. Sugiyama, S. Maeyama, S. Heun and M. Oshima, *Phys. Rev. B* **51** (1995) 14778
3. G. J. Jackson, J. Lüdecke, D. P. Woodruff, A. S. Y. Chan, N. K. Singh, J. McCombie, R. G. Jones, B. C. C. Cowie and V. Formoso, *Surf. Sci.* **441** (1999) 515.
4. G. J. Jackson, D. P. Woodruff, R. G. Jones, N. K. Singh, A. S. Y. Chan, B. C. C. Cowie and V. Formoso, *Phys. Rev. Lett.* **84** (2000) 119.
5. M. S. Kariapper, G. F. Grom, G. J. Jackson, C. F. McConville and D. P. Woodruff, *J. Phys.: Condens. Matter* **10** (1998) 8661.
6. S. M. Driver and D. P. Woodruff, *Surf. Sci.* **457** (2000) 11.

7. R. L. Toomes, M. Polcik, M. Kittel, J.-T. Hoeft, D. Sayago, M. Pascal, C. L. A. Lamont and D. P. Woodruff, *Surf. Sci.* **513** (2002) 437.
8. C. Fisher, D. P. Woodruff, R. G. Jones, B. C. C. Cowie and V. Formoso, *Surf. Sci.* **496** (2002) 73.
9. G. J. Jackson, S. M. Driver, D. P. Woodruff, N. Abrams, R. G. Jones, M. Butterfield, M. D. Crapper, B. C. C. Cowie and V. Formoso, *Surf. Sci.* **459** (2000) 231.
10. G. J. Jackson, D. P. Woodruff, A. S. Y. Chan, R. G. Jones and B. C. C. Cowie, *Surf. Sci.* **577** (2005) 31.
11. K.-U. Weiss, R. Dippel, K.-M. Schindler, P. Gardner, V. Fritzsche, A. M. Bradshaw, D. P. Woodruff, M. C. Asensi and A. R. Gonzalez-Elipe, *Phys. Rev. Lett.* **71** (1993) 581.
12. J. J. Lee, Ph.D. thesis, University of Warwick, 2003.
13. A. Hauschild, K. Karki, B. C. C. Cowie, M. Rohlfing, F. S. Tautz and M. Sokolowski, *Phys. Rev. Lett.* **95** (2005) 209601; **95** (2005) 209602.
14. T.-L. Lee, S. Warren, B. C. C. Cowie and J. Zegenhagen, *Phys. Rev. Lett.* **96** (2006) 046103.
15. J. M. Carpinelli, H. H. Weitering, M. Bartowiak, R. Stumpf and E. W. Plummer, *Phys. Rev. Lett.* **79** (1997) 2859.
16. A. Goldoni and S. Modesti, *Phys. Rev. Lett.* **79** (1997) 3266.
17. J. M. Carpinelli, H. H. Weitering, E. W. Plummer and R. Stumpf, *Nature* **381** (1996) 398.
18. H. W. Yeom, S. Takeda, E. Rotenberg, I. Matsuda, K. Horikoshi, J. Schaefer, C. M. Lee, S. D. Kevan, T. Ohta, T. Nakao and S. Hasegawa, *Phys. Rev. Lett.* **82** (1999) 4898.
19. J. Avila, A. Mascaraque, E. G. Michel, M. C. Asensio, G. LeLay, J. Ortega, R. Pérez and F. Flores, *Phys. Rev. Lett.* **82** (1999) 442.
20. E. A. Kröger, F. Allegretti, M. J. Knight, M. Polcik, D. I. Sayago, D. P. Woodruff and V. R. Dhanak, *Surf. Sci.* **600** (2006) 4813.

Chapter 26

SITE-SPECIFIC X-RAY PHOTOELECTRON SPECTRA OF TRANSITION-METAL OXIDES

JOSEPH C. WOICIK

National Institute of Standards and Technology, Gaithersburg, Maryland, 20899, USA

In the present chapter we combine the X-ray standing wave technique with photoelectron spectroscopy (PES), thus rendering PES site specific. We determine the individual contributions of atoms within the unit cell to the total valence-band emission for the two transition-metal oxide compounds rutile TiO_2 and corundum V_2O_3. Furthermore, we highlight the importance of both single-particle cross-section effects and many-body excitation effects in the interpretation of photoelectron spectra.

26.1. Introduction

In Chapter 12 we described how the X-ray standing wave (XSW) technique can be used to render photoelectron spectroscopy site specific within the unit cell of a crystal and demonstrated the method for the heteropolar crystal GaAs. In this chapter, we use the method to explore the chemical bonding in the two ionic solids rutile TiO_2 and corundum V_2O_3. We use the site-specific XPS technique to measure the individual atomic contributions to the total valence photocurrent of these two transition-metal oxides. Unlike the example of GaAs discussed in Chapter 12, the metal cations and the oxygen anions of these materials have significantly different electronic structures. Consequently, the angular-momentum dependence of their valence atomic cross-sections ("matrix-element effects") is extremely important for a quantitative comparison between theory and experiment. We will show that the site-specific XPS technique is ideally suited to study chemical hybridization and the nature of the solid state chemical bond and

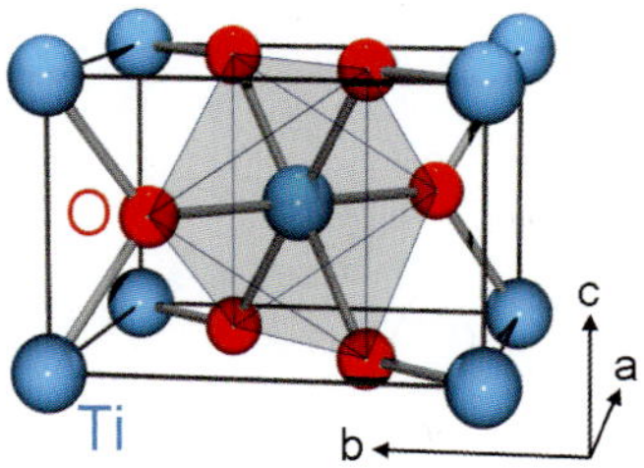

Fig. 26.1. Crystal structure of rutile TiO_2.

to test the accuracy of theoretical calculations used to describe solid state electronic structure.

26.2. Chemical Hybridization and Matrix-Element Effects in Site-Specific X-Ray Photoelectron Spectra of Rutile TiO_2

Figure 26.1 shows the crystal structure of rutile TiO_2. The rutile unit cell contains two Ti atoms and four O atoms. Six O ligands surround each Ti atom, in a slightly distorted octahedral geometry. The electronic properties of TiO_2 are therefore often described by the group theoretical treatment of the O_h point group.[1] Ti has the free-atom electronic configuration $3d^2 4s^2$. Simple electron counting therefore finds the formal ionization state of Ti in rutile to be Ti^{+4}, giving the O atoms a $2p^6$ closed-shell atomic configuration. From the crystal structure of rutile in Fig. 26.1, the valence-band density of states may be delineated by utilizing the $TiO_2(200)$ Bragg back reflection that occurs at photon energy $h\nu = 2700\,\text{eV}$.

The TiO_2 valence-photoelectron spectrum recorded from a $TiO_2(110)$ surface with photon energy 5 eV below the (200) Bragg back reflection is shown in Fig. 26.2.[2] This spectrum is compared to an *ab initio* density functional theory (DFT) calculation of the total density of states using the local density approximation (LDA).[3] (The curve labeled "*c-theory*" will be addressed below.) The curves have been normalized to equal peak heights and referenced in energy to the valence-band maximum. Additionally, all theoretical curves have been convolved with a Gaussian of width 0.4 eV to simulate the total experimental resolution. Careful inspection of the data and calculation reveals a poor energy alignment and a poor agreement on the widths of the two peaks that are characteristic of the rutile photoelectron spectrum.[4]

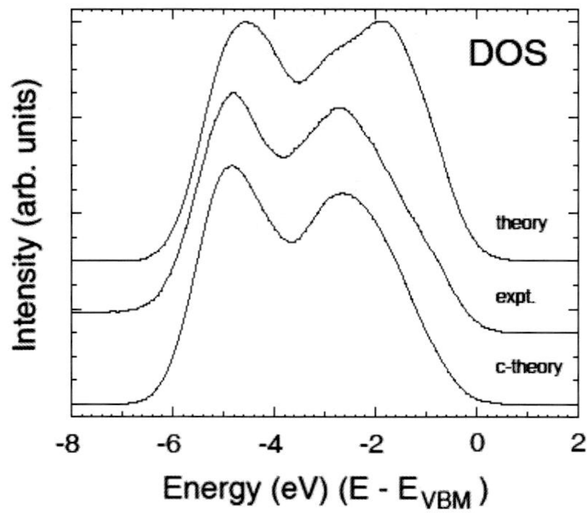

Fig. 26.2. Theoretical total electronic density of states (upper curve), valence-photoelectron spectrum recorded off of the Bragg condition (middle curve), and theoretical total electronic density of states corrected for individual Ti and O angular-momentum dependent photoelectron cross-sections (lower curve). The curves have been scaled to equal peak heights.

In order to understand the origins of these discrepancies, spectra from the rutile $TiO_2(110)$ surface were recorded at two photon energies within the photon energy width ($\Delta E = 0.38\,\mathrm{eV}$) of the $TiO_2(200)$ Bragg back-reflection condition as shown in Fig. 26.3. These photon energies were chosen to maximize the electric-field intensity on either the Ti or O atomic planes, as was done for GaAs in Chapter 12. To extract the individual Ti and O components, the curves were aligned relative to the energy position of the Ti $3p$ core line and normalized to the electric-field intensity at either the Ti or O atomic positions that was taken to be equal to the relative intensity of the Ti $3p$ or O $2s$ core line, respectively. Linear combinations of the two curves, computed according to Eq. (12.10) of Chapter 12, yielded the experimental photoelectron partial density of states curves centered on the Ti and O atoms, as shown in Fig. 26.4.

Clearly, the large contribution of Ti to the valence-band spectrum indicates significant covalent bonding between the Ti and O atoms, despite the formal Ti^{+4} charge state of Ti in rutile. Had the Ti atoms been completely ionized, there would be no Ti valence-electron emission. Although early interpretations of the rutile TiO_2 photoelectron spectrum have attributed the valence-electron emission primarily to the O $2p$ derived valence states,[4] our data support more recent resonant photoemission,[5]

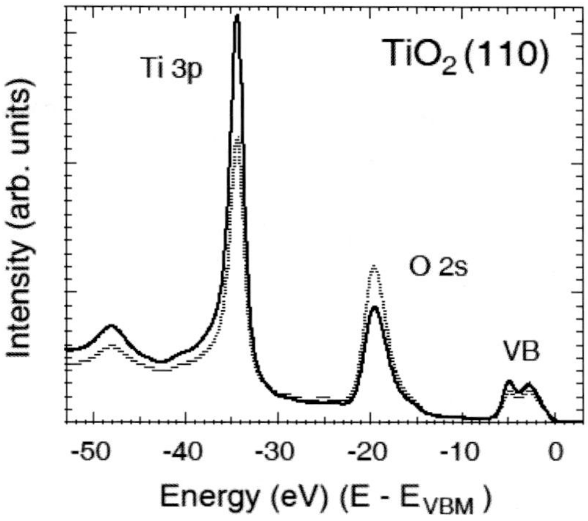

Fig. 26.3. Photoelectron spectra from the rutile $TiO_2(110)$ surface recorded within the photon energy width of the $TiO_2(200)$ Bragg back-reflection condition. The photon energies were chosen to maximize the electric-field intensity on either the Ti (solid curve) or O (dotted curve) atomic planes.

X-ray photoelectron diffraction,[6] and X-ray fluorescence studies[7] that have indicated significant Ti 3*d* admixture (*p*–*d* hybridization) in the valence band.

In order to understand the implications of these findings, theoretical Ti and O partial density of states curves, also shown in Fig. 26.4, were computed by projecting the obtained wave functions over the Ti and O valence atomic orbitals within spheres centered around the Ti and O atoms which were chosen to be equal to the known covalent radii for each species: 1.3 Å for Ti and 0.75 Å for O.

Clearly, agreement between theory and experiment is now much less than satisfactory. In particular, the second peak of the Ti valence band is nearly absent in the theory, and the triply peaked structure of the O valence band is poorly modeled. The Ti and O theoretical partial density of states curves were then further decomposed into their angular-momentum resolved components. These curves are shown in Fig. 26.5. It is clear that the second peak of the experimental Ti spectrum cannot be reproduced without inclusion of the Ti 4*p* component, and the intermediate structure of the experimental O spectrum cannot be reproduced without inclusion of the O 2*s* component.

Drawing on the physical insight gleaned from Eq. (12.8) in Chapter 12, we modeled the partial density of states curves from the weighted sums of

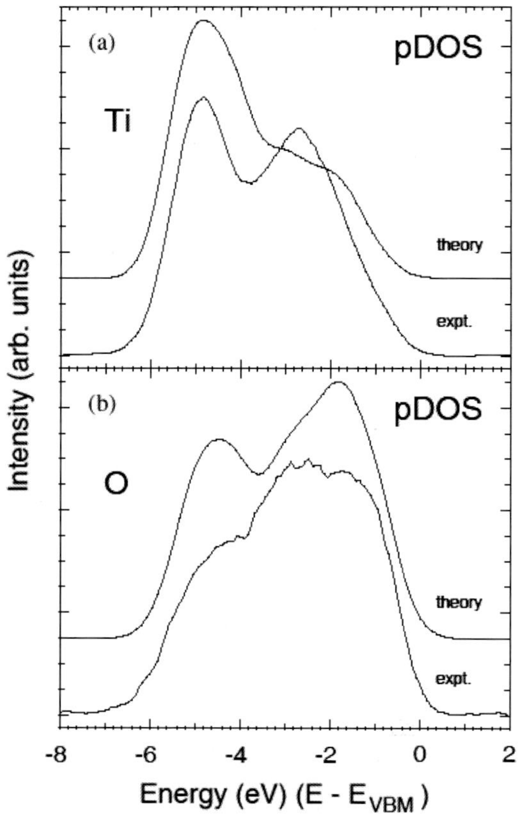

Fig. 26.4. Theoretical partial density of states and experimental site-specific valence-photoelectron spectrum: (a) Ti; (b) O. Theoretical and experimental curves have been scaled to equal peak heights. The O component has been scaled by a factor of 4 relative to the Ti component.

the different orbital components of Fig. 26.5 using the tabulated, angular-momentum dependent, theoretical atomic cross-sections (Ti $\sigma_{4s}/\sigma_{3d} = 9.9$ and O $\sigma_{2s}/\sigma_{2p} = 29$).[8] Agreement between the theoretical and experimental partial density of states curves was much improved; however, even better agreement was obtained for both the Ti and O components if the relative atomic cross-sections were scaled by an additional factor of 2. A full 1.5 of this factor of 2 could come from the choice of theoretical de-convolution radii; however, it is much more likely that additional corrections to the theoretical atomic cross-sections arise from changes in the atomic wave functions in going from the atomic to the solid state ("solid state effects").

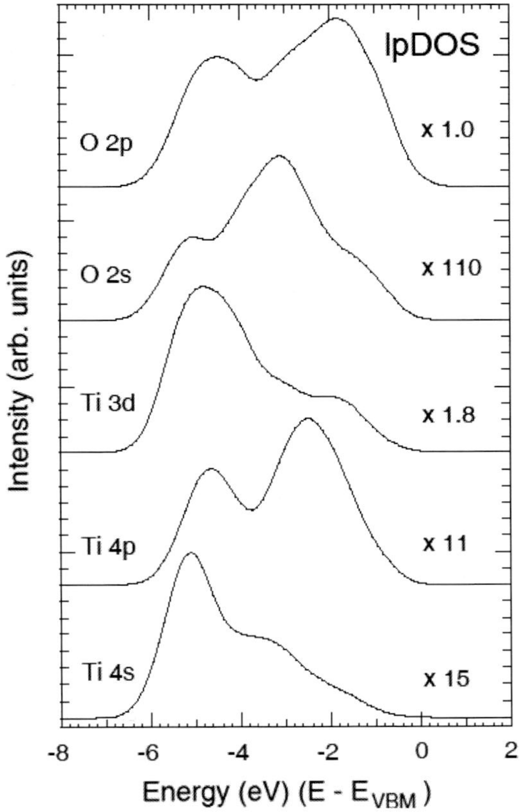

Fig. 26.5. The different angular-momentum resolved components of the theoretical Ti and O partial density of states. Multipliers relative to the O $2p$ component are indicated in each case.

The resulting theoretically *corrected* partial density of states curves are shown in Fig. 26.6. The agreement now between theory and experiment is startling, and this agreement has been achieved without any energy dependence of the cross-sections across the energy width of the valence band or deviation from the single-particle approximation of the photoelectron effect; i.e., "many-body effects" that will be addressed in the following section.

We may now recalculate the total density of states from the theoretical corrected Ti and O partial density of states using the *experimental* Ti and O *total* cross-sections determined from the areas of the individual Ti and O components of the rutile valence band ($\mathrm{Ti_{VB}/O_{VB}} = 3.4$). This curve is shown in Fig. 26.2.

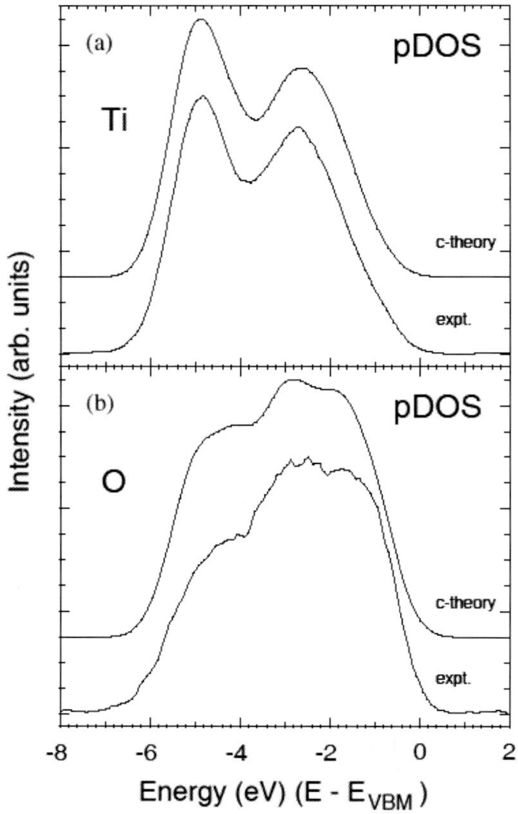

Fig. 26.6. Theoretical partial density of states corrected for individual angular-momentum dependent photoelectron cross-sections and the site-specific experimental valence-photoelectron spectrum: (a) Ti; (b) O. The curves have been scaled to equal peak heights.

At this point it is instructive to consider the molecular orbitals for an octahedral, first-row transition-metal oxide.[9] In this bonding scheme, the metal 4s orbitals bond with the ligand σ orbitals to form the $a_{1g}(\sigma^b)$ level, the metal $3d_{x^2-y^2}$ and $3d_{z^2}$ orbitals bond with the ligand σ orbitals to form the $e_g(\sigma^b)$ level, the metal 4p orbitals bond with both the ligand σ and π orbitals to form the $t_{1u}(\sigma^b)$ and $t_{1u}(\pi^b)$ levels, and the metal $3d_{xy}$, $3d_{xz}$, and $3d_{yz}$ orbitals bond with the ligand π orbitals to form the $t_{2g}(\pi^b)$ level. Additionally, there are ligand π orbitals [$t_{1g}(\pi)$ and $t_{2u}(\pi)$] that are left over and are rigorously non-bonding in O_h symmetry.

Examination of both the experimental partial density of states and the theoretical corrected partial density of states of Fig. 26.6 leads to an attractive interpretation of the electronic structure of rutile TiO_2 within this σ and π bonding scheme. The doubly-peaked Ti structure of the valence band may be attributed to the energy splitting between the σ and π groupings of the bonding states, with the σ bonds lying at the lower energy.[9] These states are mirrored in the triply-peaked O structure, and they occur at the same energy as in the Ti spectrum, indicating the sharing of electrons in a covalent bond. The O non-bonding π states then naturally compose the third peak of the O spectrum; they occur at higher energy than the O bonding π states and possess little or no electron density on the metal atoms, as expected. Close inspection of the theoretical, angular-momentum resolved components of Fig. 26.5 supports these conclusions, although the solid state electronic structure is much more complicated than the electronic structure of an isolated octahedral TiO_6 molecule. In particular, the groupings of orbitals into discrete σ and π states is not so transparent, and the effect of translational symmetry spreads the states into bands. Additionally, metal–metal and ligand–ligand interactions that are not present in isolated molecules are known to affect the valence electronic structure.[10] Interestingly, it is the fact that the last occupied states of the TiO_2 crystal lie on the O atoms with no counterpart on the Ti atoms that explains why TiO_2 is an insulator.

It is now instructive to examine the chemical hybridization of the metal and ligand orbitals within this O_h bonding scheme. As both the O $2s$ and O $2p_z$ atomic orbitals belong to the same symmetry representation of the O_h point group, the ligand σ orbitals will always contain a mixture of these states[9]:

$$|\psi_{L,\sigma}\rangle = \alpha|2s\rangle + \sqrt{1-\alpha^2}|2p_z\rangle. \tag{26.1}$$

(The ligand π orbitals are constructed solely from the O $2p_x$ and $2p_y$ atomic orbitals.) The hybridization of these orbitals orients the ligand-charge density towards the metal atoms, leading to an increased overlap between the metal and ligand wave functions. It has been stated by Mulliken that "a little hybridization goes a long way" to stabilize a chemical bond,[9] and, from the theoretical calculations of Fig. 26.5, we find that α, the mixing coefficient, is only about 10%, even though the O $2s$ valence component accounts for as much as 30% of the experimental O valence spectrum due to the much larger cross-section of the O $2s$ versus the O $2p$ atomic orbitals. This relatively small value of α results from the relatively large energy separation between the O $2s$ and O $2p$ atomic orbitals that, as seen from the data in Fig. 26.3, is about 17 eV. The energy separation between the Ti $3d$, $4s$, and $4p$ atomic orbitals is significantly smaller,[9] thereby accounting for the much larger amount of Ti $3d$, $4s$, and $4p$ hybridization observed on the Ti sites.

Amusingly, it is the added complexity of the photoemission process; i.e., the "over representation" of orbitals with smaller angular momenta (O $2s$ versus O $2p$, and Ti $4s$ and $4p$ versus Ti $3d$) that affords this direct experimental observation of chemical hybridization in the solid state electronic structure: the observation of σ and π bonds and of oxygen non-bonding π states, and positive identification of valence-band contributions from the O $2s$ and the Ti $4s$ and $4p$ orbitals.

26.3. Many-Body Effects in Site-Specific X-Ray Photoelectron Spectra of Corundum V_2O_3

The highly correlated transition-metal oxide corundum V_2O_3 is an example in which the independent-particle approximations of both solid state band-structure theory and the photoemission process fail. In the latter, this failure manifests itself by the appearance of additional structure in both core and valence photoemission electron distribution curves (EDC's) due to electronic transitions that result from screening and/or relaxation of the final-state hole produced by the photo-excitation.[11] In the former, it manifests itself as the appearance of quasiparticle excitations that are not described by a self-consistent, weakly correlated crystal potential.[12] Despite the fact that these "satellite" or "loss" features may involve charge transfer between different atoms within the crystal ("many-body effects"), the probability of creating the final-state hole is still well described by the dipole approximation. Consequently, even for highly correlated materials the valence XSW technique maintains its ability to attain site specificity as these excitations are intrinsic to the photoelectron process. It is therefore of interest to examine how LDA theory, which produces a single-particle density of states, compares with the site-specific photoelectron density of states so measured.

V_2O_3 has been studied intensively for over three decades due to its interesting electronic properties. At room temperature, it is a paramagnetic metal possessing a trigonal (corundum) crystal structure. Upon cooling to 160 K, it undergoes a first order phase transition to an insulating antiferromagnet with monoclinic symmetry.[13] This transition

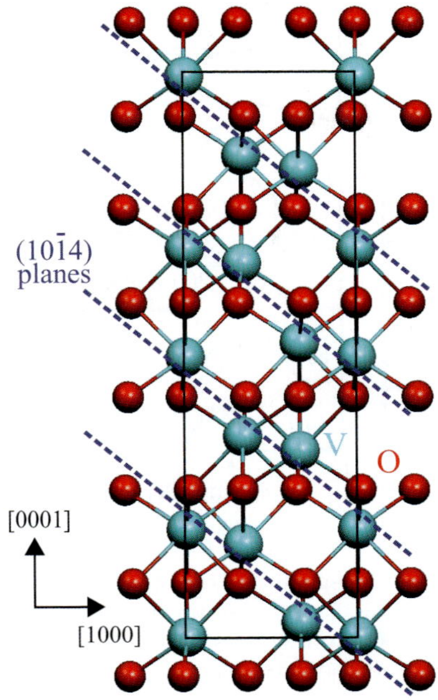

Fig. 26.7. Crystal structure of corundum V_2O_3.

is often considered to be a prototypical Mott–Hubbard metal-insulator transition[14,15]; however, more recent work suggests a more refined model to account for covalent bonding between the V 3*d* and O 2*p* states.[16,17] The crystal structure of V_2O_3 is shown in Fig. 26.7.

Photoelectron spectroscopy (PES) of the paramagnetic V_2O_3 phase has been used intensively in an attempt to understand its underlying electronic structure.[17–25] Unfortunately, interpretation of the PES data is not straightforward. Ultraviolet photoelectron spectroscopy (UPS) is highly surface sensitive, and its results may deviate from the true bulk electronic structure. X-ray photoelectron spectroscopy (XPS) is more sensitive to the bulk electronic structure owing to the larger escape depth of the photoelectrons, and, indeed, XPS studies[24,25] have detected a prominent near-Fermi-level peak that was not revealed by UPS.[18,20,23] The interpretation of XPS data is, however, also complicated. First, as we have seen from our discussion of TiO_2, matrix-element effects are pronounced in XPS. Second, the analysis of XPS data may be more complicated than suggested by a single-particle description due to many-body effects in the photoemission process[11,26] as we shall now illustrate.

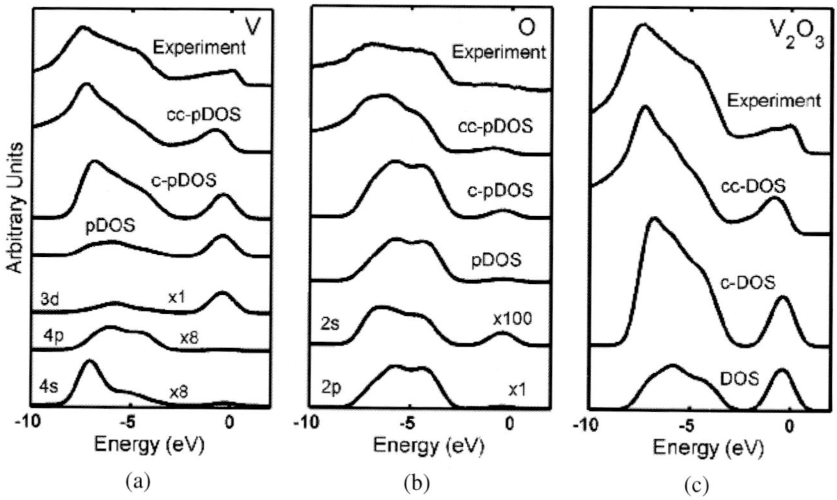

Fig. 26.8. Experimental valence-photoelectron spectra of V_2O_3, theoretical density of states curves (DOS), theoretical density of states curves corrected for individual angular-momentum dependent photoelectron cross-sections (c-DOS), and theoretical density of states curves corrected for both individual angular-momentum dependent photoelectron cross-sections and the Doniach–Šunjić effect (cc-DOS) for: (a) V, (b) O, and (c) total spectrum. "p" indicates partial DOS. Figures (a) and (b) additionally show partial angular-momentum-resolved DOS curves for V $3d$, V $4s$, and V $4p$ and O $2s$ and O $2p$, respectively. Curves have been scaled to equal peak heights.

Partial V and O photoelectron spectra of V_2O_3 are shown in Figs. 26.8(a) and 26.8(b), respectively.[27] These spectra were obtained by collecting spectra from a $V_2O_3(0001)$ surface within the photon energy width ($\Delta E = 0.6\,\text{eV}$) of the $V_2O_3(10\text{–}14)$ Bragg back reflection that occurs at photon energy $h\nu = 2288\,\text{eV}$ and applying similar analysis as before. (The observed energy width of the reflection is wider than predicted theoretically due to the large mosaic spread of the crystal.) The spectra were recorded with the photon beam incident normal to the (10–14) crystal planes and incident to the (0001) surface at 51°, as shown in Fig. 26.7. The total (usual) off-Bragg XPS valence-electron spectrum is given in Fig. 26.8(c).

Each experimental spectrum is compared with several theoretical spectra representing the filled-state portion of the DOS curve convolved with a Gaussian of width 0.47 eV to mimic the total experimental resolution. The curves denoted by "DOS" and "pDOS" are simply the total and partial valence DOS, respectively (partial in Figs. 26.8(a) and 26.8(b), total in Fig. 26.8(c)), as obtained from the LDA calculation. Also shown in Figs. 26.8(a) and 26.8(b) are the theoretical *per-orbital* partial valence DOS curves of V $3d$, $4s$, and $4p$ and O $2s$ and $2p$, respectively. The curves denoted by "c-DOS" and "cc-DOS" are explained below.

Both theoretical and experimental spectra in Fig. 26.8 feature two "lobes" — a less intense, narrower lobe extending from the Fermi level to 3 eV below it (denoted as the "upper lobe") and a more intense, wider lobe extending from 3 eV to 9 eV below the Fermi level (denoted as the "lower lobe"). The theoretical per-orbital partial DOS curves show a dominating contribution from the V $3d$ and O $2p$ orbitals to the upper and lower lobe, respectively, which agrees with the usual assignment of these features to V $3d$ and O $2p$ electron bands.[24] Simple electron counting beginning with the V free-electron configuration $3d^3 4s^2$ and assuming a formal 3^+ charge on the V ion shows that each V atom is "left" with two valence electrons beyond what is necessary for bonding the O atoms. These electrons make up most of the contribution to the high-energy upper lobe and are, of course, related to the metallic property of the phase.

Importantly, the partial XPS data of Figs. 26.8(a) and 26.8(b) clearly show that both V and O electrons contribute non-negligibly to the overall electronic structure of the valence band at all energies. This is an immediate consequence of the sharing of electrons in a covalent bond; i.e., electrons shared between the V and the O atoms have a finite probability of being photo-emitted from either. This complete hybridization of features between the partial DOS of the metal atoms and the ligand atoms has been observed previously by site-specific X-ray photoelectron spectroscopy for both α-Fe_2O_3[28] and NiO.[29] The direct identification of non-negligible covalent bonding effects observed here experimentally is important, especially for the upper-lobe electrons that are related to the Mott–Hubbard transition. It means that, despite being localized on the metal ions, these electrons are significantly p–d hybridized with the O atoms. A complete theoretical treatment must therefore take charge sharing between the V and O orbitals into account, rather than settling for a simplified picture where the V $3d$ electrons are merely split by the ligand crystal field.

LDA-based DOS calculations qualitatively support the above picture, but they are not quantitatively adequate concerning both the line shape and relative intensities of the two lobes. Using Eq. (12.10) from Chapter 12 to compute the theoretical V and O partial photoelectron curves as the weighted sum of the different orbital components (curves denoted by "c-pDOS" and "c-DOS" are the cross-section *corrected* partial and total theoretical DOS curves, respectively) improves agreement with experiment. We used weighting coefficients that were empirically found to best mimic the experimental spectrum: (O 2s)/(O 2p) = 7.3 and (V 4s)/(V 3d) = 40.4. As for TiO_2, these weighting coefficients agree within a factor of two with computed *atomic* photoelectron cross-sections.[8]

For the theoretical V partial DOS curve, improvement is in both the line shape of the lower lobe and in the relative intensity between the upper and lower lobes. This is because the experimental V spectrum is dominated by V 4s (and 4p) electrons due to their higher atomic cross-sections, while the uncorrected theoretical V partial DOS is dominated by V 3d electrons. However, for the theoretical O partial DOS curve, although noticeable, improvement between theory and experiment is not as significant. Clearly then, the discrepancy between theory and experiment for the O spectrum must not be due to matrix-element effects. This is because the contribution of the O 2s electrons is relatively small even when their enhanced cross-section is taken into account.

To obtain further insight into the cause of the remaining differences between theory and experiment for the upper lobe (cf. Fig. 26.8(c)), Fig. 26.9 shows a high-resolution XPS spectrum of the near-Fermi-edge region, compared with the matrix-element corrected, theoretical c-DOS curve broadened by 0.2 eV. Both experiment and the LDA calculation feature two peaks: one near the Fermi edge and one over an eV lower in energy. A group-theoretical analysis of the LDA peaks shows that they correspond primarily to a splitting of the V 3d derived t_{2g} states into a_{1g} (lower peak) and e_g^{π} (upper peak) states.[30] This splitting results from distortion of the octahedral geometry surrounding the V atoms within the D_{3d} point group. A similar splitting has recently been demonstrated experimentally for tetragonally distorted $SrTiO_3$ within C_{4v} symmetry by X-ray absorption measurements.[31] Clearly, the c-DOS curve predicts neither the energy separation between the two peaks nor their relative intensity. Furthermore, a significant low-energy "tail," extending for over an eV, is observed in the experimental spectrum. This tail has no counterpart in the theoretical LDA curves.

The limitations of LDA calculations in describing the experimental XPS data for V_2O_3 can be attributed to two different physical sources, the first being the treatment of electron correlation. Within LDA, it is assumed that the exchange-correlation energy per particle at each point in space is given by its value for the homogeneous electron gas.[3] While this approximation is sufficient for many simple metals and semiconductors, it is too crude to

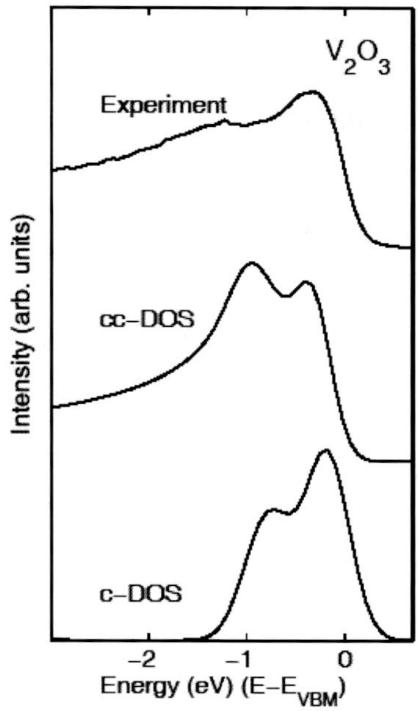

Fig. 26.9. High resolution photoelectron spectrum of the near-Fermi-edge region of V_2O_3, compared to the theoretical density of states in that region. Experimental valence-photoelectron spectrum (top), theoretical density of states curve corrected for individual angular-momentum dependent photoelectron cross-sections (c-DOS) (bottom), and theoretical density of states curve corrected for both individual angular-momentum dependent photoelectron cross-sections and the Doniach–Šunjić effect (cc-DOS) (middle). Curves have been scaled to equal peak heights.

describe a highly correlated material accurately, and, in fact, LDA fails to predict the metal-insulator transition in V_2O_3. The LDA results presented here (and elsewhere, e.g., in Refs. 17, 22, 30 and 32) are compared to the room temperature metallic phase, even though the 0 K electronic structure is computed. This approximate treatment of correlation can also lead to an inaccurate distribution of excitation energies. In many correlated materials, agreement with experiment is improved by adding an on-site Hubbard–Coulomb interaction term U for the localized d (or f) electrons,[33] but we are not aware of LDA $+ U$ calculations for the metallic, paramagnetic phase of V_2O_3, and whether or not LDA $+ U$ would yield an adequate description for V_2O_3 is controversial.[34]

The second source of error within LDA (and DFT itself) is the use of the Kohn–Sham eigenvalues. These solutions are zero-order approximations of quasiparticle excitation energies.[35] Many-electron processes of photo-excitation attributed to screening and relaxation of the valence hole beyond the quasiparticle approximation therefore will not be contained within a DFT calculation. Multiple configurations in the photoemission final state can be considered explicitly by *ad hoc* calculations within a cluster solution of a model Hamiltonian using a configuration-interaction expansion, and an empirical solution to this problem has recently rendered good agreement with experiment for V_2O_3.[16,17]

An alternative approach to the problem of strong electron correlation is dynamical mean field theory (DMFT).[36] In this approximation, the many-body spectral response function (that results from the excitation of the many-electron wave function) near the Fermi energy is computed by mapping the Hubbard model onto a self-consistent quantum impurity model. Agreement with the near-Fermi-level XPS data is found, given a reasonable choice for the many-body Coulomb repulsion parameter U,[24,25,36] and, in fact, by fitting U, the metal-insulator transition not predicted by *ab initio* LDA calculations is recovered.

In order to distinguish the effects of approximate correlation in the LDA calculation from the effects of intrinsic many-body phenomena in the photoemission process, we modeled the latter by the Doniach–Šunjić asymmetric line shape due to electron shakeup across the Fermi edge.[37] These effects are present even in the simple metals that are well described by LDA. Further motivation is provided by Fig. 26.10(a), which compares the XPS spectrum of the V $2p_{3/2}$ core level with the XPS spectrum of the V_2O_3 valence band, and by Fig. 26.10(b), which compares the XPS spectrum of the Ti $2p_{3/2}$ core level and the XPS spectrum of the TiO_2 valence band. (Core spectra have been shifted in energy to align with the dominant valence-band features.) Tailing appears in both valence and core spectra for V_2O_3, but not in valence or core spectra for TiO_2, suggesting a common physical origin associated with the metallicity of V_2O_3 and the absence of a Fermi edge for insulating TiO_2. Additionally, there appears to be a distinct correlation between the presence of a valence-band tail and the appearance of both similar and more complicated features in the core-level spectrum of V_2O_3. Similar tails in the core-level spectra of V_2O_3 have been observed previously.[38,39] This assertion also agrees with the recent data of Ref. 25, where a clear correlation between the shake-down satellites in the V $2p$ and $3p$ core-level spectra and the spectral peak measured at the Fermi level was observed.

We have incorporated the Doniach–Šunjić effect by convolving the theoretical LDA calculations with the asymmetric power-law Doniach–Šunjić line shape $E^{\alpha-1}$. This procedure neglects, among other things, a dependence of α on the energy and the contribution of additional lifetime

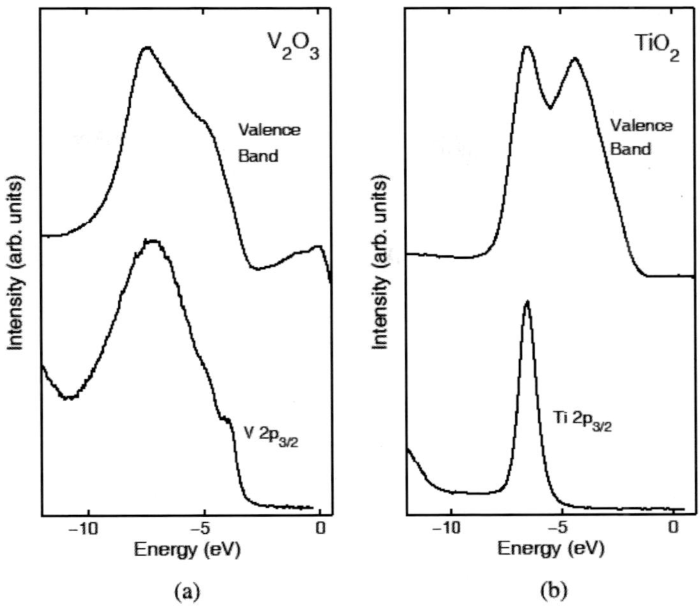

Fig. 26.10. Comparison of the $2p_{3/2}$ core level and the valence-band photoemission spectra for: (a) V_2O_3; (b) TiO_2. Curves have been scaled to equal peak heights and shifted in energy to align the dominant normalized features.

effects. A physically reasonable[37] value of $\alpha = 0.25$ was determined from a fit to the O $1s$ spectrum of V_2O_3. Figures 26.8 and 26.9 show the theoretical calculations *corrected* for the Doniach–Šunjić effect denoted by "cc." In all cases, agreement with experiment is markedly improved. In particular, the tail region, which was completely absent in the theoretical DOS and c-DOS curves, is well reproduced by the cc-DOS curve, not only for the upper lobe, but also for the lower lobe. However, the relative intensity and energy position of the two peaks of the near Fermi region are still lacking.

Based on this analysis, it is evident that the features corresponding to the two highest electron bands near the Fermi level, which are not well described even after correction for asymmetric line shape, are sensitive to the many-body correlations not properly accounted for in LDA. In fact, the presence of the higher-energy peak (and energy position of the lower peak) is dependent on whether or not V_2O_3 is in its insulating or metallic phase.[24] Lack of adequate correlation in LDA affects most significantly the relative intensity and energy separation of these peaks.

26.4. Conclusions

In conclusion, we have applied the site-specific XPS technique to measure the individual atomic contributions to the total valence photocurrent of the two transition-metal oxides rutile TiO_2 and corundum V_2O_3. The spectra so obtained are directly related to the individual, single-particle partial density of states of the metal and ligand atoms. It is clear that the site-specific XPS technique is ideally suited to study chemical hybridization and the nature of the solid state chemical bond and to test the accuracy of theoretical calculations that model solid state electronic structure. Additionally, we have highlighted the importance of both single-particle cross-section effects and many-body excitation effects in the interpretation of photoemission data, both of which are intrinsic to the photoelectron process. These effects offer both an added complexity and an intriguing opportunity towards the study of solid state electronic structure.

Acknowledgments

This research was performed at the National Synchrotron Light Source which is supported by the United States Department of Energy. The author is indebted to Dr. Leeor Kronik and Dr. Erik Nelson for their contributions to all aspects of this work.

References

1. L. A. Grunes, R. D. Leapman, C. N. Wilker, R. Hoffmann and A. B. Kunz, *Phys. Rev. B* **25** (1982) 7157.
2. J. C. Woicik, E. J. Nelson, L. Kronik, M. Jain, J. R. Chelikowsky, D. Heskett, L. E. Berman and G. S. Herman, *Phys. Rev. Lett.* **89** (2002) 077401.
3. N. Troullier and J. L. Martins, *Phys. Rev. B* **43** (1991) 1993; D. M. Ceperly and B. J. Alder, *Phys. Rev. Lett.* **45** (1980) 566; J. P. Perdew and Y. Wang, *Phys. Rev. B* **45** (1992) 13244.
4. S. Hufner and G. K. Wertheim, *Phys. Rev. B* **8** (1973) 4857.
5. Z. Zhang, S.-P. Jeng and V. E. Henrich, *Phys. Rev. B* **43** (1991) 12004.
6. R. Heise, R. Courths and S. Witzel, *Solid State Commun.* **84** (1992) 599.
7. L. D. Finkelstein, E. Z. Kurmaev, M. A. Korotin, A. Moewes, B. Schneider, S. M. Butorin, J.-H. Guo, J. Nordgren, D. Hartmann, M. Neumann and D. L. Ederer, *Phys. Rev. B* **60** (1999) 2212.
8. M. B. Trzhaskovskaya, V. I. Nefedov and V. G. Yarzhemsky, *Atom. Data Nucl. Data Tables* **77** (2001) 97. As the Ti $4p$ orbitals are unoccupied in the free Ti atom, the Ti $4p$ atomic cross section is not tabulated; consequently,

we took it to be equal to the Ti 4s one. For the same reason, we took the V 4p cross section to be equal the V 4s one.

9. C. J. Ballhausen and H. B. Gray, *Molecular Orbital Theory* (Benjamin, New York, 1964). The ligand z axis is directed towards the metal ion.
10. C. Castellani, C. R. Natoli and J. Ranninger, *Phys. Rev. B* **18** (1978) 4945.
11. S. Hufner, *Photoelectron Spectroscopy: Principles and Applications* (Springer-Verlag, Berlin, 2nd edn., 1996).
12. O. Madelung, *Introduction to Solid-State Theory* (Springer-Verlag, Berlin, 2nd printing, 1981).
13. P. D. Dernier and M. Marezio, *Phys. Rev. B* **2** (1970) 3771.
14. D. B. McWhan, T. M. Rice and J. P. Remeika, *Phys. Rev. Lett.* **23** (1969) 1384.
15. J. Zaanen, G. A. Sawatzky and J. W. Allen, *Phys. Rev. Lett.* **55** (1985) 418.
16. R. J. O. Mossanek and M. Abbate, *Phys. Rev. B* **75** (2007) 115110.
17. R. Zimmermann, R. Claessen, F. Reinert, P. Steiner and S. Hüfner, *J. Phys.: Condens. Matter* **10** (1998) 5697; R. Zimmermann, P. Steiner, R. Claessen, F. Reinert, S. Hüfner, P. Blaha and P. Dufek, *J. Phys.: Condens. Matter* **11** (1999) 1657.
18. G. A. Sawatzky and D. Post, *Phys. Rev. B* **20** (1979) 1546.
19. K. E. Smith and V. E. Henrich, *Phys. Rev. B* **38** (1988) 5965; **38** (1988) 9571.
20. S. Shin, S. Suga, M. Taniguchi, M. Fujisawa, H. Kanzaki, A. Fujimori, H. Daimon, Y. Ueda, K. Kosuge and S. Kachi, *Phys. Rev. B* **41** (1990) 4993.
21. M. Imada, A. Fujimori and Y. Tokura, *Rev. Mod. Phys.* **70** (1998) 1039.
22. H.-D. Kim, H. Kumigashira, A. Ashihara, T. Takahashi and Y. Ueda, *Phys. Rev. B* **57** (1998) 1316.
23. M. Schramme, Ph.D. thesis, Universität Augsburg, 2000; some of Schramme's experimental results are shown in K. Held, G. Keller, V. Eyert, D. Vollhardt and V. I. Anisimov, *Phys. Rev. Lett.* **86** (2001) 5345.
24. S.-K. Mo, J. D. Denlinger, H.-D. Kim, J.-H. Park, J. W. Allen, A. Sekiyama, A. Yamasaki, K. Kadono, S. Suga, Y. Saitoh, T. Muro, P. Metcalf, G. Keller, K. Held, V. Eyert, V. I. Anisimov and D. Vollhardt, *Phys. Rev. Lett.* **90** (2003) 186403.
25. G. Panaccione, M. Altarelli, A. Fondacaro, A. Georges, S. Huotari, P. Lacovig, A. Lichtenstein, P. Metcalf, G. Monaco, F. Offi, L. Paolasini, A. Poteryaev, M. Sacchi and O. Tjernberg, *Phys. Rev. Lett.* **97** (2006) 116401.
26. P. A. Cox, *Transition Metal Oxides* (Clarendon, Oxford, 1992).
27. J. C. Woicik, M. Yekutiel, E. J. Nelson, N. Jacobson, P. Pfalzer, M. Klemm, S. Horn and L. Kronik, *Phys. Rev. B* **76** (2007) 165101.
28. C.-Y. Kim, M. Bedzyk, E. J. Nelson, J. C. Woicik and L. E. Berman, *Phys. Rev. B* **66** (2002) 085115.
29. T. M. Schuler, D. L. Ederer, S. Itza-Ortiz, G. T. Woods, T. A. Calcott and J. C. Woicik, *Phys. Rev. B* **71** (2005) 115113.
30. V. Eyert, U. Schwingenschlögl and U. Eckern, *Europhys. Lett.* **70** (2005) 782.
31. J. C. Woicik, E. L. Shirley, C. S. Hellberg, K. E. Andersen, S. Sambasivan, D. A. Fischer, B. D. Chapman, E. A. Stern, P. Ryan, D. L. Ederer and H. Li, *Phys. Rev. B* **75** (2007) 140103(R).

32. L. F. Mattheiss, *J. Phys.: Condens. Matter* **6** (1994) 6471.
33. V. I. Anisimov, F. Aryasetiawan and A. I. Lichtenstein, *J. Phys.: Condens. Matter* **9** (1997) 767.
34. S. Di Matteo, *Phys. Scripta* **71** (2005) CC1 and references therein.
35. A. Görling, *Phys. Rev. A* **54** (1996) 3912; C. Filippi, C. J. Umrigar and X. Gonze, *J. Chem. Phys.* **107** (1997) 9994.
36. G. Kotliar, S. Y. Savrasov, K. Haule, V. S. Oudovenko, O. Parcollet and C. A. Marianetti, *Rev. Mod. Phys.* **78** (2006) 865; G. Kotliar and D. Vollhardt, *Phys. Today* **57**(3) (2004) 53; K. Held, I. A. Nekrasov, G. Keller, V. Eyert, N. Blümer, A. K. McMahan, R. T. Scalettar, Th. Pruschke, V. I. Anisimov and D. Vollhardt, *Psi-k Newsletter* **56** (2003) 65.
37. S. Doniach and M. Šunjić, *J. Phys. C: Solid State Phys.* **3** (1970) 285.
38. R. L. Kurtz and V. E. Henrich, *Phys. Rev. B* **28** (1983) 6699.
39. K. E. Smith and V. E. Henrich, *Phys. Rev. B* **50** (1994) 1382.

Chapter 27

PROBING MULTILAYER NANOSTRUCTURES WITH PHOTOELECTRON AND X-RAY EMISSION SPECTROSCOPIES EXCITED BY X-RAY STANDING WAVES

S.-H. YANG[*], B. C. SELL[†,‡,§], B. S. MUN[‡,¶,||] and C. S. FADLEY[†,‡,**]

[*]IBM Almaden Research Center,
San Jose, CA 95120, USA

[†]Physics Department,
University of California Davis, Davis, CA 95616, USA

[‡]Materials Sciences Division,
Lawrence Berkeley National Laboratory, Berkeley,
CA 94720, USA

[§]Physics Department, Otterbein College,
Westerville, OH 43081, USA

[¶]Advanced Light Source,
Lawrence Berkeley National Laboratory,
Berkeley, CA 94720, USA

[||]Department of Applied Physics,
Hanyang University, Ansan, Kyeonggi, 426-791, Korea

[**]fadley@physics.ucdavis.edu

We discuss a newly developed X-ray standing wave/wedge method for probing the composition, magnetization, and electronic densities of states in buried layers and interfaces of spintronic and other nanostructures. In work based on photoemission, this method has permitted determining concentration and magnetization profiles through giant magnetoresistive and magnetic tunnel junction structures, as well as individual layer densities of states near the Fermi level in a tunnel junction . Using X-ray emission and resonant inelastic X-ray scattering for detection has permitted probing deeper layers and interfaces in a giant magnetoresistance structure. Various future applications of this method in nanomagnetism and other fields of nanoscience are suggested, including using more energetic hard X-ray standing waves so as to probe more deeply below a surface and standing wave excitation in spectromicroscopy to provide depth sensitivity.

27.1. Introduction

Multilayer nanometer-scale structures are ubiquitous in current magnetic devices, and the detailed characteristics of the layers making them up, including the interfaces between layers, are often decisive as to their ultimate functional properties. Buried layers and interfaces are thus crucial elements in such devices, as well as many other nanoscale structures of potential interest in technology, but characterizing them fully presents unique challenges. Various microscopic techniques such as scanning tunneling (STM) and atomic force (AFM), low-energy electron without and with spin resolution (LEEM and SPLEEM, respectively), scanning transmission X-ray (STXM), zone-plate focused X-ray (XM), and photoelectron emission without and with spin resolution (PEEM or SP-PEEM) can provide high in-plane lateral resolution, and for XM and PEEM also element-specific resolution of structures, but most of these techniques tend to be highly surface sensitive, and all have limited depth-resolving ability for buried structures. Soft X-ray scattering on and off resonant conditions provides the ability to vary the probing depth, but the scattered intensities must still be fit to X-ray optical simulations to derive depth-resolved information. Transmission electron microscopy with electron energy loss spectroscopy is perhaps the most direct method for looking in an element-specific way at layered structures and interfaces but it is destructive in requiring the sectioning of the sample and is more limited in the range of chemical, magnetic, and bonding information available than a technique which could in some way make use of photoelectron or X-ray emission as the probing spectroscopies. Angle-resolved X-ray photoelectron spectroscopy (ARXPS), in which the takeoff angle of the photoelectrons is varied so as to change the relative degree of surface sensitivity, is by now a well-established spectroscopic technique for obtaining depth profiles of concentration, but it can be difficult to derive an unambiguous profile from such measurements.[1] The relatively newly developed X-ray standing-wave/wedge (swedge) method, which is the primary focus of this paper, is a spectroscopic technique that goes beyond ARXPS in its capabilities, and we here discuss its applications to a few systems of interest in spintronics, and summarize its advantages and disadvantages.[2–10]

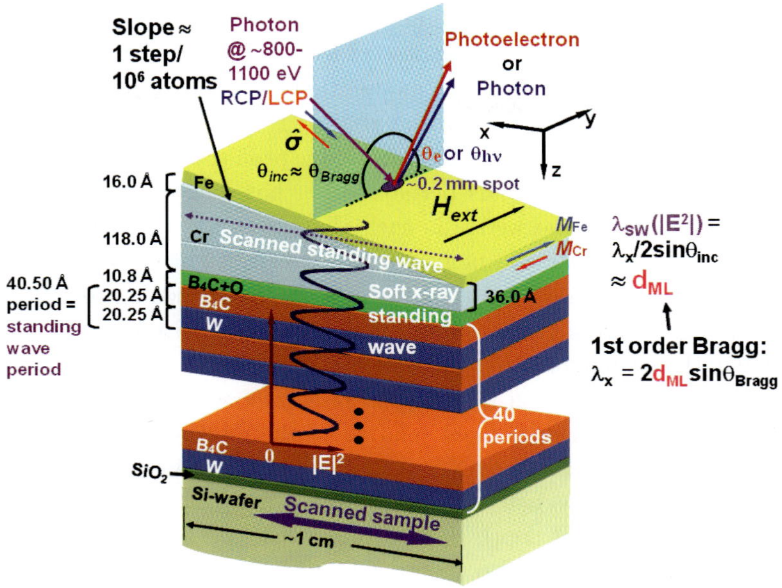

Fig. 27.1. The basic geometry of the standing-wave/wedge ("swedge") method for probing buried layers and interfaces via photoelectron emission or X-ray emission, with some specific numbers relevant to the study in Ref. 4 indicated. Soft X-rays are incident at the first-order Bragg angle for a multilayer mirror substrate of period d_{ML}, thus creating a strong standing wave above the mirror whose period is also d_{ML}. On top of the multilayer, the first layer of the sample is grown with a wedge profile, so that scanning the sample along the wedge slope (the x-direction) results in scanning the standing wave through the thickness of any layer(s) subsequently grown on top of the wedge. By changing the circular polarization of the X-ray beam from left to right, element-specific magnetic circular dichroism (MCD) measurements can be carried out. (From Ref. 4)

The basic principle of the method is illustrated in Fig. 27.1, which includes some specific parameters for the first case studied: the Fe/Cr interface, a prototype system exhibiting giant magnetoresistance (GMR).[4] A well-focused soft X-ray synchrotron radiation (SR) beam at about 1000 eV energy, is incident on a synthetic multilayer mirror at its first-order Bragg angle. This leads to a high reflectivity and a strong standing wave (SW) above the mirror. If the bilayers making up the mirror (in this example composed of B_4C and W) have a thickness d_{ML}, then the period of the SW, as judged by the square of its electric field, also has a period of d_{ML}.

Beyond this, the fact that the SW modulation is the result of interference between the incident and reflected beams implies that its intensity will range over maximum limits set by $1 + R \pm 2\sqrt{R}$, if R is the reflectivity. Thus, even a modest reflectivity at the Bragg condition of 0.01 will yield an overall SW modulation of ±20% via the last term in this expression. The sample to be studied is then grown on top of the mirror, with its base layer (here Cr) in a wedge profile, and another constant thickness layer (here Fe), plus perhaps other layers, grown on top of the wedge. The slope of the wedge is such that, over the full sample length along the x-direction in the figure, it changes in height z by a few times the standing wave period d_{ML}. Since the X-ray beam size is ~0.1 mm and much smaller than the typical sample length of ~1 cm, scanning the sample relative to the beam along the wedge slope (the x-direction) effectively scans the standing wave through the sample. It is important in this context to note that the SW phase is fixed relative to the multilayer during such a scan.[4] Thus, photoelectron or X-ray emission signals from different atoms will exhibit oscillatory behavior that can, in a direct-space manner, be interpreted in terms of depth distributions with the aid of X-ray optical (XRO) calculations[10] (*cf.* also Chapter 7 by M. J. Bedzyk in this book) in what we believe is a more direct way than is possible in the reciprocal-space scattering measurements mentioned above.

In practice, the swedge method is also combined with more standard XSW methods for determining depth-resolved information perpendicular to a set of reflecting planes, as discussed in various other chapters of this book. These are: scanning the incidence angle over the Bragg reflection condition for a given fixed photon energy so as to generate a rocking curve, and scanning the photon energy over the Bragg condition for a given fixed incidence angle. In both these types of scans, the SW modulation is negligibly small at the outset well off the Bragg condition, then grows in to a maximum at the Bragg angle, and then decreases to a small value again. Simultaneously, the phase of the SW moves vertically by about 1/2 of the SW period, thus causing significant changes in photoelectron or X-ray emission intensities. Both of these measurements, combined with appropriate XRO simulations, can be used to determine the Bragg angle at the outset of a swedge experiment, and they also provide complementary depth-resolving information that has been used together with x-coordinate scans to finally determine the thickness of the wedge for a given x-coordinate setting, as well as final depth profiles.[4] A distinct advantage of the swedge approach, however, is that several full periods of the SW can be scanned through the sample, and the resulting very nearly sinusoidal oscillations more quantitatively analyzed to determine depth profiles. One feature of such oscillations that is particularly useful is the phase shift between them for different species, which can directly be read as an approximate indicator of position with respect to the surface of the sample. These features will be illustrated below in a few examples. We thus believe that such swedge scans possess inherently greater information content than a rocking curve or a photon energy scan.

All of the experimental data reported here have been obtained on bend-magnet beamline 9.3.2 or elliptically polarized undulator beamline 4.0.2 of the Advanced Light Source, using a multi-technique spectrometer/ diffractometer which incorporates a Scienta SES 2002 electron spectrometer and a Scienta XES 300 soft X-ray spectrometer. This system also includes a custom-built specimen goniometer that can scan both X-ray incidence angle and sample vertical position under computer control.

27.2. Applications Using Standing Wave Excited Photoelectron Emission

Multilayers involving Fe and Cr were the first to exhibit the GMR effect that has been used in magnetic read heads since the early 1990s. The interface between Fe and Cr is crucial to such devices, and we thus show in Fig. 27.2 as a first example of the kind of results that can be obtained from the swedge method both the depth profile of Fe and Cr concentrations and the depth profile of element-specific magnetizations for Fe and Cr.

These profiles have been derived by measuring the Fe $2p$ and Fe $3p$, as well as the Cr $2p$ and Cr $3p$, photoelectron intensities and magnetic circular dichrosim (MCD) effects.[3,4] MCD is a common measurement in synchrotron radiation studies in which the intensity of a given photoelectron or X-ray emission peak is measured with right circular polarized (RCP) excitation and then with left circular polarized excitation, and the MCD is then the normalized intensity difference $I_{\mathrm{MCD}} = 2(I_{\mathrm{LCP}} - I_{\mathrm{RCP}})/(I_{\mathrm{LCP}} + I_{\mathrm{RCP}})$. In these measurements, which were performed at bend-magnet beamline 9.3.2 of the ALS, the polarization was varied by accepting radiation from slightly above and below the plane of the electron orbit in the storage ring, but there are also elliptically polarized undulators that can achieve this variation. Comparing the raw intensities $I_{\mathrm{LCP}} + I_{\mathrm{RCP}}$ with XRO calculations[10] permitted deriving the concentration profiles at the left of Fig. 27.2, and comparing the dichroism curves I_{MCD} with corresponding XRO calculations permitted deriving the magnetization profiles at right,

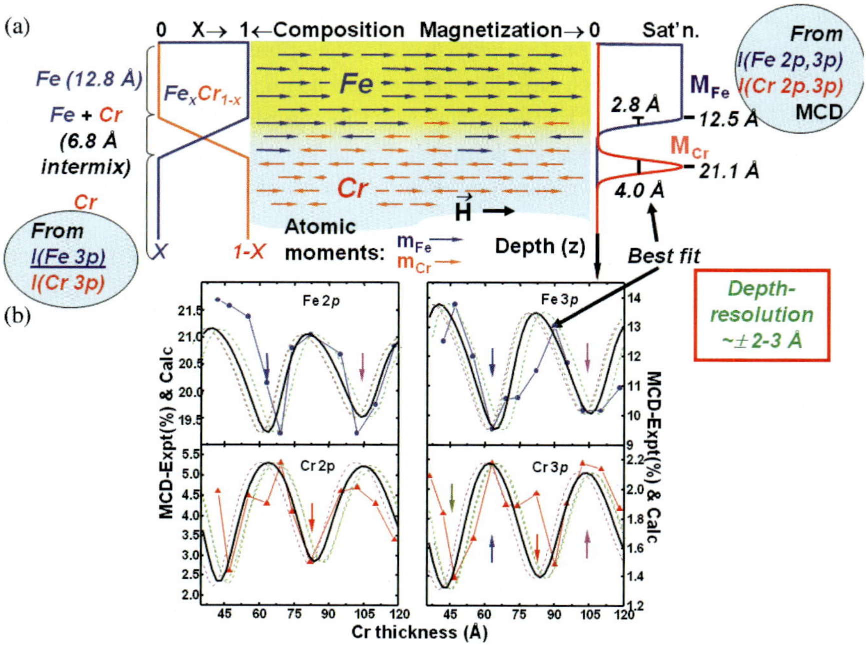

Fig. 27.2. (a) The concentration profiles and element-specific magnetization profiles of Fe and Cr through the Fe/Cr interface of the sample shown in Fig. 27.1 that have been derived from Fe $2p$ and Fe $3p$ intensity and MCD measurements, respectively. These data were obtained with a photon energy of 825 eV. (b) A summary of the Fe and Cr MCD data used to derive the magnetization profiles shown in (a). (From Ref. 4.)

with accuracies in the parameters involved that are estimated to be $\sim\pm 2$–3 Å. From the relative magnitudes and signs of the MCD effects, it could also be concluded that Cr is weakly ferromagnetic just under the interface, but that it is antiferromagnetically coupled to the Fe. Thus, a complete picture of this interface has been derived in a nondestructive manner.

As another example related to spintronics, we consider the magnetic tunnel junction, in which two ferromagnet layers (e.g., CoFe) are separated by an insulating layer (e.g., Al_2O_3 or MgO), and spin-dependent tunneling interactions can produce a large tunnel magnetoresistance (TMR). Such magnetic tunnel junction (MTJ) devices have now essentially replaced GMR in magnetic read heads. Figure 27.3 summarizes photoemission data from a sample consisting of an Al_2O_3 wedge varying in thickness from 100 Å to 55 Å, a layer of CoFe of 25 Å thickness, a layer of CoFeB of 15 Å thickness, and a final protective cap of Al_2O_3 of 10 Å thickness.[9] In Fig. 27.3(a) is shown the B 1s spectrum, which is split into two components by a large

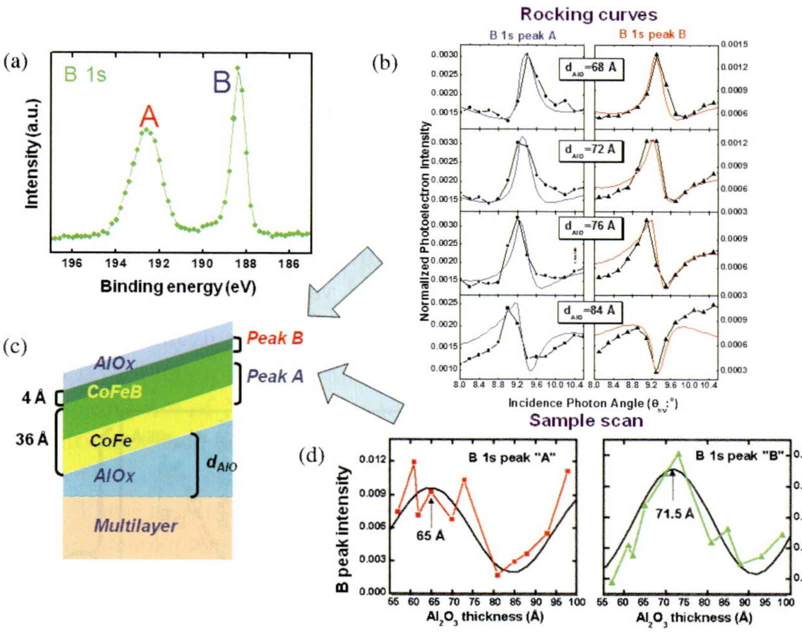

Fig. 27.3. The boron 1*s* spectrum from a CoFeB layer on top of a sample with the configuration shown in (d) and layer thicknesses as described in the text. (b) The intensity of the two components *A* and *B* in (a), obtained by rocking the sample, i.e., by scanning the X-ray incidence angle through the first-order Bragg reflection of the multilayer. (c) The distribution of the two types of boron in the sample, as derived from a fit of X-ray optical calculations to the data, with the smooth curves in (b) and (c) representing best fits. The photon energy was 1000 eV. (d) The variation of the intensities of peaks *A* and *B* as the standing wave is scanned through the sample by moving the sample in the x direction (a swedge scan). From Ref. 11.

chemical shift. These two components *A* and *B* can be verified as two chemically and spatially distinct species by either doing a rocking-curve scan and monitoring the two intensities *A* and *B* (Fig. 27.3(b)) or a scan along the wedge slope (Fig. 27.3(c)) in which the x position is fixed and the angle of incidence is varied, thereby changing the SW position. The two components *A* and *B* have markedly different behavior as a function of SW position. Analysis of the scans shown in Figs. 27.3(b) and 27.3(c), but particularly the phase shift between the oscillations in Fig. 27.3(c), reveals that their mean depths are different by about 7 Å and that peak *B* originates from atoms closer to the surface. A precise analysis of both sets of data yields the concentration profiles responsible for these two peaks indicated in Fig. 27.3(d), and the conclusion that the boron of type *B* in

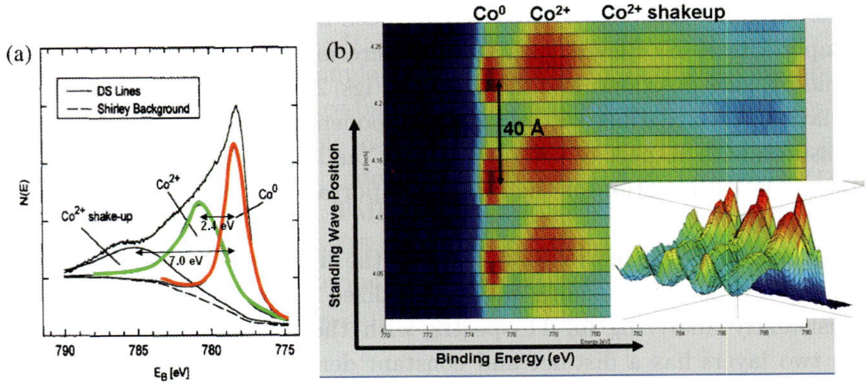

Fig. 27.5. (a) A Co $2p$ spectrum from oxidized Co, as obtained from the literature (Ref. 9), indicating the three features expected: one from Co^0 and two from Co^{2+} (a main peak and a broad shake-up or screening satellite). (b) The effect of scanning the standing wave through a sample consisting of an Al_2O_3 wedge, a Co layer, and a Ru cap, on the Co $2p$ spectrum. The photon energy was 1100 eV. Note the obvious phase shift between the Co^0 and Co^{2+} peaks. From Refs. 9 and 11.

turn in a direction indicating that the oxide is nearer the surface. Beyond this, the oscillatory patterns seen for the various core level intensities of different atoms from this sample, as plotted in Fig. 27.6, yield a family of phase shifts which can be analyzed to determine depth distributions.

For example, O $1s$ is split into what appears to be two metal-oxide components, one that is in phase with Co oxide and nearer to the surface, and one connected with Ru that is below the surface. The metallic Co signal also seems to come from not very far below the Ru on average. These results thus point to a very strong intermixing and/or island formation in the Co and Ru layers, with the relative weakness of the Ru oscillations also suggesting that it has distributed itself over depths that must be approaching the wavelength of the standing wave, which was in this case 40 Å. An approximate picture of the sample profile is shown in the inset of Fig. 27.6, and it is very different from what might have been supposed from the synthetic recipe.

27.3. Applications Using Standing Wave Excited X-Ray Emission

As another alternative with the swedge method, we consider detecting photons emitted from the sample, either as normal X-ray emission

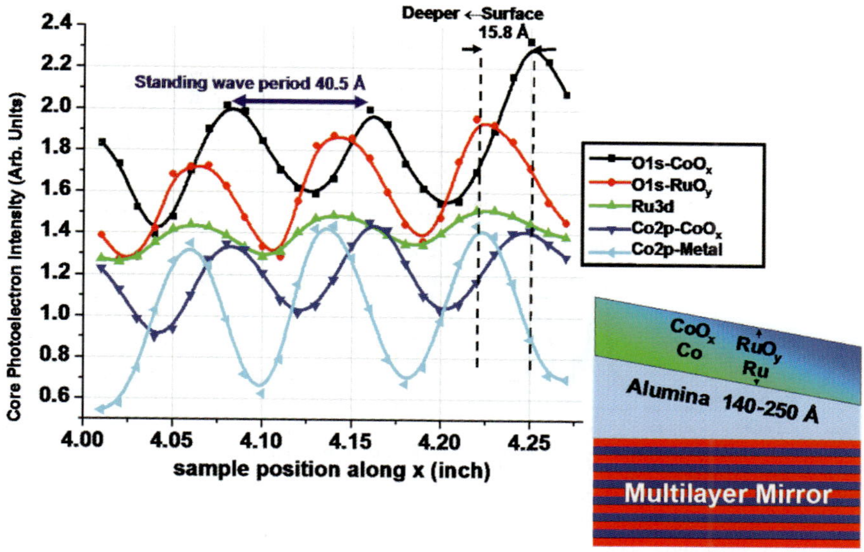

Fig. 27.6. (a) The oscillatory intensity variations of different core-level photoelectron intensities as the standing wave is scanned through a sample with the configuration shown in (b). The different peaks involved are indicated. From Refs. 9 and 11.

spectroscopy (XES), or as the closely related experiment of resonant inelastic X-ray scattering (RIXS) in which the incoming photon energy is tuned to a strong absorption resonance, and the outgoing photons, at lower energies than the excitation, are measured. The fact that photons have much greater penetration and escape depths than electrons of comparable energies makes this type of measurement capable of looking much more deeply into multilayer structures, and we will show that it thus also permits characterizing both the top *and* bottom interfaces of a given layer. In photoemission measurements, the strong attenuation of the emission from a given layer due to inelastic electron scattering during escape tends to bias the data strongly towards seeing only the top interface of a given layer.

This additional type of photon-out measurement is again illustrated for the case of the Fe/Cr system of relevance to GMR, but in this case, the sample consisted of a Cr wedge varying from 120 Å to 240 Å in thickness, an Fe layer of 16 Å thickness, and a final Al_2O_3 capping layer of 13 Å thickness.[8] There are thus two interfaces involved, Al_2O_3/Fe on top and Fe/Cr on the bottom. We have in particular looked at the Cr and Fe $L_{\alpha,\beta}$ X-ray spectra, as excited by photons at the Fe $2p_{1/2}$ absorption edge (722 eV), which yields a typical wide-energy-range spectrum as shown in

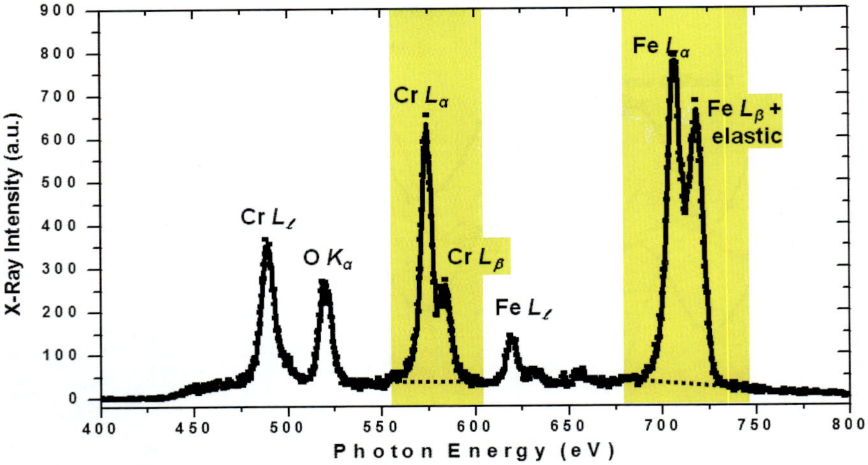

Fig. 27.7. A scan of emitted X-ray intensity over a broad energy range encompassing O, Cr, and Fe soft X-ray emission processes. The exciting X-ray, which is elastically scattered into the spectrometer, is very close to the energy of Fe L_β, and contributes about 40% of the intensity above the dashed linear background of the combined Fe L_α + Fe L_β that is plotted in Fig. 27.8. From Ref. 8.

Fig. 27.7. The Fe L spectra are thus strictly speaking RIXS emission, but the Cr L spectra would be normal non-resonant XES. Other X-rays are also emitted from Cr, Fe, and O, as indicated in Fig. 27.8.

Figure 27.8 shows the oscillatory patterns associated with Cr L and Fe L emission, in (a) for Cr L, as summed over excitation with left and right circularly polarized (LCP and RCP) radiation, in (b) the same sum for Fe L, and in (c) the Cr L/Fe L ratio. The Cr L results in (a) show the expected increase in intensity as the Cr thickness increases, since the X-rays can escape from depths well into the Cr wedge, by contrast with photoelectrons. The Cr oscillations are relatively weak, due to the fact that the emitted X-rays sample a considerable depth into the wedge, but they are well predicted by XRO theory. The Fe layer, which is only about 40% of the standing wave period in thickness, shows much stronger oscillations, which are very well predicted by XRO theory that of course takes into account the morphology of the sample. Finally, in (c) the Cr L/Fe L ratio is shown, and it also is very well predicted by theory. These intensity data permit determining the concentration profiles through both the top and bottom interfaces of Fe.

In Fig. 27.8(d)–27.8(e) is shown results of relevance to magnetic circular dichroism in Fe: in (d) is the difference between LCP and

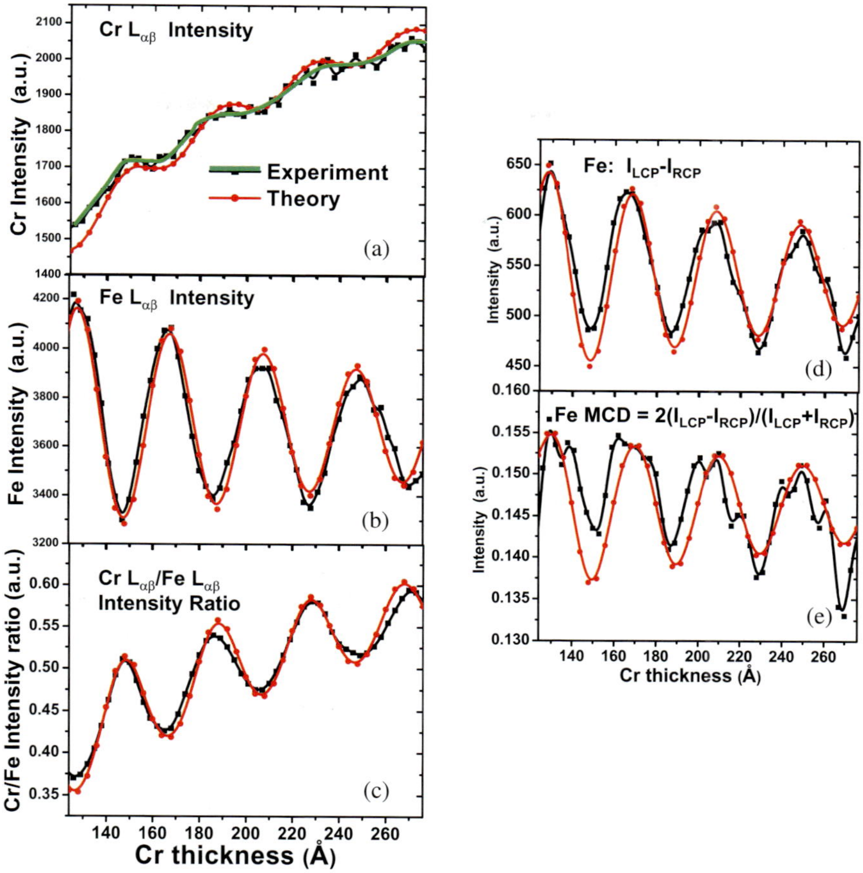

Fig. 27.8. The effect of scanning the standing wave through an Al_2O_3/Fe/Cr wedge trilayer, at a photon energy of 722 eV resonant at the Fe $L2$ edge on: (a) and (b) the individual Cr $L_{\alpha+\beta}$ and Fe $L_{\alpha+\beta}$ intensities (as summed over LCP and RCP), (c) the ratio of these intensities, Cr $L_{\alpha+\beta}$/Fe $L_{\alpha+\beta}$, (d) for Fe $L_{\alpha+\beta}$, the quantity $\mathrm{I_{LCP}} + \mathrm{I_{RCP}}$ that is proportional to Fe magnetization, and (e) the final Fe MCD as defined in the inset. (From Ref. 8)

RCP, and in (d) is twice the ratio of these two quantities, which is the MCD signal. Over three cycles of the standing wave have passed through the Fe layer, and all three of the quantities in these panels exhibit clear oscillations of the order of 25%, 15%, and 10% around their mean values, respectively. As a first conclusion connected with this data, the fact that there is a non-constant and oscillatory MCD signal can be shown rigorously[8] to require that the per-atom contribution of Fe to MCD decreases through the two interfaces. That is, if the per-atom contribution were constant, such that MCD is simply proportional to the Fe concentration at a given depth, then the intensity normalization in the MCD expression would render it constant as the SW scans through the Fe layer. Also, the slight phase shift between the summed intensity in (a) and the MCD signal in (c), which is about 3–4 Å in vertical standing wave position, and the greater asymmetry and width of the peaks in the MCD oscillations indicates that the two Fe interfaces do not have the same magnetization profile.[8] The red curves in Fig. 27.8 are the result of an analysis involving a large number of X-ray optical calculations for different layer and interface geometries. The calculations were performed for various choices of parameters between two layers i and j: the constant-composition layer thicknesses t_{ij} after allowing for a linear-gradient concentration interface thicknesses w_{ij}, which leads to a normalized concentration profile over a given interface of the form $\rho_i(z) = 1 - z/w_{ij}$, and normalized per-atom magnetization profiles of a Gaussian form $m(z) = \exp(-z^2/2\sigma_{ij}^2)$, with decay half-widths g_{ij}, where $g_{ij} = \sqrt{(2\ln 2\sigma_{ij})} = 1.177\sigma_{ij}$. The final best-fit interface parameters are shown in Fig. 27.9: black for concentration and red for per-atom magnetization. Thus, the properties of both top and bottom interfaces have been determined via this analysis, something that would not be possible with photoelectron emission.

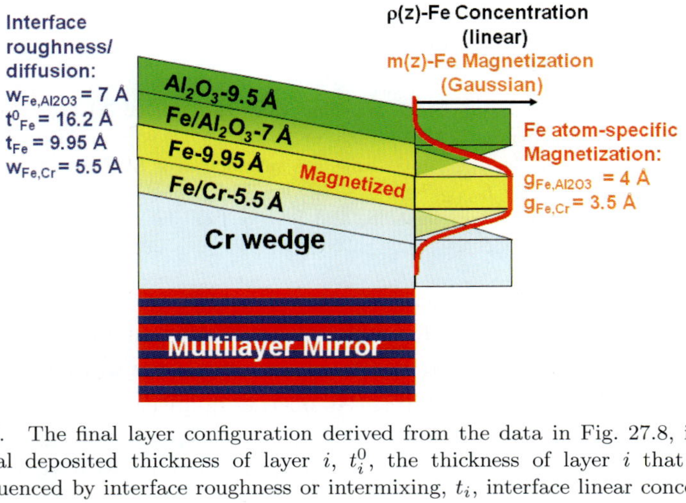

Fig. 27.9. The final layer configuration derived from the data in Fig. 27.8, including the initial deposited thickness of layer i, t_i^0, the thickness of layer i that has not been influenced by interface roughness or intermixing, t_i, interface linear concentration parameters, w_{ij}, and Gaussian atom-specific magnetization halfwidths, g_{ij}, where i and j refer to the two layers involved. From Ref. 7.

27.4. Future Applications — Hard X-Rays and Microscopy

Finally we comment briefly on some additional interesting developments that could permit the swedge method to study additional aspects of spintronic and other nanostructures, with several of these possibilities being schematically illustrated in Fig. 27.10.

Although a limitation of this method is the necessity of growing, or somehow mechanically placing, the sample on a suitable multilayer mirror + wedge, there are many possible multilayer configurations that can be used, with B_4C/W and Si/Mo being two very popular ones for X-ray mirrors, but many other pairs having been characterized.[14] Some of these would, in addition, permit growing epitaxial sample layers (e.g., AlAs/GaAs, $SrTiO_3/La_{1-x}Sr_xMnO_3$, Pt/Co), and data have already been obtained of this type for the first two systems mentioned here.[15] It is also possible to grow multilayers with smaller periods than the 40 Å discussed here, down to 25–30 Å, thus increasing the resolution for sample layers in the 10–15 Å range in thickness.

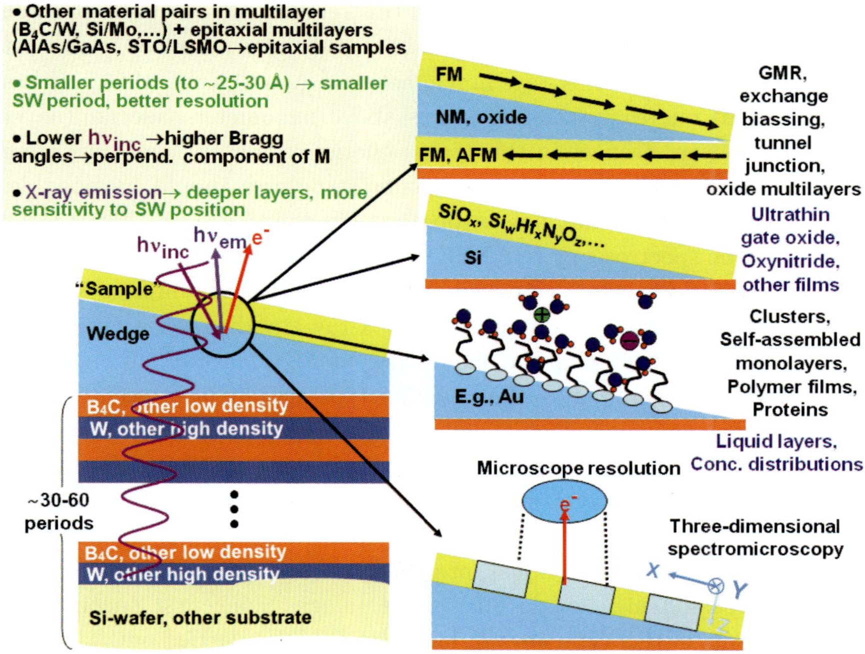

Fig. 27.10. Illustration of various possible future applications of the standing-wave/wedge (swedge) method. (From Ref. 5)

Lowering the photon energy from the *ca.* 500–1000 eV range used to date yields higher incidence angles, and thus greater sensitivity to the perpendicular component of magnetization in MCD studies, although the reflectivities also decrease as energy is lowered. However, lower energies would also yield higher cross-sections for studying, e.g., C 1*s* emission from organic or polymeric nanostructures.

Looking at X-ray emission rather than photoelectron emission, as we have already illustrated, yields greater depth penetration, and sensitivity to top and bottom interfaces, as well as deeper buried layers.

Other applications areas, as illustrated in Fig. 27.10, would be semiconductor-related nanolayer structures, self-assembled monolayers, or even high-pressure gases or liquid solutions in interactions with surfaces, with the last-mentioned being possible via the use of a windowed X-ray emission cell.[16]

Another interesting area for future development would be in using much harder X-rays for excitation of photoelectrons, going from soft X-rays in the 500–1000 eV regime up to 5 or 10 keV, in order to penetrate multilayer structures more deeply. There is presently growing activity in Europe and Japan in carrying out photoemission in this regime,[17,18] and some of these possibilities for studies of magnetism have already been discussed in a recent comprehensive review of X-rays in magnetism.[7] Among other things, it has been pointed out that standing waves above nanometer-scale multilayer mirrors are even stronger in this higher-energy regime,[7,19] and thus more accurate characterizations of even deeper structures should be possible. In related work that did not involve standing waves, such hard X-ray photoelectron spectroscopy (HAXPES) has in fact been used in connection with chemical shifts and local density calculations to study interface mixing in the Ni/Cu interface.[20] By using hard X-ray standing wave excitation, this type of characterization should be more quantitatively possible, and some very encouraging data of this type has in fact recently been obtained.[15]

As a final possibility for the future, illustrated at lower left in Fig. 27.10, carrying out soft X-ray-excited XM or PEEM studies with standing-wave excitation could provide a type of direct depth sensitivity to these laterally resolving SR techniques, provided that a few standing wave cycles can encompassed in a single microscope image. Some first encouraging measurements of this type have in fact recently been carried out.[21]

27.5. Concluding Remarks

The standing-wave/wedge method discussed here thus has demonstrated the ability to non-destructively determine buried-interface concentration profiles and element-specific magnetization profiles, as well as layer-specific densities of states, in a variety of multilayer nanostructures of interest in spintronics and other areas of current interest. A limitation of the technique is that a suitable sample structure must be grown on a multilayer mirror of sufficient reflectivity, but there nonetheless seem to be a variety of systems for which this should be possible, and a wide range of applications areas, both in magnetism and other fields.[5] Both, photoelectron emission and X-ray emission/inelastic scattering can be used as probes, with the latter providing greater bulk sensitivity and the ability to look at both the top and bottom interfaces of different layers. Using more energetic X-rays will permit studying deeper layers and interfaces, and using soft X-ray standing waves for imaging with photoelectron microscopy should also permit adding depth resolution to this family of techniques.

Acknowledgments

This work was supported by the Director, Office of Science, Office of Basic Energy Sciences, Materials Science and Engineering Division, U.S. Department of Energy, under Contracts Nos. DE-AC03-76SF00098 and DE-AC02-05CH11231. C.S.F. also gratefully acknowledges the support of the Alexander von Humboldt Foundation and the Helmholtz Association during part of this work.

References

1. C. S. Fadley, *Prog. Surf. Sci.* **16** (1984) 275; S. Oswald, M. Zier, R. Reiche and K. Wetzig, *Surf. Interface Anal.* **38** (2006) 590; Summary of the *47th IUVSTA Workshop on Angle-Resolved XPS*, (Riviera Maya, Mexico, 2007) in *Surf. Interface Anal.* **40**, 1579 (2008).
2. S.-H. Yang, B. S. Mun, A. W. Kay, S.-K. Kim, J. B. Kortright, J. H. Underwood, Z. Hussain and C. S. Fadley, *Surf. Sci. Lett.* **461** (2000) L557.
3. S.-H. Yang, B. S. Mun, A. W. Kay, S. K. Kim, J. B. Kortright, J. H. Underwood, Z. Hussain and C. S. Fadley, *J. Electron Spectrosc.* **114** (2001) 1089.
4. S.-H. Yang, B. S. Mun, N. Mannella, S.-K. Kim, J. B. Kortright, J. Underwood, F. Salmassi, E. Arenholz, A. Young, Z. Hussain, M. A. Van Hove and C. S. Fadley, *J. Phys.: Condens. Matter* **14** (2002) L406.

5. S.-H. Yang, B. S. Mun and C. S. Fadley, *Synchrotron Rad. News* **17**(3) (2004) 24.
6. S.-H. Yang, B. S. Mun, N. Mannella, A. Nambu, B. C. Sell, S. B. Ritchey, F. Salmassi, S. S. P. Parkin and C. S. Fadley, *J. Phys.: Condens. Matter* **18** (2006) L259.
7. G. Srajer, L. H. Lewis, S. D. Bader C. S. Fadley, E. E. Fullerton, A. Hoffmann, J. B. Kortright, K. M. Krishnan, S. A. Majetich, C. A. Ross, M. B. Salamon, I. K. Schuller and T. C. Schulthess, *J. Magn. Magn. Mater.* **307** (2006) 1.
8. B. C. Sell, S.-H. Yang, M. Watanabe, B. S. Mun, L. Plucinski, N. Mannella, S. B. Ritchey, A. Nambu, J. Guo, M. W. West, F. Salmassi, J. B. Kortright, S. S. P. Parkin and C. S. Fadley, *J. Appl. Phys.* **103** (2008) 083515.
9. B. C. Sell, Ph.D. thesis, University of California, Davis, 2007.
10. C. S. Fadley, S.-H. Yang, B. S. Mun, J. Garcia de Abajo, Chapter discussing X-ray optical calculation methodology, in *Solid-State Photoemission and Related Methods: Theory and Experiment*, eds. W. Schattke and M. A. Van Hove (Wiley-VCH Verlag, Berlin, 2003); S.-H. Yang, A. X. Gray, A. M. Kaiser, B. S. Mun, J. B. Kortright and C. S. Fadley, Journal of Applied Physics.
11. S. H. Yang, B.C. Sell and C. S. Fadley, *J. Appl. Phys.* **103**, 07C519 (2008); C. S. Fadley, *J. Electron Spectrosc.* 178–179 (2010).
12. M. Julliere, *Phys. Lett. A* **54** (1975) 225.
13. B. Klingenberg, F. Grellner, D. Borgmann and G. Wedler, *Surf. Sci.* **383** (1997) 13–24.
14. A summary of data obtained over many multilayer pairs. Available at http://henke.lbl.gov/multilayer/survey.html.
15. S. Doring, M. Gorgoi, C. Papp, B. Balke, S. Ueda, K. Kobayashi *et al.*, to be published; A. X. Gray, C. Papp, B. Balke, S.-H. Yang, M. Huijben, E. Rotenberg, A. Bostwick, S. Ueda, Y. Yamashita, K. Kobayashi, E. M. Gullikson, J. B. Kortright, F. M. F. de Groot, G. Rijnders, D. H. A. Blank, R. Ramesh and C. S. Fadley, *Phys. Rev. B* **82**, 205116 (2010).
16. J. Guo, T. Tong, L. Svec, J. Go, C. Dong and J.-W. Chiou, *J. Vac. Sci. Technol. A* **25** (2007) 1232.
17. *Proc. Workshop on hard X-ray photoelectron spectroscopy, HAXPES, Nucl. Inst. and Meth. A* **547** (2005), J. Zegenhagen and C. Kunz (eds.).
18. Program and abstracts of the Fourth International Workshop on Hard X-Ray Photoelectron Spectroscopy, HAXPES2011, at DESY, Hamburg, Germany, available at: http://haxpes2011.desy.de.
19. C. S. Fadley, *Nucl. Instrum. Meth. A* **547** (2005) 24–41.
20. E. Holmstrom, W. Olovsson, I. A. Abrikosov, A. M. N. Niklasson, B. Johansson, M. Gorgoi, O. Karis, S. Svensson, F. Schafers, W. Braun, G. Ohrwall, G. Andersson, M. Marcellini and W. Eberhardt, *Phys. Rev. Lett.* **97** (2006) 266106.
21. F. Kronast, R. Ovsyannikov, A. Kaiser, C. Wiemann, S.-H. Yang, D. E. Bürgler, R. Schreiber, F. Salmassi, P. Fischer, H.A. Dürr, C. M. Schneider, W. Eberhardt and C. S. Fadley, *Appl. Phys. Lett.* **93** (2008) 243116.

EPILOGUE

I am enormously grateful to my friend and colleague Dr. Alexander Kazimirov who slightly twisted my arm to start on this volume on the X-ray standing wave technique. His perseverance and dedication was the motor behind this book project. Alexander was born in Glubokiy village in Russia on July 29th, 1952 and he died, much too early, on August 12th, 2011 after having completed his last ascent, the 1559 m high peak of Algonquin Mountain in the Adirondacks in New York State. Alexander had obtained his PhD degree in physics in Moscow where he continued his work and where we met in 1993. He worked together with me during several extended stays at the Max-Planck-Institute for solid state research in Stuttgart, Germany. We continued collaborating when he moved to Northwestern University in Evanston IL, and then further to Cornell High Energy Synchrotron Source, Ithaca, NY. To his last minute, Alexander had kept on learning and exploring, and building new plans. With his premature death the scientific community lost a highly respected colleague and I am missing a good friend.

Jörg Zegenhagen, Grenoble, September 2012

Appendix 1

X-RAY STANDING WAVES — EARLY REMINISCENSES

Boris W. Batterman†

It is a pleasure to contribute to this book which summarizes work on standing waves over the last 40 years.

I first heard about Dynamical X-ray Diffraction when I took Prof. Warren's course at MIT. The Darwin curve describing perfect crystal Bragg reflection was presented as in Darwin's original work. It was very intuitive and involved simply considering partial reflection and transmission through successive planes of a perfect crystal. It gave the correct results for a specific problem but it did not encompass the larger problem of how X-rays are diffracted when one takes into account the influence of the diffracted beams within the crystal. Beams scattered from atoms and planes of atoms could interact in a dynamical way with the incident wave and the many diffracted waves.

These interactions remarkably affect the ultimate diffracted beams from the crystal as a whole. The kinematical theory which was used to interpret diffraction for structural studies ignored further interactions between diffracted and incident beams.

When I came to Bell Laboratories in 1956 the semiconductor revolution was well underway. Germanium was the working material. When I started to do some X-ray experiments on germanium, I was surprised to find that

†Walter S. Carpenter Jr. Prof. Applied and Engineering Physics, Cornell University, Emeritus; Emeritus Director, Cornell High Energy Synchrotron Source, CHESS, Ithaca, NY, USA.

in the Bragg case, the diffracted beam was nearly as strong as the primary beam. Warren had nurtured a pair of calcite crystals that were nearly perfect from the X-ray point of view and these were taken out of their secret storage place once a year when he had his students to do an experiment to measure the Darwin curve.

I found myself surrounded by a world of perfect crystals of germanium, where the worst of them was probably better than Warren's Calcite crystals.

To study these, Dynamical Theory was necessary. We were all astounded to hear of the Borrmann effect whereby, under certain diffraction conditions, perfect crystals could behave as if photoelectric absorption of X-rays would be turned off, and the crystal would transmit substantial intensities even if the normal absorption factor could be as large as exp (-100). Understanding this would demand a full knowledge of Dynamical Diffraction Theory, originally created by P.P. Ewald and generalized by Max von Laue. In the course of working with germanium and silicon, I delved into this theory.

About this time 1959, Prof Warren suggested that I consider writing a review paper on Dynamical Diffraction. My colleague Henderson Cole and I joined forces to bring this to fruition.[1] In the middle of this effort, I remember specifically a figure in R.W. James' work[2] which showed the X-ray wavefield intensity in a crystal as the reflection curve of a Bragg peak was scanned through the range of Darwin total reflection. At the low angle end of the range, the intensity at the atom planes was zero, and as one progressed across the range of total reflection by increasing the incident angle, the intensity at the planes increased four times the value an atom would experience when the incident traveled through the crystal when no diffraction took place.

This allowed me to immediately understand the peculiar features of the Darwin curve from an absorbing crystal. For a crystal with no absorption, the Darwin curve was flat-topped with unity reflectivity. With absorption, (the Darwin–Prins curve) was no longer flat topped and a strong asymmetry developed, giving a diffracted intensity substantially lower at the high angle side of the curve.

The superposition of the internal incident and diffracted waves produced a standing wave whose nodal planes moved with respect to the crystal planes as the incident beam traversed the range of total reflection. At the low angle side, the interaction of the wave field with the atoms of the Bragg planes would be minimal, but at the high angle side an antinode would be at the atomic sites and the absorption of X-rays would be greater, leading to a loss of diffracted beam and hence an asymmetry in the Darwin curve.

At this time, there was much discussion of the recently discovered Borrmann effect, and it was clear that a standing wave was set up in the crystal that could reduce photoelectric absorption in the same manner as the beams at the low angle end of the Darwin total reflection curve and lead to enhanced transmission. Cole and I discussed a neutron experiment that had been performed. The crystal was one in which neutrons could be absorbed by a neutron/gamma process, the signature of which would be an emitted gamma ray. It was verified that when the diffraction condition in transmission satisfied the Bragg condition, the gamma ray emission was reduced.

The idea of a secondary process providing a signature of a diffraction event was very appealing to me. It was at this time that I came up with the idea of doing this with X-rays. The simplest way to do this was to use a germanium crystal. The secondary radiation would be the Ge K_α fluorescence and the incident radiation would be Mo K_α, a conventional X-ray tube source whose energy was high enough to excite the Ge host atoms. The set up was almost trivial since I had a good working double crystal diffraction apparatus. I set up a slow scan to run a Darwin peak overnight. I used two scintillation counters, one to record the diffracted Darwin curve, the other was placed in the plane of incidence between the incident and diffracted beams as close as possible to the germanium crystal. For stability reasons, over many hours of the run at seconds of arc precision, I started the scan at the end of the work day and let it run all night.

When I came in the next morning, I was flabbergasted at what I saw. On the trace of one chart recorder was a very good peak showing the expected asymmetric Darwin–Prins curve. The second recorder showed the Ge K-fluorescence coming from the crystal. For the first few hours of the run, the intensity was quite constant representing the absorption of the incident beam. Then, the intensity started to dip below the average value and kept continuously getting weaker as the Bragg condition was approached.

The remarkable thing to me at this time was that the lowering of the fluorescence was observable some 40 or so Darwin widths before the actual reflected beam could be observed. The decrease intensified until the reflected beam became strong and the fluorescence dropped to nearly zero in the range of total reflection. As the high angle side of the range of total reflection was approached, the Ge fluorescence signal rose until it was considerably higher than the average fluorescence observed when off the Bragg condition and it gradually subsided to this value again, some 40 Darwin widths on the high angle side.

At first look, the large dip to nearly zero of the fluorescence seemed obvious. When in the region of near total reflection, very little X-rays penetrated the crystal and the fluorescence would have to drop dramatically. The reasons for the long tails on the fluorescence were not so obvious.[a]

It occurred to me at this time that the asymmetry of the tails was a measure of the standing wave intensity at the host atoms, changing from nodes at the atoms on the low angle side to antinodes at the high angle side.

Then, the most significant conclusion was that the motion of the standing wavefield was in fact showing the location of the germanium atoms in the lattice. This in one sense was a solution to the phase problem in crystallography because it allowed one to locate an atom using intensity data only.

In our case, this was somewhat trivial since the location of germanium atoms in germanium was never in doubt. I then got quite excited about determining foreign atom locations. It was clear that one could look at the fluorescence curve and tell whether a foreign atom was substitution or interstitial by merely observing the asymmetry of the fluorescence with respect to the Darwin curve. It appealed to me that a simple observation of the asymmetry could locate an atom. You only had to ask what the asymmetry left/right or right/left was to determine whether an atom was substitutional or interstitial. This work[3] was published in 1964.

This of course was a bit too simple, because even if an atom was interstitial, on certain Bragg planes it could appear substitutional. So, to pin things down between these extreme positions, one would have to measure the fluorescence using several Bragg planes to triangulate the atom position exactly.

[a]The reason that the asymmetry started so far from the angle at which the Bragg reflection was strongest is as follows: When the incident and reflected beams superpose in the crystal, the modulation of the intensity at the atoms depends on E_h/E_o the ratio of the amplitudes of the reflected and incident waves. The Darwin curve intensity is equal to this quantity squared, $(E_h/E_o)^2$. Lets say the angle of incidence and is such that tails of the Darwin curve is at 1%. That means that (Eh/Eo)2 equals .01 (Eh/Eo) is 0.1. Thus the modulation (change of intensity of the wave at the atom) is changed by 10% even if the Darwin tail is at only 1% of the peak reflectivity, ((Eh/Eo) is unity in the total reflection range) however the flouresence will have changed by 10%.

It was clear that creating a standing wave in the Bragg case allowed one to locate atoms by using the angle of incidence as a tuning knob to move standing waves in a controlled manner and by observing the fluorescence, locate the fluorescing atom!

The next step was to put foreign atoms in a crystal and use the standing wave technique to locate its position. Towards this end, I approached my Bell Lab colleague Dr. J.R. Patel, and he pointed out to me that silicon could dissolve a rather high amount of Arsenic atoms, up to about 10^{20} atoms/cc and still maintain its crystal perfection.

This was a good combination since the silicon host would not produce detectable fluorescence and arsenic could be readily excited with Mo K_α radiation. This turned out to be a very straightforward demonstration of the standing wave technique[4] and showed that arsenic was indeed a substitutional impurity. This work was published in 1969.

Several years after the Arsenic work was published, I received a communication from Jene Golovchenko who was quite interested in the standing wave technique because it had similarity to his earlier work on particle channeling in solids. I invited him to come to Cornell to perform an experiment he designed on a silicon crystal which had a shallow layer of arsenic diffused in through the crystal surface. He wanted to further develop the ability to locate atoms at arbitrary interstitial locations. The experiment was done at Cornell with Walter Brown as collaborator. That paper was published in 1974.[b]

With my student Lonni Berman and my colleague Jack Blakely from Cornell we used standing waves to investigate the structure of monolayer Au atoms on silicon in a UHV environment. The success of this experiment[5] showed that standing waves existed in the vacuum above the crystal surface and could be effectively used to probe the structure.

An exciting innovation was developed by Mike Bedzyk when he was at Cornell. He realized that X-ray standing waves would exist above the surface of a totally reflecting mirror for X-rays at low glancing angles. The standing wave period depended on the incidence angle and therefore could be effectively used to probe the details of large spacing structures that ranged over many tens of Angstroms. He was awarded the Warren Prize for this work.

It is particularly pleasing to me that the standing wave technique has found wide applications. The advent of synchrotron sources with their high

---

[b]Note by the editor: See next appendix written by Jene Golovchenko.

brightness and intensity made standing wave experiments easier to carry out with increased precision and sensitivity.

References

1. B. W. Batterman and H. Cole, *Rev. Mod. Phys.* **36** (1964) 681.
2. R. W. James, *The Optical Principles of the Diffraction of X-rays*, G. Bell and Sons Ltd., London, 1950.
3. B. W. Batterman, *Phys. Rev.* **133** (1964) 759.
4. B. W. Batterman, *Phys. Rev. Lett.* **22** (1969) 1.
5. L. E. Berman, B. W. Batterman and J. M. Blakely, *Phys. Rev. B* **38** (1988) 5397.

Appendix 2

REMEMBRANCES OF X-RAY STANDING WAVES DAYS

Jene Golovchenko

Harvard University, Department of Physics, 17 Oxford Street, Cambridge, Massachusetts 02138, USA

It is many years ago that I first started thinking about and tinkering with X-ray standing waves. My recollections about those days are mostly framed in the context of the few mentors and many (then) young colleagues and students I had the pleasure and good fortune to interact with.

It began for me when I was a Post Doc working part time in the Radiation Physics Department run by Walter Brown at Bell Laboratory at Murray Hill New Jersey, and part time in the Physics Department at Brookhaven National Laboratory working with Allan Goland on Long Island, New York. I did a lot of traveling back and forth that year. Those were the early seventies, positions doing research were hard to find and this was my second post doc. The first was in Aarhus Denmark. I had actually done my experimental PhD work at Bell Labs before that, working with Walter Gibson, a member of the Radiation Physics Department and his connection with the University of Aarhus helped get me a first post doc job there. All of my research work up until then had been focused on using ion beams from low energy accelerators ($\sim 1 - 2$ MeV) to study "particle solid interactions", and the hot topic that mostly occupied my thoughts until then was ion beam channeling.

The ion channeling effect involves the penetration of energetic charged particles like protons and alpha particles through crystal lattices. When the particle beam is incident on a crystal target along low index axial or planar crystal directions, the penetration through the lattice is governed by a series of highly correlated collision sequences that result in reduced energy loss rates, enhanced particle ranges, and reduced ion induced damage in the crystal. It turns out that that the channeling effect can be used under favorable conditions to determine the atomic scale location of impurity atoms in the lattice when combined with Rutherford backscattering, ion induced X-ray fluorescence or induced nuclear reactions. Sometimes impurities could be located in the unit cell to the order of 0.1 Å, but these were special circumstances. Thus, in addition to the intellectually interesting physics involved in understanding the ion channeling effect, it had found a potentially important application in solving the "impurity lattice location problem."

One day, Fred Young was visiting from Oak Ridge National Laboratory and he and John Poate, from the department, were chatting in the hallway outside the lab where my desk was located. Fred was interested in point defects in metals which he explored with diffuse X-ray scattering. They were speculating on whether one could "channel" X-rays through crystals (just like ions) to localize defects with X-ray, similarly to the way it was done with ion channeling. I jumped into the conversation claiming that I thought one could. I actually knew virtually nothing about X-rays at the time. But one of my great interests as a student had been in solid state device physics, and I had studied the books of S. Wang (Solid State Electronics) and C. Kittel (Introduction to Solid State Physics) very carefully. The clue for me was a recollection from Kittel's book of how electron diffraction created energy band gaps for electrons in crystals.

In Kittel's book (Third Edition) there was a discussion of the nearly free electron model and an explanation of the magnitude of the energy gap at a Brillouin zone boundary in terms of the two Schrödinger equation eigenfunctions, one at the bottom and one at the top edges of the energy band gap. These two eigenmodes were completely modulated standing wave solutions (displayed in Figure 3 of Chapter 9 in that book). They had periodic intensities with nodes and antinodes at the maximum and minimum of the Fourier component of the crystal potential responsible for the coupling of plane wave components that formed the eigenfunctions. In simple cases, these standing waves had nodes and antinodes that coincided with the physical crystal planes. Kittel actually referred to the connection between energy gaps for electrons in a crystal and frequency gaps for X-rays in a crystal due to Bragg reflection in a footnote. His Appendix A had an introduction to the dynamical theory of X-ray diffraction, but I remembered neither of these at the time. But it was clear to me that the fundamental optical physics had to be very similar for X-rays and electrons and that using it, one could modulate and control the X-ray intensities in a crystal on an atomic scale just like the channeling effect controlled the flux of ions in the crystal.

One thing led to another and after discussions with my mentor Walter Brown at Bell and a fellow named Dave Keating at Brookhaven we contacted Bob Batterman at Cornell, to try an experiment to see if one could indeed really locate atoms in the Bragg band gap using X-ray standing waves. Batterman was an expert on dynamical diffraction and he invited Walter and me up to try an experiment. We used the Bragg geometry which easily allowed the continuous spatial translation of a single fully modulated standing wave mode with incidence angle. Scanning the angle of incidence of the X-ray beam within an X-ray band gap continuously moved the antinode of the standing waves from lying on a set of atomic planes to lying between them, just like in the electron case described by Kittel when scanning in energy between energy band edges.

The main problem is that accompanying the standing wave motion in the angular Bragg band gap region is a very strong extinction effect that dramatically modulates the overall penetration depth of the X-rays in the crystal. In beautiful experiments, Bob had already studied X-ray fluorescence signals from an angular Bragg scan of a germanium crystal. The main signal in the Bragg band gap was a giant dip in germanium fluorescence intensity due to the extinction effect but there were also slightly asymmetric wings outside the band gap region that were correctly ascribed to standing wave effects. To accurately obtain information by which one could locate impurity atoms in the crystal required observing the strong standing wave effects inside the Bragg band gap without being dominated by these extinction effects there.

We needed to find a way to study fluorescence signals from impurities that were atomically well localized in the crystal, but they all had to be close enough to the surface so extinction effect would be eliminated. Doing this without distorting the crystal with too many impurity atoms meant that sensitivity would be a problem (these were pre-synchrotron days). Arsenic in silicon was important at the time because it is a dopant in silicon based devices. Arsenic had a very high solid state solubility in silicon, and even at very high concentrations it mostly goes into substitutional sites. There were some issues because all arsenic dopant atoms are not electrically active donors at high concentrations. Walter Brown knew Richard Fair at Bell

Labs who was studying the problem and Walter convinced him to diffuse a shallow, high concentration arsenic layer into a silicon crystal for us. Walter also found a lithium drifted silicon X-ray detector and associated counting electronics and together we drove to Cornell to perform the experiment with Bob Batterman on his scanning double crystal monochromator and GE X-ray machine. I recall how the goniometer was driven by a clock motor and the data was taken in single slow angular scans.

Things looked good from the start and we returned to Bell with data that looked promising. Specifically, the big extinction effect in the fluorescence signal was gone and there were giant standing wave effects to be seen in the raw data. Together with help from a wonderful computer support person named Marilyn Robbins I plotted up all the possible results one could obtain for different potential locations of impurity atoms in the lattice from the dynamical theory and compared and fitted them to the experimental results. Indeed it looked like one could "channel" X-rays in a crystal and perform precision lattice location experiments using X-ray standing waves. I was terribly excited and had the feeling I was witnessing the birth of a new piece of science with real applications.

I wrote up a paper on the experiment and after inputs from Walter and Bob, it was sent off to the Physical Review and published in 1974 (*The Observation of Internal X-Ray Wave Fields during Bragg Diffraction with an Application to Impurity Lattice Location*[1]). I was convinced that with improvements in beam collimation, detector sensitivity, geometry and angular stability, impurities could be located in crystals with a few hundredths of an Ångstrøm accuracy. I was unfortunately not able to find a way to continue this project here in the US. So I wrote to Jens Lindhard back in Aarhus Denmark to enquire whether they would be interested in having me return to Denmark once again but this time to pursue this new X-ray field. He invited me back for a second stay and I began building an apparatus from scratch dedicated to exploring these X-ray standing wave effects, with a new X-ray generator, fine focus molybdenum X-ray tube, homemade flex bearing double crystal monochromator, asymmetric monochromator crystal, SiLi detector, a nuclear physics style mutichannel, multiscaling pulse analyzer, with a piezo electrically driven angular scanning system. It worked great and the next standing wave papers came from Denmark.

I had a great student named Stig K. Anderson who did his thesis work with me on this project along with Georg Mair, a Post Doc from Austria. We studied impurities that were ion implanted in a sample and even figured

Arsenic Atom Location on Passivated Silicon(111) Surfaces[20]
Synchrotron X-Ray Standing Wave Study of Arsenic on Si(100)[21]
X-Ray Standing Wave and Tunneling Microscope Location of Gallium Atoms on a Silicon Surface[22]
Arsenic and Gallium Atom Location on Silicon(111)[23]
Giant Vibrations of Impurity Atoms on a Crystal Surface[24]
Thermal Vibration Amplitudes and Structure of As on Si(001)[25]

References

1. J. A. Golovchenko, B. W. Batterman and W. L. Brown, *Phys. Rev. B* **10** (1974) 4239.
2. S. K. Andersen, J. A. Golovchenko and G. Mair, *Phys. Rev. Lett.* **37** (1976) 1141.
3. G. L. Miller, R. A. Boie, P. L. Cowan, J. A. Golovchenko, R. W. Kerr and D. A. H. Robinson, *Rev. Sci. Instrum.* **50** (1979) 1062.
4. C. Liu and J. A. Golovchenko, *Phys. Rev. Lett.* **79** (1997) 788.
5. C. Liu and J. A. Golovchenko, *Optics Letters* **24**(9) (1999) 587–589.
6. J. A. Golovchenko and Chien Liu, in *X-ray and Inner-Shell Processes*, AIP Conference Proceedings **506**, edited by Donald S. Gemmell, Stephen H. Southworth, R. W. Dunford, E. P. Kanter and Linda Young, American Institute of Physics, pa. (2000) 621–637.
7. P. L. Cowan, J. A. Golovchenko and M. F. Robbins, *Phys. Rev. Lett.* **44** (1980) 1680.
8. J. A. Golovchenko, D. Kaplan, B. M. Kincaid, R. A. Levesque, A. E. Meizner, M. F. Robbins and J. Felsteiner, *Phys. Rev. Lett.* **46** (1981) 1454.
9. J. A. Golovchenko, R. Levesque and P. L. Cowan, *Rev. Sci. Instrum.* **52** (1981) 509.
10. J. M. Bedzyk, W. N. Gibson and J. A. Golovchenko, *J. Vac. Sci. Tech.* **20** (1982) 634.
11. P. L. Cowan, J. A. Golovchenko and M. F. Robbins, *Phys. Rev. Lett.* **44** (1980) 1680.
12. J. A. Golovchenko, *Bull. Amer. Phys. Soc.* **25** (1980) 426.
13. J. A. Golovchenko, J. R. Patel, D. R. Kaplan, P. L. Cowan and M. J. Bedzyk, *Phys. Rev. Lett.* **49** (1982) 560.
14. J. A. Golovchenko and J. R. Patel, *Phys. Rev. Lett.* **50** (1983) 153.
15. J. R. Patel and J. A. Golovchenko, *Phys. Rev. Lett.* **50** (1983) 1858.
16. J. R. Patel and J. A. Golovchenko, *Proc. of the Microscopy of Semiconducting Materials Conf.*, Oxford, England, **349** (1983) 347.
17. J. R. Patel, J. A. Golovchenko, J. C. Bean and R. J. Morris, *Phys. Rev. B* **31** (1985) 6884.
18. J. A. Golovchenko, B. M. Kincaid, R. A. Levesque, A. E. Meixner and D. R. Kaplan, *Phys. Rev. Lett.* **57** (1986) 202.

19. J. R. Patel, P. E. Freeland, M. S. Hybertsen, D. C. Jacobson and J. A. Golovchenko, *Phys. Rev. Lett.* **59** (1987) 2180.
20. J. R. Patel, J. A. Golovchenko, P. E. Freeland and H. J. Gossman, *Phys. Rev. B* **36** (1987) 7715–7717.
21. J. Zegenhagen, J. R. Patel, B. M. Kincaid, J. A. Golovchenko, J. B. Mock, P. E. Freeland, R. J. Malik and K.-G. Huang, *Appl. Phys. Lett.* **53** (1988) 252.
22. J. Zegenhagen, J. R. Patel, P. Freeland, D. M. Chen, J. A. Golovchenko, P. Bedrossian and J. E. Northrup, *Phys. Rev. B* **39** (1989) 1298.
23. J. R. Patel, J. Zegenhagen, P. E. Freeland, M. S. Hybertsen, J. A. Golovchenko and D. M. Chen, *J. Vacuum Science & Technology B* **7** (1989) 894.
24. R. E. Martinez, E. E. Fontes, J. A. Golovchenko and J. R. Patel, *Phys. Rev. Lett.* **69** (1992) 1061.
25. G. E. Franklin, E. Fontes, Y. Qian, M. J. Bedzyk, J. A. Golovchenko and J. R. Patel, *Phys. Rev. B* **50** (1994) 7483.

Appendix 3

X-RAY STANDING WAVE WORK AT SUNY ALBANY: A PERSONAL SUMMARY

Walter M. Gibson[†]

As usual, new directions in my research or new insights from our research often come from students or research associates. My involvement in X-ray standing wave (XSW) research and applications is no different. In this case, the initiator was Jene Golovchenko who was working with me at Bell Laboratories, Murray Hill, NJ on his dissertation research for a PhD from Rensselaer Polytechnic Institute in Troy, NY. Jene's thesis topic was related to determination of interatomic potential distributions in thin single crystals through the use of transmission channeling of helium ion beams.

Beginning in the mid 1960s and continuing to 1976 when I left Bell Labs to become Chair of Physics at the State University of New York (SUNY) at Albany, NY, I had been involved with study and application of ion channeling in single crystals together with about a dozen or so PhD students and collaborators from Rutgers, Princeton, RPI, City University of NY, Aarhus University, University of Paris, Lyon University, Brookhaven Nat. Lab., Argonne Nat. Lab. and Oak Ridge Nat. Lab. Much of this work in the later years of this period ('70–'76), centered on the use of ion channeling to determine and study various aspects of dilute impurity atom location inside and on the surfaces of single crystals. This was the context for my strong interest in the new potential tool for precise atom location measurements inside or on the surface of crystals. The origins of Jene's interest in the possibility to use X-ray standing waves to determine atomic

[†]Distinguished Service Professor Emeritus of Physics, State University of New York at Albany, NY 12222, USA.

positions are best related by him but I recall that he was stimulated by a paper by Batterman and Cole and earlier papers by Batterman from Cornell University. In any case, Jene, in his inimitable way, got lots of people excited about X-ray standing waves.

I remember at one point, Jene went to Cornell to talk with Bob Batterman about his ideas and came back really excited because while there, he had the opportunity to talk with P. Ewald who was then rather old and retired, spending his last years with his daughter and son-in-law Hans Bethe. As I recall, Jene said that Ewald got more and more excited as he described his analysis of how the reflectivity and the fluorescence of lattice and impurity atoms change as the incident direction of a highly parallel and monochromatic incident X-ray beam is scanned in angle across a highly perfect single crystal sample. He related how Ewald, in his excitement, got out an old bound notebook in which he had worked out the same dynamical diffraction analysis but said that he had never published or even shown it to anyone because he didn't think that it would ever be possible to measure. It is typical of Jene that rather than being disappointed that this analysis had been done a very long time before, he was excited to find a kindred spirit with whom he could discuss the details of the physics of X-ray standing waves and their potential applications. In fact, I may have the whole thing wrong, but it is pleasant to remember this way. I ask Jene about his visit with Ewald.

Following completion of his thesis work, Jene spent a few more years at Bell, after which he went to the University of Aarhus (as part of a remarkably productive exchange between Aarhus, Bell, and Albany which was initiated by a year that I spent at Aarhus in 1966–67 and continued for more than two decades). While I was settling in at Albany, an enthusiastic graduate student Mike Bedzyk came to talk to me about possible PhD research topics. I probably tried to sell him on some ion channeling projects but when I told him about Jene and his XSW ideas, Mike said that is what he wanted to do. Not dissuaded by my observation that we had no money to support such work, no experience in this area, and no equipment aside from some ancient junked X-ray equipment in the basement, Mike asked me how he could contact Jene and went ahead full speed. The rest, as they say, is history. XSW work at SUNY Albany petered out about 1989, largely because I became distracted and derailed by a succession of administrative posts at the University.

The years between '76 and '89 were, at Albany, a remarkably active and productive XSW period. There were five PhD theses from SUNY

Albany, including Mike Bedzyk's (1982), which was one of the first two XSW based theses (contemporary with a Licentiate thesis by Niels Hertel at the University of Aarhus,[a] where Jene went following three post doctoral years at Bell Labs). Other XSW PhD theses were those of Bhupendra Dev (1985), Thomas Thundat (1987), Ke-gang Huang (1988) and Mohammed Kayed (1992).[b] Important guiding and supporting roles were provided by research associates, Niels Hertel (University of Aarhus), Viktor Aristov (Institute of Solid State Physics, Moscow), Jörg Zegenhagen (Hamburg University) and a visitor Kenneth Thygesen (State Univ. College, Plattsburg, NY). There were also three theoretical theses stimulated by the experimental XSW program and directed by Professor Tara Das, on the electronic and hyperfine structure of atoms adsorbed on Silicon and Germanium single crystal surfaces. These were by M.S. Mohapatra, S.N. Mishra, and N. Sahoo.

Since most XSW work that has been reported uses highly sophisticated and expensive Synchrotron facilities, often with ultra-high-vacuum chambers and extensive computer support, it may be of interest to describe the XSW system at Albany developed initially by Mike Bedzyk and used (and improved) by him and others for many years. From the remains of two old GE X-ray generators, purchased in the '60s when the physics building was constructed, Mike put together an operating W anode X-ray source. He also resurrected a diffractometer table, had some stuff built in the machine shop and borrowed an old sodium iodide detector. By looking under rocks, I came up with enough funds to buy some asymmetrically cut crystals, an analog to digital converter and maybe even a CRT monitor. No funds were available to buy a computer to control the measurement and to analyze the data. The best that I could do was buy a couple of CPU chips (costing as I remember less than $100) which I presented to Mike with the instruction that he could get resistors, capacitors, wire, circuit boards, other components and perhaps some advice from the Physics Dept. electronic shop. A couple of years later when we had managed to get some NSF funding, I went to Mike and Niels Hertel who had joined us by then to tell them that we could finally buy a computer, only to be met with a somewhat indignant response that "What do we need a computer for?

[a]N Hertel, "Standing X-Ray Waves Interpretation and Application to Lattice Location and Relaxation Studies" Licentiate Thesis, University of Aarhus, Denmark, 1981.

[b]Editors note: For the PhD theses related to XSW conducted at the State University of New York Albany, please see the appendix.

We have a perfectly good computer". I retreated with a weak "Must have lost my head". This system was, in 1988, taken by Ken Thygesen to the State University College at Plattsburg, NY where it was used for another decade for undergraduate instruction and research.

In addition to demonstrating that meaningful adsorbed atom lattice positions and structures can be made at atmospheric pressure and ambient temperature with a simple, inexpensive X-ray standing wave analysis system, an impressive number of measurements were made at SUNY Albany. The first studies showed that for Si(111) samples, for which the residual oxide layer was removed with dilute HF solution from which the samples were moved slowly through a benzene overlayer containing dissolved bromine, Br atoms were adsorbed in positions well correlated with Si(111) lattice atomic positions.[1] Furthermore, it was shown that the 0.24 ML of adsorbed bromine effectively blocked oxidation of the silicon for days when the sample was exposed to air (if the sample was stored in a closed chamber with silica gel desiccant.) For Si(111) surfaces, cleaved while submerged in benzene containing dissolved Br, the bromine was found to be in two different locations, one over the first monolayer silicon atoms (atop position), the second, more loosely bound is in the interstitial hollow (subsurface) over the fourth layer Si atoms.[2,3] The surface silicon atoms were found to be relaxed outward by 0.13 ± 0.06 Å. The maximum coverage was found to be 0.24 monolayer and 1.0 monolayer in the atop and subsurface position respectively. This much detail is given only to illustrate the kind of information that could be obtained from these open air measurements. Adsorbed atom studies were also carried out for Si(111), and Si (220) on chemically cleaned and cleaved surfaces for adsorption of Se[4] and Ga.[5] Electrochemically deposited Ni[6,7] and Ga were studied on Si(111) and Ge(111) surfaces. Large coherent fractions were observed at room temperature. XSW was also used to study UHV deposited epitaxial nickel silicide films on Si(111).[8]

It should be noted that the silicon surfaces prepared by chemisorption of Br either after removal of the oxide with HF or on cleaved surfaces may have more natural structures for interface studies than the highly distorted surfaces prepared by ultra-high-vacuum techniques. Indeed the ability of Br to strongly adsorb and block air oxidation was also used to study atomic layer interfaces by other techniques such as ion channeling.[9]

In order to enter the modern world, an XSW facility was built as part of a SUNY beam line that was developed at the National Synchrotron Light Source (NSLS) at Brookhaven National Laboratory.[10] This facility was used for XWS analysis of $NiSi_2$ epitaxial films on Si(111)[11] and to study high reflection order and near-normal incidence X-ray standing waves.[12]

An important part of the SUNY Albany contribution to XSW studies have been the extensive theoretical investigations stimulated by the XSW experimental program.[13–21] Atomic and electronic structure calculations were carried out by a self-consistent field Hartree-Fock LCAO-MO cluster approach. The atomic structure was determined by the minimum total energy criterion. Electronic local density of states, hyperfine structure, chemical shifts, vibrational frequency and atomic vibrational amplitudes were also calculated.

Appendix: PhD theses related to XSW conducted at the State University of New York at Albany

M. J. Bedzyk, *X-Ray Standing Wave Analysis for Bromine Chemisorbed on Silicon* Ph.D. Thesis Dept. of Physics, State University of New York at Albany, NY (1982).

B. N. Dev, *Structural Studies of Chemisorption on Silicon Surfaces*, Ph.D. Thesis, Dept. of Physics, State University of New York at Albany, NY (1985).

T. G. Thundat, *Structural Investigations of Chemically and Electrochemically Deposited Metals on Silicon and Germanium Surfaces*, Ph.D. Thesis, Dept. of Physics, State University of New York at Albany, NY (1987).

K.-G. Huang, *Application of Synchrotron X-Ray Standing Waves, in Surface and Interface Studies*, Ph.D. Thesis, Dept. of Physics, State University of New York at Albany, NY (1988).

M. A. Kayed, *Photoionization and Photoemission by the Standing-Wave Intensity of Dynamical X-Ray Diffraction*, Ph.D. Thesis, Dept. of Physics, State University of New York at Albany, NY (1992).

References

1. M. J. Bedzyk, W. M. Gibson and J. A. Golochenko, *J. Vac. Sci. Technol.* **20** (1982) 634.
2. B. N. Dev. V. Aristov, N. Hertel, T. Thundat and W. M. Gibson, *J. Vac. Sci. Technol. A* **3** (1985) 975.
3. B. N. Dev, V. Aristov, N. Hertel T. Thundat and W. M. Gibson, *Surface Science* **163** (1985) 457.
4. B. N. Dev. T. Thundat and W. M. Gibson, J. *Vac. Sci. Technol. A* **3** (1985) 946.
5. T. Thundat, S. M. Mohapatra, B. N. Dev, W. M. Gibson and T. P. Das, *J. Vac. Sci. Technol. A* **6** (1988) 681.
6. T. Thundat, J. Zegenhagen, K. Thygesen and W. M. Gibson, *Surface Science* **230** (1990) 205.
7. T. Thundat, J. Zegenhagen and W. M. Gibson, *J. Vac. Sci. Technol. A* **5** (1987) 1484.

8. J. Zegenhagen, M. A. Kayed, K.-G. Huang, W. M. Gibson, J. C. Phillips, L. J. Schowalter and B. D. Hunt, *J. Appl. Phys. A* **44** (1987) 365.
9. H. S. Cheng, L. Luo, M. Okamoto, T. Thundat, S. Hashimoto and W. M. Gibson, *J. Vac. Sci. Technol. A* **5** (1987) 607.
10. P. J. Eng, Lorrie A. Krebs, Ki-Bong Lee, Peter W. Stephens, Steven Woronick, K.-G. Huang, Mohammed A. Kayed, Jorg Zegenhagen, Walter M. Gibson, David A. Hansen, John B. Hudson and James C. Phillips, *Nucl. Instr. Meth. A* **266** (1988) 210.
11. J. Zegenhagen, K.-G. Huang, W. M. Gibson, B. D. Hunt and L. J. Schowalter, *Phys. Rev. B* **39** (1989) 10254.
12. K.-G. Huang, W. M. Gibson and J. Zegenhagen, *Phys. Rev. B* **40** (1989) 4216.
13. K. C. Mishra, B. N. Dev, S. M. Mohapatra, W. M. Gibson and T. P. Das, *Hyperfine Interactions* **15/16** (1983) 997.
14. B. N. Dev, K. C. Mishra, W. M. Gibson and T. P. Das, *Phys. Rev. B* **29** (1984) 1101.
15. B. N. Dev. S. M. Mohapatra, K. C. Mishra, W. M. Gibson and T. P. Das, *J. Vac. Sci. Technol.*
16. B. N. Dev, S. M. Mohapatra, K. C. Mishra, W. M. Gibson and T. P. Das, *Phys. Rev. B* **36** (1987) 2666.
17. S. M. Mohapatra, N. Sahoo, K. C. Mishra, B. N. Dev, W. M. Gibson and T. P. Das, *Hyperfine Interactions* **34** (1987) 581.
18. S. M. Mohapatra, B. N. Dev, K. C. Mishra, N. Sahoo, W. M. Gibson and T. P. Das, *Phys. Rev. B* **38** (1988) 12556.
19. S. M. Mohapatra, B. N. Dev, K. C. Mishra, W. M. Gibson and T. P. Das, *Phys. Rev. B* **38** (1988) 13335.
20. S. M. Mohapatra, N. Sahoo, B. N. Dev, T. P. Das and W. M. Gibson, *Phys. Rev. B.*
21. S. M. Mohapatra, B. N. Dev, L. Luo, T. Thundat, W. M. Gibson, K. C. Mishra, N. Sahoo and T. P. Das, *Reviews of Solid State Science* **4** (1990) 473 (World Scientific Pub. Co.)

Appendix 4

PERSONAL RECOLLECTIONS ABOUT RESEARCH ACTIVITIES RELATED TO THE X-RAY STANDING WAVE METHOD

Seishi Kikuta

Japan Synchrotron Radiation Research Institute/SPring-8
1-1-1 Kouto, Sayo-cho, Sayo-gun, Hyogo, 679-5198, Japan

In this note, I describe the early stage of our research activities involving the X-ray standing wave method and then briefly summarize the activities made in Japan.

In the early 1960's, Kohra at the University of Tokyo designed and constructed a high-precision X-ray double/triple crystal diffractometer. I was involved in studying X-ray dynamical diffraction phenomena by using this instrument. In particular, we intended to observe intrinsic rocking curves of a nearly perfect Si single crystal by using a highly parallel and highly monochromatic incident beam. Annaka joined us and we studied the effect of temperature on the width of the rocking curve.[1] Just then, Batterman reported that the intensity of fluorescence X-rays from a Ge crystal excited by Mo K_α radiation under the Bragg case diffraction condition changes anomalously, and pointed out that it represents evidence for the successive formation of the two types of X-ray standing wave (XSW) fields, i.e. two Bloch waves corresponding to the two branches of the dispersion surface.[2] This stimulated us and we expected that the Compton and thermal diffuse scatterings would also be influenced by the formation of the characteristic wave fields in the diffraction process. Annaka, Kohra and me have observed the intensity change of these scattering signals during the diffraction process in the Bragg case, for a Si crystal with Cu K_α radiation, and for a Ge crystal with Mo K_α and Cu K_α radiation.[3,4] Since the complete separation of the Compton and thermal diffuse scattering signals was difficult, we realized a condition under which either Compton scattering or thermal diffuse scattering is the stronger signal, by choosing a proper combination of X-ray wavelength and type of crystal. The observed features of the recorded intensity curves were explained by the difference in mechanism of the scattering as follows: The Compton scattering is caused mainly by the outer electrons of atoms, and since they are largely delocalized, there is no appreciable difference in the scattering when the position of the XSW field changes. On the other hand, the thermal diffuse scattering is caused by the whole electrons containing the inner electrons and its cross section is larger when the wave field is with the antinode on the lattice planes. Worthy of mention, the localization of the scattering event is still not as strict as for the photoabsorption and subsequent fluorescence emission process. In addition, Annaka analyzed the directional dependence of the thermal diffuse scattering accompanying the Bragg case diffraction which occurs when a certain reciprocal lattice point exists near the Ewald sphere.[5] He also analyzed the intensity anomaly of the fluorescence emission accompanying the Laue case diffraction.[6] A little later, Annaka *et al.* measured the intensity of the K-series fluorescence from GaAs and the photocurrent in CdS modulated by an XSW field.[7]

From 1971, I started research concerning electron emission accompanying X-ray dynamical diffraction. For this purpose, I designed and constructed a high-precision X-ray double-crystal diffractometer installed in a high-vacuum chamber.[8] The rotation mechanism of the second crystal based on torsion of a cylinder provided an accuracy of 0.1 sec of arc. Emitted electrons were energy-analyzed by a 127° cylindrical electrostatic analyzer. We observed the change of the yield of K- and L-photoelectrons and KLL Auger electrons emitted from a Si crystal under the Bragg-case (220) diffraction condition with Cu K_α radiation. Its profile has a peak on the high angle side of the Bragg angle and a shallow dip on the low angle side.[9,10] In contrast to the X-ray fluorescence curve observed by Batterman, which resembles a mirror reflection of the diffraction curve, the electron emission curve closely resembles the XSW field intensity at the atom on the surface. This is because the absorption coefficient of electrons is much larger than the dynamical absorption coefficient of X-rays. Thus, electrons are observed from a shallow depth and are not affected by the extinction effect. Takahashi *et al.* observed angular changes in the yield of Si K-photoelectrons under the two conditions of asymmetric Bragg-case X-ray diffraction, i.e. grazing incidence, non-grazing exit and vice versa,[11,12] with remarkably different

15. E. Vlieg, A. E. M. J. Fischer, J. F. Van Der Veen, B. N. Dev and G. Materlik, *Surf. Sci.* **178** (1986) 36.
16. A. Iida, T. Matsushita and T. Ishikawa, *Jpn. J. Appl. Phys.* **24** (1985) L675.
17. T. Kawamura, Y. Fukuda, M. Oshima, Y. Ohmachi, K. Izumi, K. Hirano, T. Ishikawa and S. Kikuta, *Surf. Sci.* **251/252** (1991) 185.
18. T. Niwa, M. Sugiyama, T. Nakahata, O. Sakata and H. Hashizume, *Surf. Sci.* **282** (1993) 342.
19. A. Saito, K. Matoba, T. Kurata, J. Maruyama, Y. Kuwahara, K. Miki and K. M. Aono, *Jpn. J. Appl. Phys.* **42** (2003) 2408.
20. O. Sakata and H. Hashizume, *Acta Cryst.* **A51** (1995) 375.
21. A. Yu. Kazimirov, M. V. Kovalchuk, I. Yu. Kharitonov, L. V. Samoilova, T. Ishikawa and S. Kikuta, *Rev. Sci. Instrum.* **63** (1992) 1019.
22. T. Takahashi and M. Takahashi, *Jpn. J. Appl. Phys.* **32** (1993) 5159.
23. A. Saito, K. Izumi, T. Takahashi and S. Kikuta, *Phys. Rev. B* **58** (1998) 3541.
24. T. Ohta, H. Sekiyama, Y. Kitajima, H. Kuroda, T. Takahashi and S. Kikuta, *Jpn. J. Appl. Phys.* **24** (1985) L475.
25. T. Ohta, Y. Kitajima, H. Kuroda, T. Takahashi and S. Kikuta, *Nucl. Instrum. & Methods A* **246** (1986) 760.
26. T. Yokoyama, Y. Takata, T. Ohta, M. Funabashi, Y. Kitajima and H. Kuroda, *Phys. Rev. B* **42** (1990) 7000.
27. S. Takenaka, T. Yokoyama, S. Terada, M, Sakano, Y. Kitajima and T. Ohta, *Surf. Sci.* **372** (1997) 300.
28. M. Sugiyama, S. Maeyama and M. Oshima, *Phys. Rev. Lett.* **71** (1993) 2611.
29. M. Sugiyama, S. Maeyama, S. Heun and M. Oshima, *Phys. Rev. B* **51** (1995) 14778.

Appendix 5

X-RAY STANDING WAVES — THE EARLY DAYS IN HAMBURG

GERHARD MATERLIK

Diamond Light Source Limited, Diamond House, Didcot, Oxfordshire OX11 0DE, UK

This paper describes the pioneering implementation and development of the X-ray standing wave method with synchrotron X-radiation at HASYLAB in Hamburg. Many technical challenges had to be overcome and step by step, one application after the other was successfully implemented to bring this method, a powerful combination of spectroscopy and diffraction, to its present mature status.

I learnt the basics of X-ray dynamical diffraction as a student with U. Bonse in Münster to understand the principle of X-ray interferometry. Of course, Max von Laue's book *Röntgenstrahlinterferenzen*[1] was a wonderful introduction. Unfortunately, it was never translated into English, but for all diffraction scientists who were able to read it, it was a remarkable text book. Standing waves or *stehende Wellenfelder* were debated in great detail. They were the basis to explain the principle of a Bonse–Hart Interferometer using the Borrmann effect as illustration.

It was only when I was contacted in early 1975 by Jene Golovchenko from Aarhus I recognized that the movement of the planes when rotating an ideal crystal through a Bragg case reflection curve could be used to measure positions of foreign atoms within a crystal. He told me that he had just done this kind of experiment for an implanted (thus getting rid of extinction effects in earlier measurements) layer of As in Si. He had used an X-ray tube as source and wanted to follow up with synchrotron X-radiation (SXR) to increase the sensitivity and signal to noise ratio in on the EXAFS part. I wanted to continue the XANES work that I had started at Cornell, and understand and apply the various physical properties that determine the near-edge structure such as polarization and crystal symmetry, valence states, and atomic contributions, spin dependence, etc.

In addition, I wanted to get XSW working for measurements on bulk, surfaces and interfaces. Jörg Zegenhagen was the first Ph.D. student tasked to build an XSW instrument and test it on metal crystals. We were always faced with the comments of some colleagues that XSW were great but would probably just work for Si and Ge. The group of W. Űhlhoff at KFA (*Kernforschunganlage*) in Jülich, Germany, was best known to be able to grow, cut, and polish perfect Cu single crystals and I was able to win his support for this challenging task.

XSW on Cu single crystals meant: do not even dare to breathe when you are handling them! But to make life even more exciting (!) we later thought that the solid–liquid interface would be an excellent problem to study. No structural details were really known about the Helmholtz layer since Helmholtz had described it almost 100 years before. D. M. Kolb from Fritz-Haber Institute in Berlin gave us the chemical equipment and taught us the preparation methods. Jörg did a marvelous job to get all this working for his Ph.D. It needs to be mentioned that just when we had finally measured our first Tl/Cu(111) XSW spectrum in 1982, DORIS was turned off for an energy upgrade to find a new HEP particle and we lost at least one year before we were back again. Bad luck again for XSW. We were really looking forward to having a dedicated SXR facility! What huge impact would photon science be able to make if we could built our own light sources! I worked very hard and many extra nights and weekends with many colleagues who had also realized this opportunity to make the proposal for the ESRF a success.

In 1984, we published our first XSW paper with SXR[4] and it proved that XSW would really work extremely well with SXR. The work with Cu[5–7] showed that XSW could be used not only for Si and Ge but also to study *in situ* the solid–liquid interface under controlled potential conditions. It also demonstrated that the atomic structure of the inner Helmholtz layer was accessible and that this structure was controlled on the atomic scale via the potential and that it very sensitively depended on the concentration of all the chemical substances inside and above the liquid. We later on also used Au single crystal[8,9] with H. Abruna, who joined us from Cornell for a sabbatical.

During this early period, Niels Hertel came from Aarhus and stayed with us for several months. He introduced us to, and further developed with us, the concept of fluorescence-selected Fourier components of the electron density function.[10] Mike Bedzyk joined the group as post-doctoral researcher after he had finished his degree in Albany. The initial success gave us some money to rebuild an old UHV chamber for XSW, and Peter Funke started this project and got surface studies going in a UHV environment.[11,12] So far it had been Br chemically deposited on Si and Ge using the Bell Lab preparation recipe in which specially Mike was a real crack. Together with A. Frahm and Mike's skill in preparing the Br layer on the analyzer crystal of an LLL X-ray interferometer we were finally able to do the triangulation of the Br atoms.[13]

Also Mikhail Kovalchuk visited from Moscow with an electron detector, where the sample was placed inside the detector as source of emitted electrons that were registered at the wire. This gave us a great chance to understand electron emission from wave fields and their energy dependence, which Mike worked out in great detail, and we demonstrated their behavior in non-centrosymmetric crystals as a function of energy.[14] Just by changing the energy from the Ga- to the As K-absorption edge in GaAs we could move the (200) wave field from one atomic site to the other. This led to the basic definition of what a diffraction plane really meant in physical terms and where they were positioned within a unit cell.[15] Applying the relation to the structure factor provided a new understanding of how to measure the phase of the structure factor with XSW.[16]

We were eagerly looking forward to using the first high-resolution electron detector in UHV. This opened the way to other halogenides on Si and on Ge and comparing them with the chemically prepared adsorption sites and with other UHV methods. Peter Funke's set-up took the first results with fluorescence in 1985. Of course, we now did not scan the angle of the sample in UHV but the photon energy via the monochromator. The dream of high resolution came true when a wiggler became available after rebuilding DORIS for synchrotron radiation work. With Alex Lessmann and Wolfgang Drube, we managed to get the first results,[17] and this work also led to establishing the technique of high-energy photoemission.[18]

In 1986, Paul Cowan came along for a sabbatical to do Dexter[19] with us under grazing incidence (together with S. Brennan and T. Jach) and when Mike left to join the CHESS activity, Bhupen Dev came from Walter Gibson and observed XSW in $LiNbO_3$. We wanted to study a very defective crystal to see how XSW works for such crystals, which were even more challenging in terms of lattice perfection than metal crystals. The different dopants Ti, Er, and Fe in this system generate various different optical effects so that the resulting material can be used as optical storage medium or light amplifier or waveguide. Together with Thomas Gog, we showed that in all three cases, the impurity preferred the Li site contradicting some channeling measurements and chemical intuition.[20,21]

Bhupen also got measurements with an ultra-high vacuum baby chamber, going where the sample was prepared in a different chamber (usually prepared by R. L. Johnson from Hamburg University or F. Grey from Risø Institute), transferred under a Be dome and measured with XSW — now mostly at the first HASYLAB wiggler beamline. We wanted to measure XSW during a phase transition and at higher temperature, and Pb on Ge(111) was the proper system, which gave puzzling results with other methods. The measurements revealed that local order remained even above the Pb melting transition.[22]

We had always wondered about the influence of thicker surface layers. When will the phase induced by the layer shift the substrate XSW position? In $NiSi_2$/Si(111) we found a good system to check it. E. Vlieg and F. van der Veen from NIKHEF (*Nationaal instituut voor subatomaire fysica*) Amsterdam, had just studied it with low-energy ion scattering, and it was very good to compare these methods as well and determine the structure precisely with XSW. For thicknesses of several layers it worked perfectly.[23] The layer problem was finally put to rest years later during my research stay at SSRL in 1993–1994. Along with Sean Brennan's group, we studied thick AlGaAs layers on GaAs. We observed the XSW formation from grazing incidence up to the substrate lattice reflections and showed how they lock into the various different scales of the various sub-lattices.[24–26]

Norbert Greiser became attracted to XSW, and we decided to measure several Fourier components of the distribution function simultaneously by using a three-beam case geometry. Although the central part of the overlap of both excitation conditions was not used, we learnt that already the wings of the curves are sufficient to determine the position of atoms simultaneously in several directions.[27]

Having seen the three-beam case paper, H. Spalt from Darmstadt brought a new challenge along: the direct measurement of *Eigenvectors* of phonons by using XSW in a three-beam case.[28,29]

In UHV, several systems were studied applying the set-ups at HASYLAB very much routinely. V. Etalienaemi from Helsinki as Ph.D. student, Enrique Michel from Madrid as Humboldt Fellow, C. Sanchez-Hanke from Madrid as Ph.D. student, J. Falta as post-doctoral researcher, and many more formed an exciting team and I greatly enjoyed the privilege of working with them. In UHV, we used electron spectroscopy, fluorescence, and photon-stimulated desorption as detection channels. Needless to say, each of these methods has specific pros and cons and it has been important to understand these differences and apply them in measurements in the most advantageous way. The emphasis of all these studies was now on understanding growth and structures of materials on surfaces.[30]

Since the 1990s, we expanded the XSW methodology and observed other much more weakly modulated interference fields. We looked at the reversed wave field geometry using the Kossel effect[31,32] and developed atomic resolution X-ray holography[33] and direct imaging of atoms using a white incident beam, where all interference fringes from different wavelengths are being washed out but not in forward scattering direction[34]; and discovered photon interference X-ray absorption fine structure (PIXAFS),[35,36] where the incoming and the elastically scattered waves form an interference field inside the sample, which is swept across the atoms causing very weak modulation of the absorption signal by changing the wavelength over a large range, the Fourier transform of the absorption spectrum reveals the various coordination shells just like in EXAFS where this modulation originates from photoelectron interferences.

This later development added a wonderful twist to the story of XSW: starting out with strongly modulated fields and ending up with weakly modulated fields leaving a strong signal if the conditions are right.

I do not think that this story is over yet, and it would be fun to speculate about future developments such as the work on speckle interferometry, which revealed again interference patterns.[37] But this is not the intention of this summary.

Developing XSW with synchrotron light has been a great time and I would like to thank all the above-mentioned colleagues (and all the others whose work I could not mention here, but which has also been essential for the progress made over the last 30 years) for giving me the great privilege to work with them in a very constructive and open spirit.

At the end, SXR and XSW really worked together and proved to be a perfect match.

References

1. M. V. Laue, *Röntgenstrahl-Interferenzen* (Akademische Verlagsanstalt, Frankfurt am Main, 1960).
2. U. Bonse and G. Materlik, *Z. Phys. B* **24** (1976) 189.

3. P. L. Cowan, J. A. Golovchenko and M. F. Robbins, *Phys. Rev. Lett.* **44** (1980) 1680.
4. G. Materlik and J. Zegenhagen, *Phys. Lett. A* **104** (1984) 47.
5. G. Materlik, J. Zegenhagen and W. Uelhoff, *Phys. Rev. B* **32** (1985) 5502.
6. G. Materlik, M. Schmäh, J. Zegenhagen and W. Uelhoff, *Phys. Chem.* **91** (1987) 2921.
7. J. Zegenhagen, G. Materlik and W. Uelhoff, *J. X-Ray Sci. Technol.* **2** (1990) 214.
8. H. D. Abruna, T. Gog, G. Materlik and W. Uelhoff, *J. Electroanal. Chem.* **360** (1993) 315.
9. G. M. Bommarito, D. Acevedo, J. F. Rodriguez, H. D. Abruna, T. Gog and G. Materlik, X-Ray Standing Wave Studies of Underpotentially Deposited Metal Monolayers, in *Synchrotron Techniques in Interfacial Electrochemistry*, C. A. Melendres and A. Tadjeddine (eds.) (Kluwer Academic Publishers, Netherland, 1994), p. 371.
10. N. Hertel, G. Materlik and J. Zegenhagen, *Z. Phys. B* **58** (1985) 199.
11. P. Funke and G. Materlik, *Solid State Commum.* **54** (1985) 921.
12. P. Funke, G. Materlik and A. Reimann, *Nucl. Instrum. Meth. A* **246** (1986) 763.
13. G. Materlik, A. Frahm and M. J. Bedzyk, *Phys. Rev. Lett.* **52** (1984) 441.
14. M. J. Bedzyk, G. Materlik and M. V. Kovalchuk, *Phys. Rev. B* **30** (1984) 2453.
15. M. J. Bedzyk and G. Materlik, *Phys. Rev. B* **32** (1985) 6456.
16. B. N. Dev and G. Materlik, X-Ray Standing Waves in Noncentrosymmetric Crystals and the Phase Problem in Crystallography, in *Resonant Anomalous X-Ray Scattering*, G. Materlik, C. J. Sparks and K. Fischer (eds.) (North Holland, Amsterdam, 1994), p. 119.
17. W. Drube, A. Lessmann and G. Materlik, *Rev. Sci. Instrum.* **63** (1992) 1138.
18. A. Lessmann, W. Drube and G. Materlik, *Surf. Sci.* **323** (1995) 109.
19. P. L. Cowan, S. Brennan, T. Jach, M. J. Bedzyk and G. Materlik, *Phys. Rev. Lett.* **57** (1986) 2399.
20. T. Gog, T. Harasimowicz, B. N. Dev and G. Materlik, *Europhys. Lett.* **25** (1994) 253.
21. T. Gog and G. Materlik, in *Insulating Materials for Optoelectronics: New Developments*, F. Agulló-López (ed.) (World Scientific, Singapore, New Jersey, London, Hong Kong, 1995), p. 201.
22. B. N. Dev, F. Grey, R. L. Johnson and G. Materlik, *Europhys. Lett.* **6** (1988) 311.
23. E. Vlieg, A. E. M. J. Fischer, J. F. van der Veen, B. N. Dev and G. Materlik, *Surf. Sci.* **178** (1986) 36.
24. A. Lessmann, M. Schuster, S. Brennan, G. Materlik and C. Riechert, *Rev. Sci. Instrum.* **66** (1995) 1428.
25. M. Schuster, A. Lessmann, A. Munkholm, S. Brennan, G. Materlik and H. Riechert, *J. Phys. D: Appl. Phys.* **28** (1995) A206.
26. S. A. Stepanov, E. A. Kondrashkina, R. Köhler, D. V. Novikov, G. Materlik and S. M. Durbin, *Phys. Rev. B* **57** (1998) 4829.

27. N. Greiser and G. Materlik, *Z. Phys. B: Condens. Matter* **66** (1987) 83.
28. H. Spalt, A. Zounek, B. N. Dev and G. Materlik, *Phys. Rev. Lett.* **60** (1988) 1868.
29. A. Zounek, H. Spalt, G. Materlik, *Z. Phys. B* **92** (1993) 21.
30. V. Etelaeniemi, E. G. Michel, G. Materlik, *Surf. Sci.* **251/252** (1991) 483.
31. Th. Gog, D. Novikov, J. Falta, A. Hille, G. Materlik, *J. Phys. IV*, **C9** (1994) 449.
32. T. Gog, D. Bahr and G. Materlik, *Phys. Rev. B* **51** (1995) 676.
33. T. Gog, P. M. Len, G. Materlik, D. Bahr, C. S. Fadley and C. Sanchez-Hanke, *Phys. Rev. Lett.* **76** (1996) 3132.
34. P. Korecki and G. Materlik, *Phys. Rev. Lett.* **86** (2002) 2333.
35. Y. Nishino and G. Materlik, *Phys. Rev. B* **60** (1999) 15 074.
36. Y. Nishino, L. Tröger, P. Korecki and G. Materlik, *Phys. Rev. B* **64** (2001) 201101.
37. R. Eisenhower and G. Materlik, W. Meyer-Ilse, T. Warwick and D. Attwood (eds.), in *AIP Conf. Proc.* **507** (2000), 488.